The Ecology of Resource Degradation and Renewal

Mine spoil and the mine water discharging from an adit at Cwm Rheidol lead mine in the River Rheidol catchment, Cardiganshire, Dyfed, Wales.

The Ecology of Resource Degradation and Renewal

The 15th Symposium of
The British Ecological Society
10–12 July 1973

edited by
M. J. Chadwick
and
G. T. Goodman

A HALSTED PRESS BOOK

JOHN WILEY & SONS
NEW YORK

Osney Mead, Oxford,
85 Marylebone High Street, London W1M 3OE,
9 Forrest Road, Edinburgh,
P.O. Box 9, North Balwyn, Victoria, Australia.

First published 1975

Library of Congress Cataloging in Publication Data
Main entry under title:

The Ecology of resource degradation and renewal.

(The 15th Symposium of the British Ecological Society)
Held July 10–12, 1973.
'A Halstead Press book.'
1. Pollution—Congresses. 2. Conservation of natural resources—Congresses. 3. Environmental protection—Congresses. I. Chadwick, Michael J. II. Goodman, Gordon T. III. Series: British Ecological Society. Symposium; no. 15.
QK911.B7A35 no. 15 [QH545.A1], 574.5'08s, [301.31]
ISBN 0-470-14295-2 75-5776

Printed in Great Britain

Contents

Preface

In 1965 the British Ecological Society published a Symposium volume dealing with man-induced environmental changes and some of the problems that may arise in their wake (*Ecology and the Industrial Society*, edited by G. T. Goodman, R. W. Edwards and J. M. Lambert, Blackwell). The present volume contains the papers contributed to a second Symposium organized by the Society to discuss similar questions from a biological standpoint. Coming ten years after the first attempt was made, its publication offers an opportunity to look back and compare the states of awareness and knowledge of such questions displayed then and now by the world scientific community and by national and intergovernmental agencies responsible for dealing with these problems.

Much has happened during this ten-year interval. We have seen the emergence of the great international environmental debate, and many would add, its vociferous climax and irresolute attenuation.

In any event, prior to the 1960s there was only a very limited awareness of the special environmental problems created by 'urban and industrial development'. The perception of such questions in terms of economic and population growth, food and resource depletion, and pollution effects was limited to a relatively few far-sighted individuals. At the intergovernmental level, the specialized U.N. agencies had done valuable pioneering work in measuring changes in population, food and land-resources and in one outstanding instance (UNSCEAR), was providing a *modus operandi* for the assessment of global radionuclide hazards from atomic-weapons testing.

At the national level, in the developed countries, pollution was seen as one of the chief problems. The scientific and management approaches to this tended to be rather rigidly compartmentalized into the various sectors covered by the industrial chemist, engineer, and health authority. The role of the biologist as a manager was barely recognized despite the considerable amount of ecological thinking already in use and further required to deal adequately with air, water and land pollution.

Looking back, the Society may justly claim that its 1965 Symposium acted as a useful nodal point, demonstrating and perhaps establishing the claim of the ecologist to be able to contribute meaningfully to this developing field. Reading through the 1965 papers now, one is struck by the way they demonstrated how organisms could be used as indicators of environmental quality, in particular as a kind of early warning of the appearance of an incipient problem which if left to develop, might have significantly adverse consequences for man in terms of health, resource supply or amenity.

Equally striking is the dominance of a 'problem-recognition' and 'problem-posing' rather than a 'problem-solving' approach. Apart from the papers on sewage treatment and derelict land, 'environmental management' or the abatement or control of pollution are not seen too clearly as the ultimate objectives which motivated the original concern. But perhaps this is expecting too much by hindsight from a document which can, in perspective, be seen to be well in advance of its time.

Since its appearance, we have witnessed the rapid growth of environmental awareness all over the world. A somewhat depressing picture has been built up from a series of by now familiar, disquieting environmental episodes and accidents, like the unplanned side-effects of organochlorine compounds in Europe and North America or methyl mercury in Japan and Iraq or oil-spillages at sea and in coastal waters.

This environmental concern has culminated at the International level in the creation of the U.N. Environment Programme by the nations who contributed to the U.N. Conference on the Human Environment at Stockholm in 1972. The Governing Council of U.N.E.P. now has proposals for the development of a Global Environmental Monitoring System and an International Referral System for the co-ordination of environmental information.

During this time, the U.N. agencies have intensified their own environmental efforts and in addition, several regional programmes have been developed by mutual agreement between various collaborating governments (for example via O.E.C.D. or E.E.C.). The scientific community as represented by the various Unions and Scientific Committees of the I.C.S.U. family have co-operated in several important environmental projects.

Environmental concern at the national level has been made visible by the setting up of environmental protection agencies by many governments either *de novo* or by the restructuring of the existing environmental groups to form more integrated agencies. New environmental legislation is commonplace.

In the United Kingdom, we have seen the creation of the Department of the Environment, the Central Unit for Environmental Pollution, a permanent Royal Commission for Environmental Pollution and a new 'Control of Pollution Act' 1974, with wide ranging powers covering air, water and land pollution.

All this executive action at national and international levels, coupled with a positive response from some large industries has transformed the possibilities of carrying out effective environmental management. Indeed it may be wondered whether the executive machinery has not outstripped our scientific and technical powers to produce cost-effective environmental management solutions?

The realization is growing that it is not enough to have an awareness or clear definition of an environmental problem. Scientific and technical advances in environmental study must be interpreted and presented in the form of policy options which are usable by often hard-pressed administrators who have the final responsibility for actually doing something about an environmental problem. This will involve a blending of the traditional skills of mathematicians, physicists, chemists, biologists, economists, planners, sociologists, legal experts and many others. The present Symposium volume reflects this management oriented inter-disciplinary approach very clearly. Papers such as that contributed by Dr. A. D. Christie indicate the range of thinking needed to help solve management problems.

We believe that the 'easy' phase of environmental science is over. The simpler, quicker, 'problem-posing' phase is giving way to the 'long-haul' of 'problem-solving', and real solutions demand a great deal of relevant quantitative data on which to base policy-options. We must now enter a much less dramatic phase, steadily building a solid foundation of ecological knowledge in the appropriate multi-disciplinary context, which will serve as a basis for environmental management and control in the future. If as some believe, there is an irresolute attenuation of the environmental debate at the present time, it is not because the problems have 'gone away' but because we are failing to move from the relatively easy and qualitative 'problem-posing' phase to this much more difficult and quantitative 'problem-solving' phase of environmental science. The future quality of the environment depends upon the development of environmental management as a science by making this difficult transition. We hope that the present volume constitutes a step in this direction.

Acknowledgements

The 15th Symposium meeting of the British Ecological Society on *Degraded Environments and Resource Renewal* was held at Bodington Hall of the University of Leeds. The Society expresses its gratitude to the staff of the Hall for their part in the success of the meeting. Our thanks are also due to Dr. D. R. Hodgson who took on the burden of acting as Local Organiser for the Symposium. The smooth running of the meeting was entirely due to his efficient handling of all the local arrangements. In this he was ably supported by Dr. G. P. Buckley.

The University of Leeds extended its hospitality to the participants of the Symposium during the meeting and our thanks are due to the Vice-Chancellor and the University for this.

The Sessions were chaired by Professor H. W. Woolhouse of the University of Leeds, Mr. Grant Davis of the United States Department of Agriculture (Forest Service), Dr. W. Berg of the State University of Colorado and Professor P. B. Tinker of the University of Leeds. We are most grateful to them for giving their time and services during the meeting. In addition, an afternoon was spent visiting opencast coal gaining operations and restoration at Snydale and pulverized fuel ash disposal and revegetation sites at Thorpe Marsh, Gale Common and Ferrybridge. We are particularly indebted to the National Coal Board Opencast Executive (Mr. S. M. Hoy and Mr. B. W. Tait) and the Central Electricity Generating Board (Mr. M. G. Masterson and Mr. P. M. Owens) respectively for arranging these excursions and providing expert guidance around the sites.

Before, and during the meeting Mrs. I. Fletcher provided secretarial assistance which was invaluable. Subsequently Mrs. L. Wainhouse helped with the preparation of manuscripts. Mrs. J. Chadwick prepared the author and subject indices. We would like to express our appreciation of the work they carried out on our behalf.

Finally we would like to thank the contributors to the Symposium for their papers and expert discussion in which they participated during the sessions.

SECTION ONE

The cycling of materials in environments

The cycling of materials in disturbed environments

M. J. CHADWICK *Department of Biology, University of York, U.K.*

Introduction

Commoner (1972) gives some cogent advice to the enthusiastic yet relatively untutored 'ecologist' that current interest in environmental matters seems to have produced. He suggests that a persistent effort to answer the question 'Where does it go?' can yield surprisingly valuable information about an ecosystem. He stresses that nothing 'goes away' when it is discarded. Materials are merely converted from one molecular form to another and become transferred from place to place. Although the concept of materials cycling has been implicit in much early ecological work, and even explicit in agriculturally based work, it is only within the last twenty-five years that the universal applicability and widespread importance of this concept has been generally recognized. This recognition, in turn, has been instrumental in calling attention to the really serious nature of environmental perturbations which otherwise might have been regarded as merely rather unfortunate local incidents.

Basic biogeochemical cycles

Ecologists recognize that functional attributes of an ecosystem are as important as the structural attributes that occupied so much early ecological work. Ecosystem function stresses the interrelationship between organisms and the non-living components of the system. Many naturally occurring elements pass from the abiotic environment to living organisms. After molecular rearrangement they are eventually released to the environment again as the products of excretion and decaying organic matter. This back-and-forth movement or flux of elements has been referred to as a *biogeochemical cycle* (Hutchinson 1944). That the cycle is driven by a one-way flow of energy has been emphasized by Odum (1963) when relating the two major functional attributes of an ecosystem.

In any biogeochemical cycle some forms of the element are far less accessible to organisms than other more available forms that are relatively easily utilized by organisms. It is usual, therefore, to distinguish between the relatively unavailable or *storage pool* of an element and the exchangeable reservoir or *active pool.* The active pool usually forms the source of an element for organisms to utilize and incorporate into their biomass, which can then be distinguished as the *utilized pool.* All three pools may exist in a range of different forms and as well as there existing fluxes between one pool and another, fluxes also exist between different forms within the same pool. All fluxes can be characterized by rate constants that depend on a whole range of environmental factors. In the diagrams that follow (Figs 1 to 5) the pools (A—storage pool; B—active pool; C—utilized pool) are distinguished as being of relatively different capacities although the sizes do not represent exact quantities.

In some cycles adjustments are made between the various pools with relative ease. These are the *gaseous cycles* involving carbon, nitrogen or oxygen, which are considered to be *cyclic* or perfect because the natural negative feedback pathways (c and d in the diagrams) balance well the positive feed back pathways (a and b in the diagrams), when subjected to some rather unnatural influences. Other biogeochemical cycles, the *sedimentary cycles*, are characterized by relatively large amounts in the unavailable storage pool. These cycles tend to be less perfect and are regarded as *acyclic*. Hutchinson (1948) has pointed out that man, acting as 'a mighty geological agent' (Vernadsky 1945) has speeded up many of these cycles. Thus man dislodges many cycles from their relatively *natural form* (1) to a considerably *perturbed form* (2) characterized by a high degree of acyclism. If the resource is to be renewed, to enable continuing reutilization, it must be established in a *recycling form* (3).

It is the purpose of this paper to examine a number of these cycles in their relatively natural and perturbed forms and to indicate the effect of recycling. This will be done using extremely simplified 'figure-of-eight' cycle diagrams for the nitrogen cycle, phosphorus cycle, mercury cycle, a cycle involving water utilization and disposal following extraction from aquifers and nutrient cycles on colliery spoil tips.

Nitrogen cycle

The level of nitrogen availability, in a form utilizable by plants, is extremely important in controlling biomass production in an ecosystem. Nitrogen is

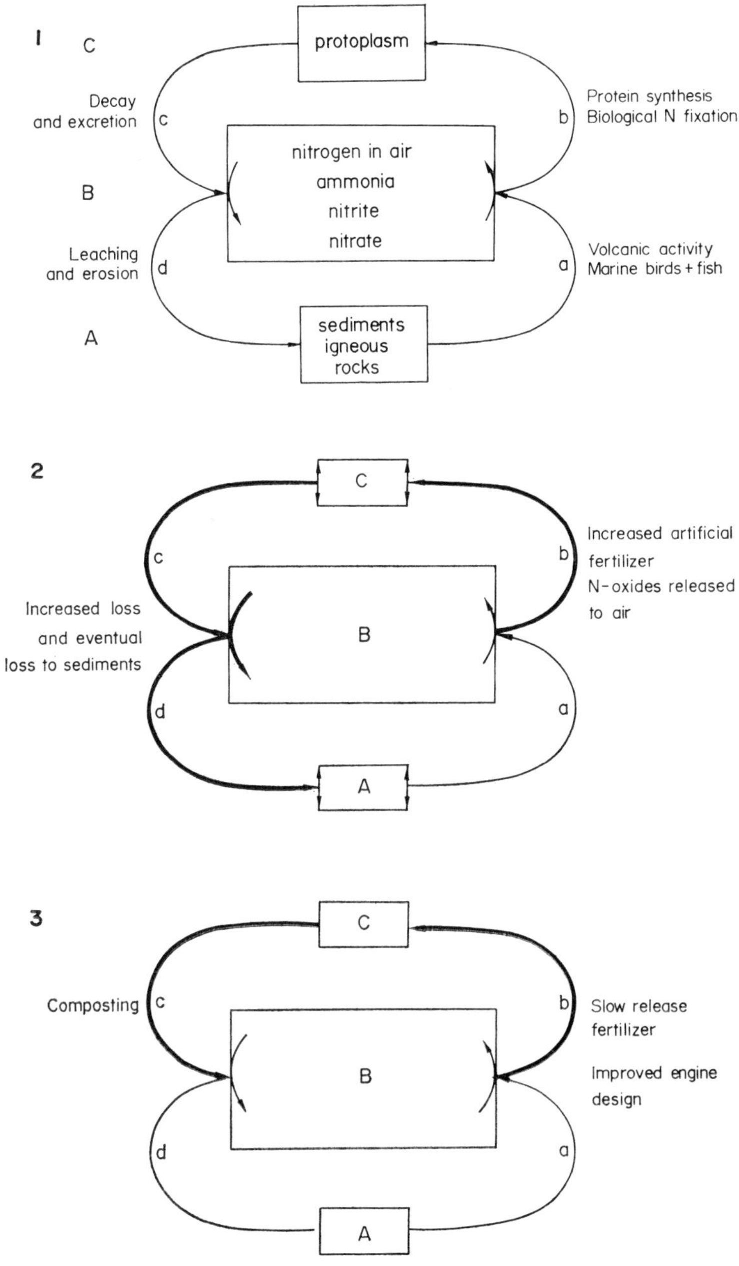

Figure 1. The nitrogen cycle. (1—'natural' cycle; 2—perturbed cycle; 3—recycling cycle; A—storage pool; B—active pool; C—utilized pool; a and b—positive feedback pathway; c and d—negative feedback pathway. Arrows in 2 indicate increasing or decreasing pool.)

required in protein synthesis and is supplied in a utilizable form by biological nitrogen fixation by microorganisms (Fig. 1 b). Under natural conditions this may occur at a rate of about 10 lb per acre per annum. Under agricultural conditions this rate can be greatly increased by the use of leguminous crops. Non-biological fixation in the atmosphere supplies only relatively small amounts. Excretion and the decay of organic matter releases fixed nitrogen (Fig. 1 c) and leaching and erosion causes losses from the active pool (B) to sediments that, with igneous rocks, make up the storage pool (A). Volcanic activity and the activities of marine birds and fish return nitrogen compounds to the active pool (Fig. 1 a).

Massive intrusion by man into the cycle is thought to be threatening the whole integrity of the nitrogen cycle (Commoner 1970). This is because nitrogen has hitherto been represented mainly in reduced forms (nitrogen gas or combined with hydrogen rather than oxygen in living things). The effects of man's intrusion on a massive scale is shown in Fig. 1–2. Technology introduces nitrogen into the environment almost entirely in oxidized forms. Power plants, automobiles and rapid increase in artificial nitrogenous fertilizer use increase the transfer represented by b, c and d in Fig. 1–2 but have little effect on a. Surface waters become polluted and as well as the disruption of biological systems nitrogen oxides in both water and air constitute potential human health hazards.

Figure 1–3 indicates that at least some improvement towards a cyclic condition can be achieved. Improved fertilizer practices (and improved engine design) can reduce the rate of loss from the active to the storage pool and the increased return of organic matter to the soil can give a constant rate of release into the active pool for subsequent uptake by plants. More balanced fluxes are established between positive and negative feedback controls leading to a more cyclic condition.

Phosphorus cycle

Phosphorus, like nitrogen, is an essential constituent of protoplasm and often limits biomass production. However the large reservoir of this element is in the storage pool, represented by phosphate rocks. Phosphate release to the active pool occurs only slowly (Fig. 2–1 a) whereas plant uptake is usually more rapid (Fig. 2–1 b). Fixation in the soil and loss to marine sediments (Fig. 2–1 d) may also be rapid. Man ahs intervened to make available in large quantities the phosphates necessary for rapid crop growth and other purposes. Phosphate deposits, that took millions of years to accumulate, are being mined extensively and much of the phosphate produced eventually finds its way into ecosystems, particularly aquatic ones, causing their altera-

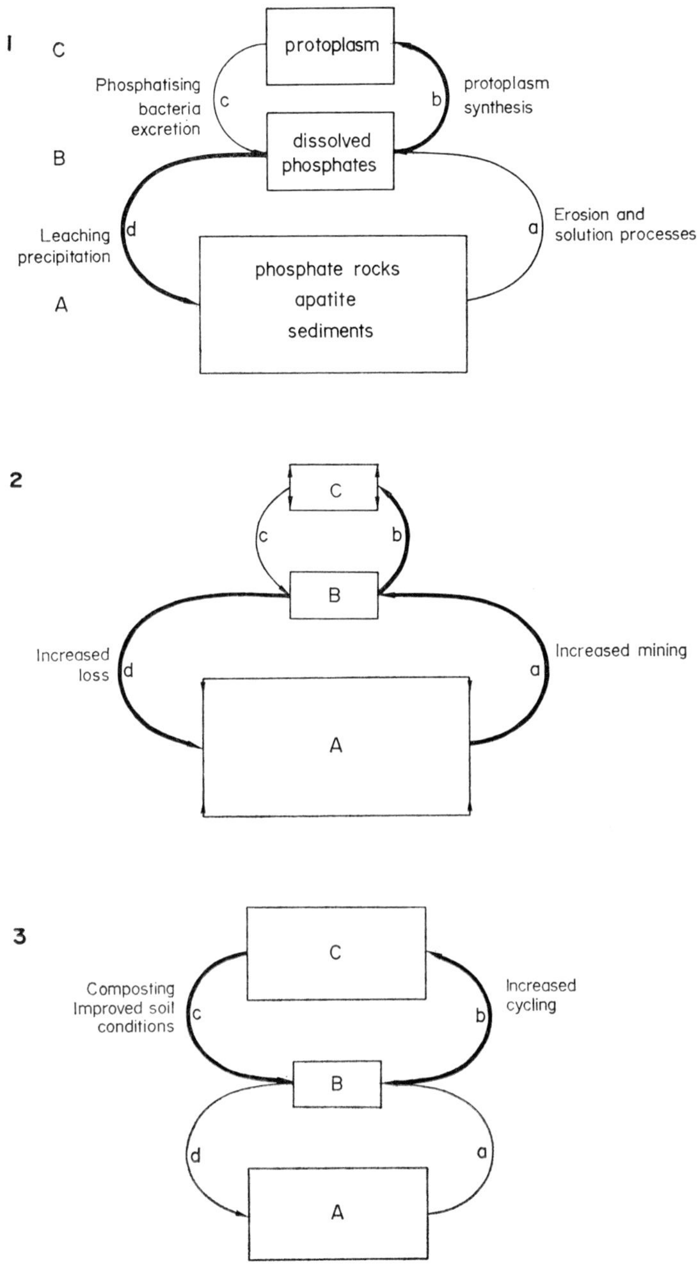

Figure 2. The phosphorus cycle. (See Fig. 1 for notation.)

tion and disruption (Hasler 1970). Eventually much of the phosphorus is lost again to the storage pool in a widely dispersed form, making it difficult to recover (Fig. 2–2). The depletion of phosphorus reserves, coupled with a largely acyclic characteristics suggest that eventually a greater degree of recycling must take place whereby composting procedures and methods of improving soil conditions to allow increased uptake of naturally released phosphate, are established (Fig. 2–3). It would be necessary to eliminate the major causes of eutrophication of acquatic systems.

Mercury cycle

Mercury is one of the less abundant elements in the Earth's crust. Under natural conditions small amounts passed from the storage pool to the active pool, mainly by weathering (Rankama & Sahama 1950). Dissolved mercury was taken up by plants and entered organisms to be returned again to the soil during decay processes (Fig. 3–1 b and c). Little information is available on natural levels and a close study of the cycling of this element has only been made since the industrial and agricultural use of mercurials has become common practice. However, mining of cinnabar and the extraction of the metal to meet industrial demand has undoubtedly caused increased rates of cycling in the biosphere. Increasing demand, coupled with the depletion of high-grade mercury mines suggests the development of a severe shortage of the element over the next thirty years (Lovering 1969). Thus the situation illustrated by the perturbed cycle (Fig. 3–2) will have to be transmuted to a recycling situation (Fig. 3–3) or alternative compounds developed.

Water utilization and renovation

The principles illustrated by general biogeochemical cycles considered on a world scale can be applied to more specific situations. In a degraded environment the object will be to move towards resource renewal by developing a recycling-type cycle out of a perturbed-type cycle. Both water utilization and disposal and vegetation establishment and maintenance on colliery spoil illustrate attempts at such management procedures.

In Britain, as elsewhere, groundwater is commonly extracted on a large scale for both public supply and industrial use. In Britain, chalk formations and the Bunter Sandstone form the principal aquifers and the rates of extraction exceed those of natural replenishment. The result is that large volumes of the aquifers are dewatered, giving rise to problems of continued extraction but also additional problems arise due to the return of the water,

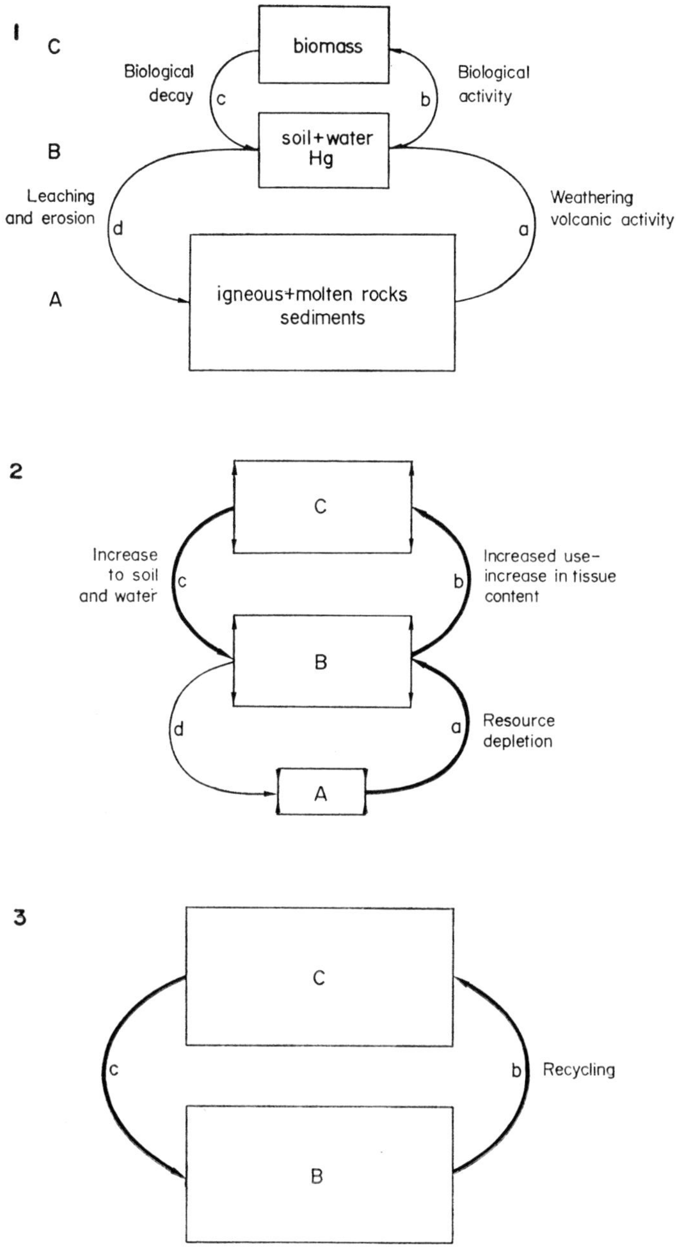

Figure 3. The mercury cycle. (See Fig. 1 for notation.)

as effluent, to surface water courses. Thus water, in an unwanted form, is discharged where it is not required and at the same time water sources, at the point they are required, are depleted. The natural and perturbed cycles are illustrated in Fig. 4–1 and 2.

It has seemed desirable to try and work out a system whereby application of the polluted water, following utilization, to a soil—vegetation system could result in renovation of the water applied, recharge of the aquifer and also possibly an increase in plant yield due to the irrigation and fertilization effect of the applied water. In a well-documented scheme in Pennsylvania (Kardos 1967; Parizek & Myers 1969; Sagmuller & Sopper 1967; Sopper 1968; Sopper & Sagmuller 1966) sewage effluent was applied to woodland and agricultural crops. These, together with the physicochemical and microbiological system represented by the soil, played a complementary role in water quality improvement. Crop yields were increased and the aquifers replenished, allowing continued water extraction. The polluting effect of the effluent on streams and rivers was considerably reduced. A similar pilot scheme in Britain, with a number of modifications, has been described by Chadwick *et al.* (1974). The modifications included the use of badly polluted river water from a tributary of a major river in place of sewage effluent and its application to natural heathland vegetation in place of tree and agricultural crops. This represents an attempt to move towards the type of recycling system shown in Fig. 4–3.

Table 1 gives an indication of the improvement in water quality obtained for three elements and also the excess of each element found in the irrigated vegetation over the amounts in similar unirrigated stands of vegetation.

Table 1. Amounts of three nutrients applied to lowland heath vegetation in polluted water, the excess of each nutrient in the irrigated vegetation (tops) and amounts in the soil water at 1·58 m depth (from Chadwick *et al.* 1974).

Nutrient	Amount applied	Amount in vegetation tops	Amount in soil water at 1·58 m
N	21·50	14·30	14·83
P	2·35	1·11	0·07
K	18·80	7·20	6·30
		(all in g. m^{-2})	

Improvement was also obtained in the alkalinity (reduction of 97%), total hardness (59%) calcium hardness (67%), conductivity (24%) sodium (5%) and anionic detergent (82%). This was using a very high rate of application (1,565 mm total, at a daily rate of 152 mm).

The attempt to renovate water by this method appears to be successful for at least some of the elements responsible for water eutrophication.

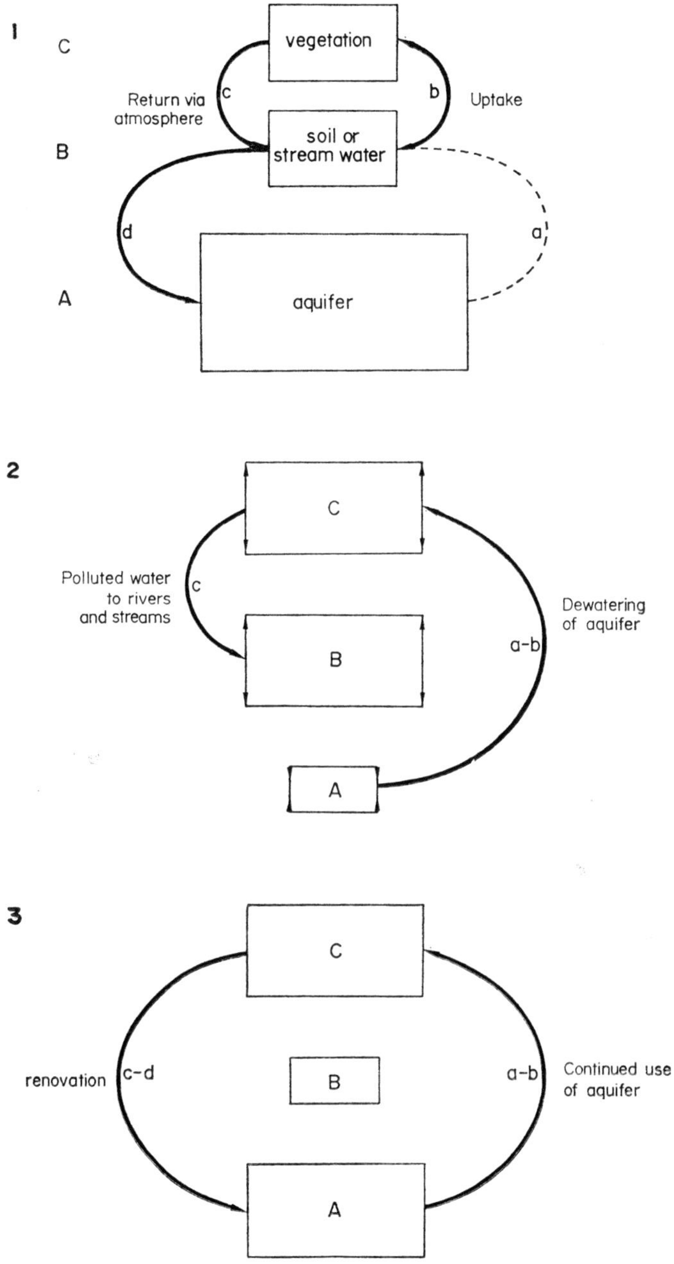

Figure 4. The cycle of water following groundwater extraction. (See Fig. 1 for notation.)

Although only partially successful in the case of nitrogen, the situation is probably better than indicated in Table 1. Most of the nitrogen applied in the ammonium form was apparently removed (96%) but much of it was probably converted to nitrate. The effect of irrigation was to increase soil pH from 3·4–4·0 to 4·7–5·4. Under higher pH conditions the ammonia oxidizing bacteria increased considerably, the nitrite oxidizers to a lesser extent and the denitrifiers hardly at all. Given faster growing plant species (conifers, grasses or arable crops) it is anticipated that nitrogen removal would have reached a higher level. Thus the indications are at least hopeful for the establishment of a recycling system.

Vegetation establishment and maintenance on colliery spoil

Studies aimed at elucidating nutrient cycles on colliery spoil heaps, in relation to reclamation work, have been carried out for some years at the University of York (Dennington, unpublished material). One of the main aims of the work has been to ascertain the input and output of nutrients from the system in an attempt to anticipate maintenance problems.

The sporadic vegetation that establishes naturally on undisturbed colliery spoil usually has available in the active pool only low levels of mineral nutrients, released only slowly from the storage pool by weathering processes (Fig. 5–1 a and b) and supplied in rainfall. Nutrients are also only slowly released from dead organic matter as the usually acid spoil conditions reduce bacterial activity so that a mat builds up on the colliery spoil surface (Fig. 5–1 c). In addition, topographical features of most tips lead to the loss of considerable amounts of nutrients in run-off water and by erosion of particulate matter (Fig. 5–1 d). Although this situation does not represent a 'natural' cycle it represents the situation before man attempts any management practices on the sites.

Table 2. Total annual nutrient input (kg.ha^{-1}) for three colliery spoil sites.

Site	Na	K	Ca	Mg	P	Total N	NO_3	NH_4	Organic N	Precipit cm yr^{-1}
Bullcroft	16·10	2·84	22·42	5·10	0·25	9·34	3·47	4·31	1·56	64·23
Mitchell	8·62	2·08	11·33	2·59	0·18	8·34	3·10	4·59	0·93	63·78
Maltby	10·67	3·72	16·44	5·35	0·67	8·71	3·01	4·13	1·57	63·29

Table 2 gives the total annual input of six nutrients in rainfall obtained by fortnightly collection and analysis at three colliery spoil sites in south Yorkshire. They represent levels capable of maintaining vegetation at a low level of productivity except that a proportion will be lost in run-off and a considerable proportion will be supplied during the winter months when

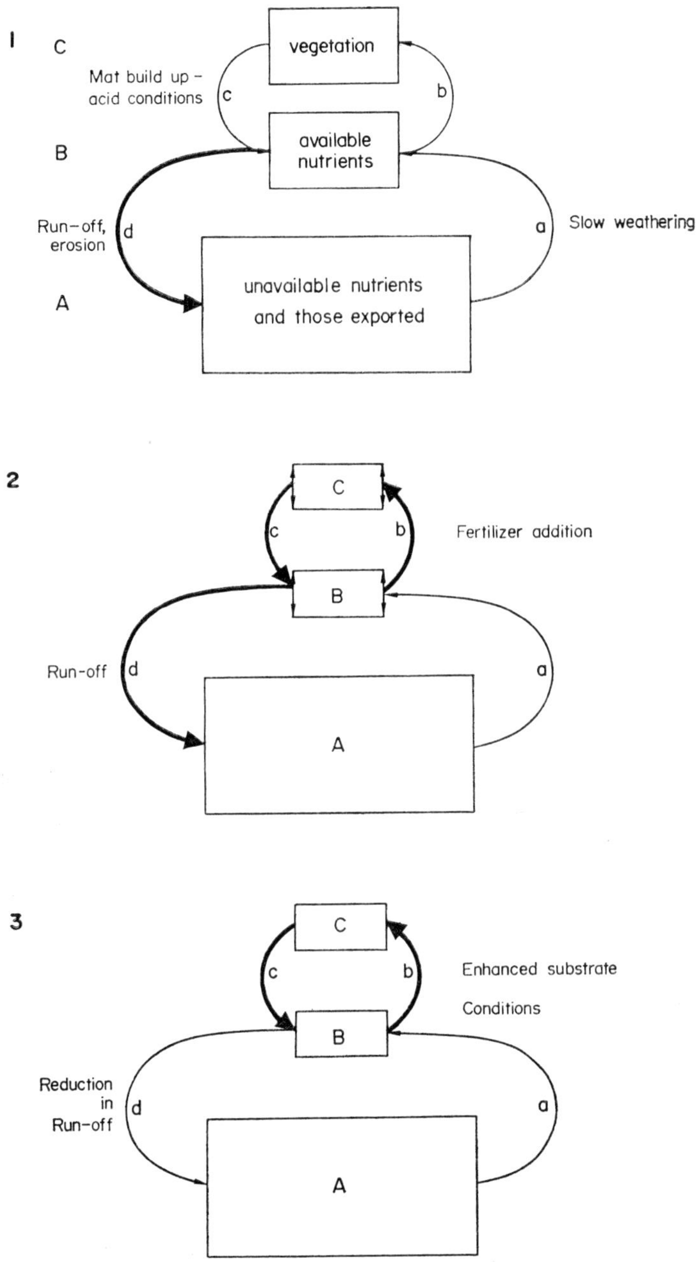

Figure 5. The nutrient cycle on colliery spoil heaps. (See Fig. 1 for notation.)

growth will be slow or have ceased entirely. For these reasons nutrients must be supplied artificially in order to establish vegetation successfully and often to maintain it. When this is done not only is the transfer from the active to the utilized pool speeded up (Fig. 5–2 b) but an improvement in substrate conditions gives increased bacterial activity and an increased rate of transfer from the utilized to the active pool (Fig. 5–2 c). Until a closed vegetational cover is established, however, it may also mean the rate of nutrient loss by run-off and erosion is still high. The effect on run-off of establishing grass species on colliery spoil is shown in Table 3. Although the effect obviously depends on both the intensity and amount of rainfall, it is not unusual to find that run-off from bare colliery spoil is three times as great as on grassed areas. It is thus possible to reduce run-off and establish something approaching a cyclic biogeochemical cycle shown in Fig. 5–3.

Table 3. Percentage run-off from grassed and bare colliery spoil sites.

Rainfall over period (cm)	Percentage run-off grassed	bare
0·203	3·35	9·85
0·584	3·42	3·42
1·727	3·94	10·42
1·854	3·56	10·00
1·941	4·74	79·27

A degraded resource can therefore become renewed but a comparison between the way in which certain nutrients are partitioned in the ecosystem (Table 4) suggests that in a relatively mature Douglas-fir ecosystem (Cole *et al.* 1969) the soil system may represent a much greater part of the active pool than for colliery spoil ecosystems that have undergone reclamation within the last ten years. Renewal to a stable state obviously requires long-term management procedures.

Table 4. Nutrient supply represented by annual precipitation supply and available soil nutrients compared with the nutrient content of the standing crop for acid colliery spoil and a second growth Douglas-fir ecosystem.

		K	Ca	P
Colliery Spoil	Precipitation	0·56	3·53	0·19
	Vegetation	1·00	1·00	1·00
	Soil	0·10	4·24	1·61
Forest	Precipitation	negligible	0·01	negligible
	Vegetation	1·00	1·00	1·00
	Soil	1·06	2·23	58·70

Conclusions

The cycling of materials is one of the two main functional attributes of any ecosystem. It is important that the cycle be understood for the widespread cycling of single nutrients. It is equally important to elucidate the cycles for specific resource management activities. The investigation of biogeochemical cycles has been responsible for demonstrating the adverse effects of massive intervention by man in some of the cycles. For many cycles, however, there are parts of the cycle that are imperfectly understood and quantitative relationships have not been established. Much information is required on residence times in the atmosphere and considerable study must be undertaken to ascertain the significant parameters that need to be determined for any cycle. For example, it has been suggested recently that particle-borne acid may be a better index of potential health hazards than the measurements of sulphur dioxide concentrations that are currently made (Brosset 1973).

Where management practices are aimed at the renewal of degraded environments it is of equal importance to determine the existing cycles and the effect of management practices on them. A positive attempt to rectify acyclic tendencies and establish balanced cycles is required rather than the following of simple rule of thumb procedures.

It is against this background that the contributions to this Symposium are made.

Summary

The importance of biogeochemical cycles in relation to disturbed environments is emphasized. Comparisons are made between the cycling of nutrients essential for plant growth and the cycling of more 'exotic' elements and compounds. The application of recycling procedures in resource renewal, in both aquatic and terrestrial situations, is stressed.

References

BROSSET C. (1973) Air-borne acid. *Ambio* **2**, 2–9.

CHADWICK M.J., EDWORTHY K.J., RUSH D. & WILLIAMS P.J. (1974) Ecosystem irrigation as a means of groundwater recharge and water quality improvement. *J. appl. Ecol.* **11**, 231–47.

COLE D.W., GESSEL S.P. & DICE S.F. (1969) Distribution and cycling of nitrogen, phosphorus, potassium and calcium in a second-growth Douglas-fir ecosystem. In *Symposium on Primary Productivity and Mineral Cycling in Natural Ecosystems*, University of Maine Press.

COMMONER B. (1970) Threats to the integrity of the nitrogen cycle: nitrogen compounds in soil, water, atmosphere and precipitation. In *Global Effects of Environmental Pollution* (Ed. by S.F. Singer), Reidel, Dordrecht, Holland.

COMMONER B. (1972) *The Closing Circle.* Cape, London.

HASLER A. D. (1970) Man-induced eutrophication of lakes. In *Global Effects of Environmental Pollution* (Ed. by S. F. Singer), Reidel, Dordrecht, Holland.

HUTCHINSON G.E. (1944) Nitrogen and the biogeochemistry of the atmosphere. *Amer. Scient.* **32**, 178–95.

HUTCHINSON G.E. (1948) On living in the biosphere. *Scient. Monthly* 67, 393–8.

KARDOS L.T. (1967) Waste water renovation by the land—a living filter. In *Agriculture and the Quality of our Environment* (Ed. by N.C. Brady), A.A.A.S. Monograph 85.

LOVERING T.S. (1969) Mineral resources from the land. In *Resources and Man.* Freeman, San Francisco.

ODUM E.P. (1963) *Ecology.* Holt, Rinehart & Winston, New York.

PARIZEK R.R. & MYERS E.A. (1969) Recharge of groundwater from renovated sewage effluent by spray irrigation. In *Proc. 4th Amer. Water Res. Conf.*

RANKAMA K. & SAHAMA T.G. (1950) *Geochemistry.* Chicago.

SAGMULLER C.J. & SOPPER W.E. (1967) Effect of municipal sewage effluent irrigation on height growth of white spruce. *J.For.* 63, 822–3.

SOPPER W.E. (1968) Waste water renovation for reuse: key to optimum use of water resources. *Water Res.* **2**, 471–80.

SOPPER W.E. & SAGMULLER C.J. (1966) Forest vegetation growth responses to irrigation with municipal sewage effluent. In *Proc. 1st Pan Amer. Soil Conserv. Cong.*

VERNADSKY W.I. (1945) The biosphere and the noosphere. *Amer. Sci.* 33, 1–12.

Arable ecosystems and the use of agrochemicals

G. R. POTTS and G. P. VICKERMAN *Game Conservancy, North Farm, Washington, Pulborough, Sussex, U.K.*

Introduction

For some time now there has been widespread concern that the use of agrochemicals may be making a substantial contribution, directly or indirectly, to the pollution of the ecosphere. Indeed, it has been claimed (Anon. 1971) that the agricultural environment itself is being degraded with the use of agrochemicals. However, this has yet to be convincingly demonstrated and the present paper describes investigations now being carried out on some of the effects of agrochemicals on cereal and grassland ecosystems. The agrochemicals considered here are the fungicides, herbicides, insecticides and molluscicides, collectively known as pesticides.

Methods

The Game Conservancy Partridge Survival Project started in 1968. The study area consists of about 300 large fields in a 65 km^2 area of the South Downs near Washington in West Sussex. There is no running water but a series of valleys run from north to south across the area and down the dip slope of the chalk. The boundaries to the west and east consist of flood plain water meadow areas. The northern boundary is formed by the scrub covered or wooded scarp slope which rises to an average height of 180 m above sea level. Woods and coastal conurbations are to the south.

Cereals are grown in two-thirds of the fields, grass in the majority of the others. The traditional crop rotation involves the use of grass and clover leys; for example, in the most simple version it is,

WHEAT→BARLEY→BARLEY (undersown)→GRASS (2 yrs)

but during the period 1967 to 1974 this was abandoned in one area (called 'modern arable') and oilseed rape and beans replaced the grass as break crops. Most pesticides were used in the latter area; all use of these is continually monitored. Some work has also been carried out on a 250 ha mixed farm near Catterick Camp in North Yorkshire.

The work on partridges (*Perdix perdix* and *Alectoris rufa*) and the importance of arthropods to their survival has already been described (Potts 1970, 1974). The detailed methods involved in the partridge work are given by Potts (1973), and the work on arthropods and agricultural ecology is described by Potts & Vickerman (1974). This latter paper describes early results of this work and documents ecological changes in the cereal ecosystem which have resulted from recent changes in the methods of farming. Particular attention is given to the composition of the field and ground layer insect fauna and to spatial variations in the relative abundance of predators and their prey and of parasites and their hosts. The aim is to account for important changes in the fauna and to predict the future course of ecological side-effects which will accompany the many pending changes in modern farming techniques.

The arthropod densities in the cereal crops are measured with the use of a 'backpack' petrol driven Dietrick vacuum insect net. Each sample consists of five randomly placed sub samples of 0·092 m^2 and generally contains a mixture of arthropods, vegetation and soil. The samples are deep frozen and then stored in alcohol for sorting under a microscope; the details being given in Potts & Vickerman (1974). Part of the routine work involves the monitoring of the use of all agrochemicals on the farms in the study areas. The information in the present paper is drawn from our study of arable ecosystems and from some early experience of the senior author in work on pesticides in the North Sea.

Increase in the use of pesticides

So far this century about 55 different pesticides have been widely used in Britain for the protection of farmland crops other than vegetables and fruit.

The number readily available has increased from about 4 in 1925 (arsenic, sulphuric acid, calcium cyanamide and rotenone) to the present total of about 45. Judging from the actual use on two farms (Fig. 1) the number widely used has increased in parallel to the total available, that is from 1 or 2 in 1930 to about 20 in 1973. In 1973 17 different active ingredients were used on North Farm (the largest of the 15 farms in the Sussex study area). As a total this is by no means atypical for an arable farm, although it is higher than that for the smaller mixed farm in north Yorkshire where only 9 active ingredients were used (Fig. 1).

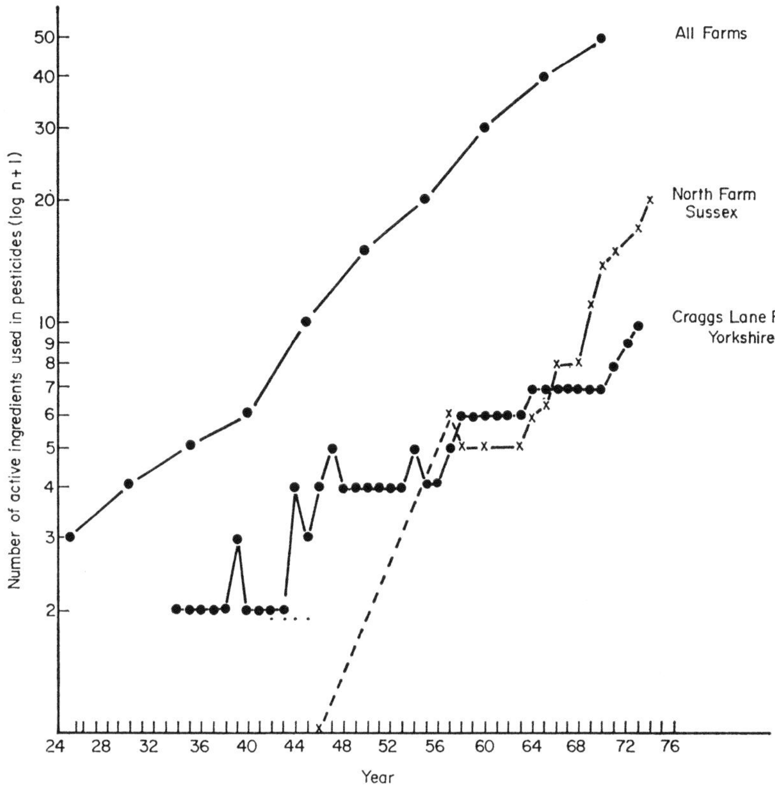

Figure 1. The number of active ingredients used by the arable departments on two farms and by all arable farms. The latter is an estimate based on general enquiries together with data given by Martin (1973).

PESTICIDE CLASSES AND THEIR USES

Fungicides

(a) to control seed-borne diseases
(b) to control leaf diseases

Organo-mercury fungicides or recent replacements such as mixtures of maneb and quinasaphate are used to control seed-borne diseases such as loose smut (*Ustilago nuda* (Jens) Rostr.). The major leaf diseases are mildew (*Erysiphe graminis* DC.), leaf blotch (*Rhynchosporium secalis* (Oudem.) J. J. Davis) and the brown and yellow rusts (*Puccinia hordei* Otth. and *P. striiformis* Westend.).

Since 1970 a number of chemicals have been introduced to control these, of which three, tridemorph, ethirimol and polyram have been widely used; a fourth (dodanil) is likely to be widely used from 1974.

Herbicides

(a) post-harvest pre-emergence control of monocotyledonous weeds
The main problems here are the rhizomatous grass weeds such as *Agropyron repens* (L.) Beauv. and *Agrostis stolonifera* L. Although they are controlled mostly by cultivation, chemicals such as TCA and dalapon can also be used. A new and more effective chemical, glyphosate, has recently been introduced. Pre-emergence chemical control of the wild oat (*Avena fatua* L.) first became feasible with the introduction of di-allate in 1961.
(b) post-emergence control of monocotyledonous weeds
Grass weeds, especially wild oats (*Avena fatua* L. and *A. ludoviciana* Dur.), blackgrass (*Alopecurus myosuroides* Huds.) and (in the study area) *Poa trivialis* L. can now be controlled by various chemicals. Two, chlortoluron and a mixture of metoxuron and simazine, introduced in 1972, are now likely to be widely used to control grass weeds in the study area.
(c) narrow spectrum post-emergence control of dicotyledonous weeds
(d) broad spectrum post-emergence control of dicotyledonous weeds
The use of chemicals for these purposes and subsequent changes in the flora have been discussed in earlier papers (Potts 1970; Potts & Vickerman 1974).
(e) to enhance straw shortening and strengthening
Cycocel is occasionally used for this purpose, especially where high levels of nitrogen have been applied to winter wheat.
(f) for desiccation (to aid harvesting of crop, or burning of stubble)
The latter includes the pre-harvest use of diquat and the very wide use of paraquat on stubbles. The effects of this and of subsequent burning will be described in a separate paper.

Insecticides

(a) to control seed and seedling pests
γ BHC seed dressings are used almost universally on cereal seeds in the U.K. mainly to control larvae of Tipulidae and Elateridae. Ecologically there appears to be much over-use of the compound because attacks of these pests are now uncommon, but the application is cheap even as an insurance. The wheat bulb fly (*Leptohylemyia coarctata* Fall.) does not occur commonly in any of the study areas and the seed has not therefore been treated with dieldrin or chlorfenvinphos.

Dieldrin (as aldrin in potato fertilizer) was introduced on the Yorkshire farm in April 1957 and its use continued (including use as a sheep dip from 1958) until 1966 when it was replaced by organophosphorus substitutes;

DDT was used on only three occasions from 1954 onwards. In Sussex dieldrin has never been used on North Farm, except as a sheep dip, but DDT was used on several occasions during the late 1960's, especially in 1969.

Most chemicals would not be developed for use unless there was a demand and unless the research could be carried out in the expectation of future sales. Pesticides have become increasingly available in recent years because of technical developments in organic chemistry. However, the use of most chemicals cannot be clearly assigned to increased demand, because the need in terms of pest 'evels has not been monitored until now. It appears that there was a clear demand for systemic mildew inhibitors for decades before they came into use. Cereal mildews are not a new threat, though many other leaf diseases are a greater problem now than formerly (Jenkins *et al.* 1972; Melville & Lanham 1972). On the other hand the anti-polygonum and anti-monocotyledon herbicides have been introduced in response to increases in these weeds.

Cereal leaf diseases began to affect yields seriously in the study area in 1965. This may be the result of changes in the host-parasite resistance relationships rather than any fundamental environmental change. The breakdown of disease resistance in many cereal varieties is probably the result of the breeding necessary for short-term yield gains.

At present the relationship between block cropping (i.e. several adjacent fields with the same crop) and the incidence of leaf diseases is being investigated. For some diseases, especially mildew, there does not appear to be an important relationship between crop area and susceptibility to disease, but for others there may be. A South East Region Agricultural Development and Advisory Service survey (R. G. Hughes, *in litt.*) in the study area suggested that a non-susceptible crop more than 180 m wide reduced the spread of brown rust. These studies are being carried out because both the size of fields and the incidence of block cropping have greatly increased over the past decade.

Cereal aphids are perhaps the most important pests of cereals in Britain. Their numbers have increased in recent years and the increase has been world wide (Baryanovits 1973; Potts & Vickerman 1974). Insecticides were first used to control cereal aphids in the study area in 1973, but in parts of Essex (and the Netherlands) such treatments are now routine.

Monocotyledonous weeds have greatly increased in recent years. Most of the increases have followed changes in crop rotation and stubble hygiene (e.g. wild oats used to be 'rogued' in the early stages of infestation and grass rhizomes were killed by prolonged cultivation). Another change is that the combine harvester now passes the weed seed on to the stubble whereas it was formerly removed with the sheaves. More dicotyledon herbicides are now necessary as the continued use of early compounds has led to the development of floras dominated by species which they did not control.

ECOLOGICAL SIDE-EFFECTS

Farmland ecosystems

Although pesticides are usually used for obvious economic reasons it is perhaps surprising that they are used on a massive scale in an ecosystem of which comparatively little is known. The natural control mechanisms already present in the system are not understood and, perhaps even more important, the long-term effects of pesticide applications are not known. These problems are perhaps best exemplified by the cereal aphids. For example, the use of the most common aphicide (dimethoate) on a crop

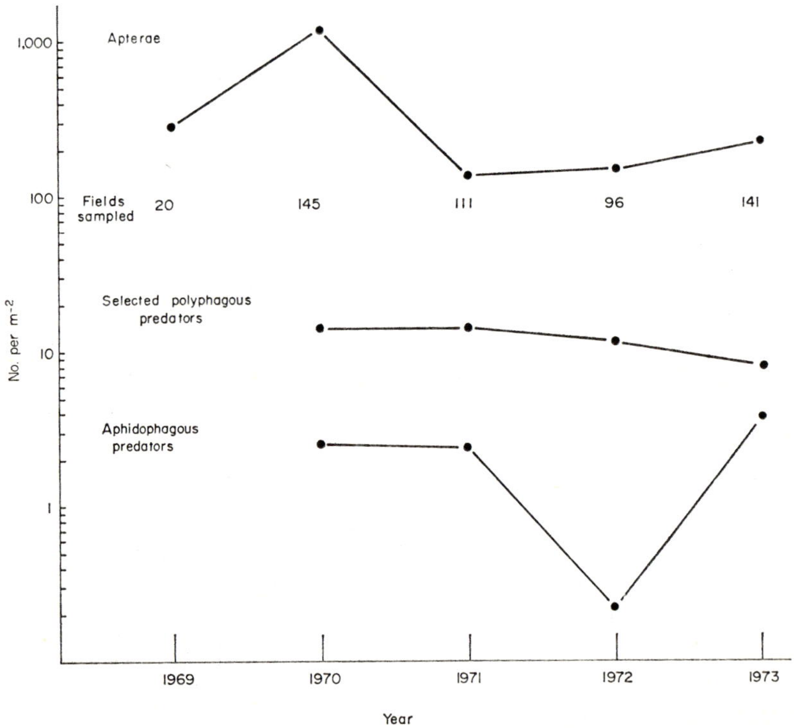

Figure 2. Trends in the late June density of apterous aphids in cereal fields in the Sussex study area from 1969 to 1973. Trends of aphid predators (see text) and polyphagous predators are given from 1970 to 1973.

of winter oats in the study area on 10 April 1973 reduced the number of apterous aphids by 97%—but also reduced the total for all other arthropods by 80%. Further cereal aphid outbreaks on the scale of 1963 and 1968 could result in the widespread aeriel application of such chemicals.

Mean leve‘s of cereal aphid populations in the study area over the period 1969–73 are given in Fig. 2. It must be emphasized that cereal aphids although comparatively scarce in most cereal fields, may vary in number (per square metre) between different fields by more than a thousand-fold. The factors which result in an outbreak are therefore clearly important.

It has been shown (Potts & Vickerman 1973) that there are significant inverse relationships between the numbers of apterous cereal aphids present in samples taken from different fields and both the diversity of the arthropod fauna of those fields and the proportion of predatory arthropods in them

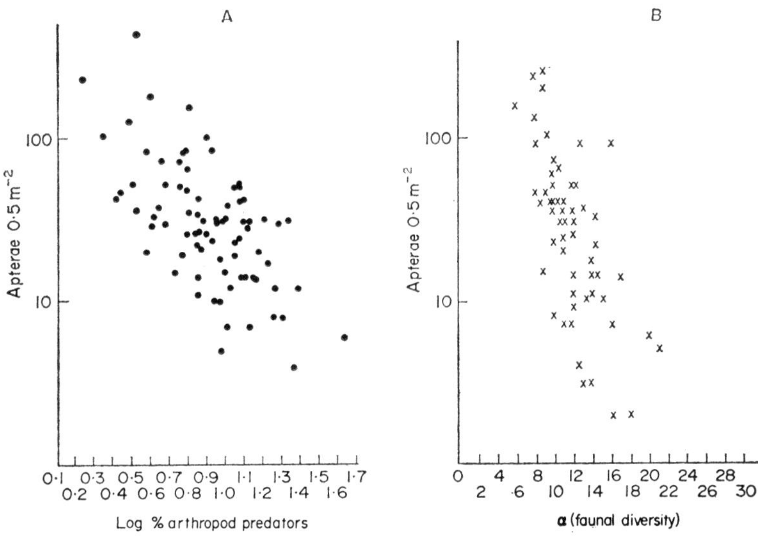

Figure 3. A. The relationship between the density of apterous aphids and the log of the percentage of arthropods which were predators in cereal fields in 1972. B. The relationship between the density of apterous aphids in cereals and the faunal diversity (= c) in 1971. Both correlations are significant (*P*< 0105).

(Fig. 3). The aphid specific predators such as the Syrphidae (mainly *Syrphus balteatus* Deg.), Coccinellidae (mainly *Propylea 14-punctata* (L.)), Cantharidae (mainly *Cantharis rufa* L.), and Neuroptera (mainly *Chrysopa carnea* Stephens) are not generally sufficiently numerous to account for this relationship. However, the role of other arthropod predators is being investigated in a series of field and laboratory experiments. The results have so far shown that 16 common predatory species readily feed on cereal aphids in the

Table 1. A comparison of the late June densities of selected insect taxa in cereal crops according to the type of farm system over the years 1970 to 1973.

System	Rest (375)* Mixed	North Farm (118)* (modern arable)	
Selected dicotyledonous feeders		Mean no. per m²	
Curculionidae	1·60	1·42	N.S.
Chrysomelidae	1·46	0·84	—
e.g.			
Crepidodera transversa (Scop.)	0·36	0·08	†
Cassida flaveola Tb.	0·06	0·06	N.S.
Chaetocnema concinna (Marsh)	0·34	0·30	N.S.
Gastrophysa polygoni (L.)	0·36	0·04	†
Selected monocotyledonous feeders			
Tenthredinidae	4·41	2·38	†
Jassidae	6·56	3·75	†
Delphacidae	3·94	2·77	N.S.
Lema melanopa (L.)	0·18	0·10	N.S.

* Number of fields in sample.
† $P \leqslant 0{\cdot}01$.

laboratory. Moreover, analyses of gut contents of predators captured in pitfall traps have shown that they do feed on them in the cereal crops. Further work, involving the identification of aphid predators by precipitin testing is being carried out so that the likely impact of these polyphagous arthropod predators can be calculated. Changes in the combined density of some of the polyphagous predators (Opiliones, *Bembidion* sp., *Trechus* sp., *Risophilus atricapillus* (L.), *Agonum dorsale* (Pont.), *Tachyporus* sp.) over the period 1970–73 are given in Fig. 2 together with similar records for the aphidophagous species.

The effects of modern farming techniques on the fauna of cereal crops have been investigated and it is already apparent that the predatory complex has been reduced on the modern farm. For example, in the spring of 1970 and 1971 the arthropod fauna emerging from cereal fields which had been undersown with a grass/legume mixture the previous year was compared with that emerging from other cereal fields. The absence of undersowing resulted in a 27% decrease in faunal diversity and a reduction in the proportion of predatory insects by 15·2%. The likely impact of this can perhaps be assessed by the fact that in East Anglia 30% of cereals were undersown in 1950 whereas by 1970 the figure had fallen to 3% (Potts 1970).

It is essential that the effect of changes such as these can be identified and quantified. By far the most widespread and consistent use of pesticides in cereals is that against the dicotyledonous weeds. However, the effect of this on the weed dwelling fauna is not a simple one. The crop rotation on the farm is more important and many weeds and niches are not utilized in cereals which are spatially and temporally separated from grass crops which had been undersown in the previous year. Many typical species of weed dwelling insect are far less common on modern arable farms (Table 1) than one would expect from the distribution of their host plants; for example, *Cassida flaveola* Tb. and *Gastrophysa polygoni* (L.) are quite rare or absent even where their favoured host plants, *Stellaria media* (L.) Vill. and *Polygonum aviculare* L. respectively, are abundant. The number of insect taxa present on modern arable farms is much less than on traditional ley farms, especially early in the

Table 2. The number of taxa recorded in 'Dietrick' vacuum insect net samples from cereals on different types of farms in early June 1972, West Sussex.

Number of taxa	Fields by type of farm Traditional ley	Intermediate	All arable	Total
26–30	—	—	3	3
31–35	—	6	3	9
36–40	1	10	7	18
41–45	4	8	4	16
46–50	6	9	4	19
51–55	7	5	—	12
56–60	3	3	—	6
61–65	1	1	—	2
	22	42	21	85

χ^2 (8 × 3) = 29·39 *P* (14 d.f.) < 0·01

season (Table 2). Although no one knew of these kinds of relationship when dicotyledonous weeds were first sprayed and the effects could not therefore bemeasured, we now expect to be able to assess the overall effects of monocotyledon sprays as they are introduced. Limited trials in 1973 suggest that the elimination of *Poa trivialis* from cereals will not change the total number of arthropods present but it will seriously reduce the density of some species which are favoured food of partridge chicks (see Vickerman 1974).

Adjacent ecosystems

The build up of dieldrin in the Shag (*Phalacrocorax aristotelis*) population in the North Sea during the 1960's was measured by a joint project of the University of Durham and Shell Chemicals Ltd and is given in Fig. 4.

The accumulation of p,p′-DDE, the main metabolite of DDT, is rather less well understood but is essentially similar; the main use of DDT in the U.K. appears to have been during the mid 60's (Coulson *et al.* 1972).

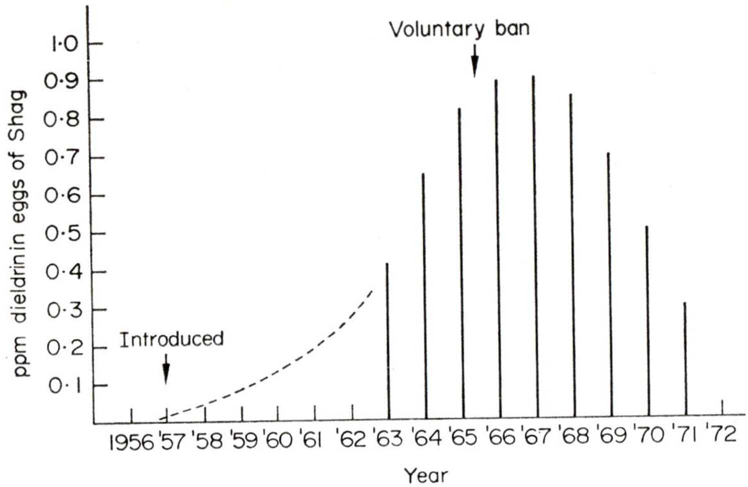

Figure 4. Dieldrin content of the eggs of a Shag population in the North Sea.

Detailed investigations of the breeding success of the Shag (Potts 1966) and adult survival (Potts 1969) indicated that the losses in the egg and early chick stages were unusually hing. For example of 1,541 eggs laid in 1963 and 1964 only 34% produced ten-day-old chicks. The earlier studies of Snow (1960), on what must have been a much less contaminated Shag population, in the early 1950's included 921 eggs; 67% of these produced ten-day-old chicks. A large part of the difference was clearly not directly attributable to the presence of pesticides (Potts 1966) but some of it may have been caused by them in view of the very high frequency of egg breakages (Appendix I) and the thinner egg shells. Shell weights were down by 17% in the sample taken by Ratcliffe (1970). Nowever, the population as a whole appeared to be unaffected, until the 'crash' of 1968 attributable to a 'red tide' (Coulson *et al.* 1968).

This example and that of the present investigation of the decline of the Grey Partridge (Potts 1974) draw attention to the difficulty of quantifying small, but nevertheless biologically significant side effects of pollutants. Indeed the effect of most pesticides on partridge survival is only through their role in the farming system as a whole; their effect can only be evaluated in relation to the rest of the modern 'technological package'. The effects vary greatly according to the ecological circumstances, such as disease, predation and to the effects of the weather.

Discussion

A basic strategy of agriculture is to maximize yield by removing crop competitors by chemical means. However, such an approach failed in the controlled environment of glasshouses so that biological control has become necessary (Gould 1971) and our general experience suggests to us that this may be the outcome on farmland. Certainly, the relationships between increases in crop hygiene and yield are no longer simple and the benefits of pesticides have become intangible to some farmers.

It is often claimed that the interests of wildlife or game conservation and good farming are opposed and that the conservation interest can only be met by compromising the farm standards. It appears, however, that the more important conflict is between those systems of farming which rely on the *ad hoc* empirical approach, 'pouring on' new technology, and those systems which might wish to consider or even harness those elements of the crop ecosystem which are sympathetic to the natural control of crop competitors. It must be accepted that in the short term the second approach will be more tedious and it will tax ecology and ecologists beyond their imagination. But, just as new technologies continue to be grafted piecemeal on to existing systems which we do not understand, so ecologists will surely have to sort out the problems some day.

In the meantime no farmer wants to use pesticides which exacerbate his problems in the long term and which, in addition, incur severe environmental costs. The only way to prevent these problems and to provide data for the conservation of various farmland species is to monitor and account for events in the farmland ecosystem; this is our aim.

Acknowledgements

We are especially grateful to the late Christopher Hynt founding the Partridge Survival Project for the Game Conservancy. Also we have been extremely fortunate to have first class cooperation and help from all those concerned with the farming in the West Sussex study area, especially at North Farm, Washington. We are particularly grateful to Mr E. B. Potts for his detailed records of agrochemical use which began in 1934 on Craggs Lane Farm, Catterick Camp, north Yorkshire and to Mr B. P. W. Morris for detailed records at North Farm starting in 1957. The work is partly supported by a grant from the Natural Environment Research Council starting in 1971 and by a further grant from the Agricultural Research Council starting in 1973, in addition to the funds of the Game Conservancy Partridge Survival Project.

Summary

The amount of pesticides used on farmland continues to increase rapidly. One expects the ecological hazards of this to be hard to identify, but the long term agricultural benefits are now also intangible. The biological significance of stability, diversity and predation in the farmland ecosystem are discussed with reference to cereal aphids.

References

ANON. (1971) Blueprint for Survival. *Ecologist* **2**, No. 1.

BARYANOVITS F. (1973) The increasing problem of aphids in agriculture and horticulture. *Outlook on Agriculture* 7, 102–8

COULSON J.C., POTTS G.R., DEANS I.R. & FRASER S.M. (1968) Mass mortality of sea birds in Northumberland due to a red tide. *Brit. Birds* **61**, 381.

COULSON, J.C., DEANS I.R., POTTS G.R., ROBINSON J. & CRABTREE A.N. (1972) Changes in organochlorine contamination of the marine environment of Eastern Britain, monitored by Shag eggs. *Nature, Lond.* **236**, 454–6.

GOULD H.J. (1971) Large scale trials of an integrated control programme for cucumber pests in commercial nurseries. *Pl. Path.* **20**, 149–56.

JENKINS J.E.E., MELVILLE S.C. & JEMMETT J.L. (1972) The effect of fungicides on leaf diseases and on yield in spring Barley in south-west England. *Pl. Path.* **21**, 49–58.

MARTIN H. (1973) *Insecticide and Fungicide Handbook for Crop Protection.* Blackwell, Oxford.

MELVILLE S.C. & LANHAM C.A. (1972) A survey of leaf diseases of spring Barley in south-west England. *Pl. Path.* **21**, 59–66.

POTTS G.R. (1966) Studies on a marked population of the Shag with special reference to the breeding biology of birds of known age. *Ph.D. Thesis.* University of Durham.

POTTS G.R. (1968) Success of eggs of the Shag on the Farne Islands, Northumberland, in relation to their content of dieldrin and pp'DDE. *Nature, Lond.* **217**, 1282–4.

POTTS G.R. (1969) The influence of eruptive movements, age, population size and other factors on the survival of the Shag (*Phalacrocorax aristotelis*). *J. Anim. Ecol.* 38, 53–102.

POTTS G.R. (1970) Recent changes in the farmland fauna with special reference to the decline of the Grey Partridge. *Bird Study*, **17**, 145–66.

POTTS G.R. (1973) Pesticides and the fertility of the Grey Partridge (*Perdix perdix*) *J. Reprod, Fert. Suppl.* **19**, 391–402.

POTTS G.R. (1974) The Grey Partridge; problems of quantifying the ecological effects of pesticides. *Proc. International Congr. of Game Biologists* **11**, 405–13.

POTTS G.R. & VICKERMAN G.P. (1974) Studies on the cereal ecosystem. *Adv. Ecol. Res.* 8, 107–97.

RATCLIFFE D.A. (1970) Changes attributable to pesticides in egg breakage frequency and eggshell thickness in some British Birds. *J. appl. Ecol.* 7, 67–107.

SNOW B.K. (1960) The breeding biology of the Shag on the Island of Lundy. *Ibis* 102, 554–75.

VICKERMAN G.P. (1974) Some effects of grass weed control on the arthropod fauna of cereals. *Proc. 12th Br. Weed Control Conf.* 3, in press.

Appendix 1

Details of broken eggs of the Shag, *Phalacrocorax aristotelis*, Farne Islands, 1963–1964.

Breakage resulting from:	*Number in category*
1. Extremely thin shell	12 } 49
2. Cracked in nest during incubation	37 }
3. Simultaneous polygyny*	58
4. Possibly accidental, found on edge or very near nest	58
5. Poor nest, laid on rock etc.	43
6. Found away from nest, possibly broken by Gulls	20 approx.
	228

Notes

a. The loss in category 3 is clearly underestimated since many eggs were lost but not found broken, also losses in category 6 exclude eggs which may have been removed from the colony.
The total number of eggs concerned was 1,541 of which 1,038 were accounted for in one way or another.
Minimum percentage of cracked eggs (*i.e.* excluding categories 3 to 6 = 49/1,038 = 4·7%.

b. The whole egg concentration of dieldrin was 0·63–1·19 (95% confidence limits).
The whole egg concentration of DDE was 0·67–1·37 (95% confidence limits)

c. Shell weights down 17% (Ratcliffe 1970).

d. There are no similar data for less contaminated populations of Shag.

* See Potts (1968)

The dispersal and persistence of p,p′-DDT*

F. MORIARTY *Monks Wood Experimental Station, Abbots Ripton, Huntingdon, U.K.*

Introduction

DDT was first synthesized by Zeidler in 1874. Its insecticidal properties were discovered by Müller in 1939, and its large-scale use as an insecticide started during the Second World War. Its major uses are for the control of insects in agriculture, forestry, public health, and in some industrial processes. In most of these applications DDT is released deliberately into the environment, unlike many other potential pollutants, which, given the will, can, at a cost, be used without escaping into the environment.

This paper will discuss where DDT occurs, attempt tentative explanations for the pattern of distribution, and, in particular, consider what is meant by the term *persistence*. These topics have an intrinsic interest, but they also have a much wider significance: many of the ideas that can be developed about DDT as a pollutant are probably applicable to other pollutants too.

Distribution in the physical environment

The first obvious question is, how much DDT is manufactured and used? It is impossible to quote firm figures for the annual production of DDT throughout the world. Manufacturers make annual returns in the U.S.A., but not in this or many other countries. It can be seen (Fig. 1) that production in the U.S.A. reached a peak in 1963, when it was $8{\cdot}13 \times 10^{10}$g. Probably about half of the world's total production comes from the U.S.A., and perhaps 70% of the American production is exported (Woodwell *et al.* 1971).

* Usually p,p′-DDT will be referred to simply as DDT in the text. The o,p′-isomer is not included in this abbreviated designation.

Comparable amounts of DDT/square mile appear to have been used in England and Wales in the mid-1960s. The principal uses have been for agriculture and horticulture, where it is estimated that in 1962–4 249 tons ($2\cdot53 \times 10^8$g) were used annually (Strickland 1965). Use was very patchy, with 81% being applied to 128,000 acres of top fruit. The total land area of England and Wales is 1·6% of that in the U.S.A., so a proportional quantity there would be $1\cdot58 \times 10^{10}$g.

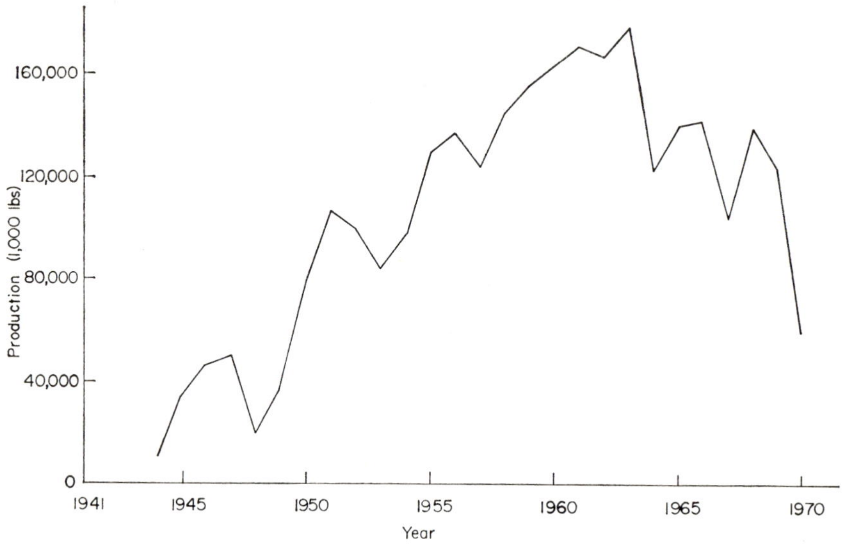

Figure 1. Figures from the annual reports of the 'United States Tariff Commission. United States Production and Sales.' Production figures for DDT before 1944 and after 1970 are included with those for other synthetic insecticides.

Future production figures are difficult to guess, but the pattern of use is changing. Political pressures have already reduced usage in parts of Europe and North America, whereas amounts are likely to increase in some less developed parts of the world.

AMOUNTS IN, AND LOSS FROM, SOIL

Much of this DDT is sprayed on plants and soil. DDT has a very low vapour pressure, of $1\cdot5 \times 10^{-7}$mm mercury at 20°C (Balson 1947), and is not very soluble in water. The solubility is 1·2 parts in 10^9 at 25°C (Bowman *et al.* 1960). One might expect therefore that most of the DDT would remain in or on the soil until either it was altered chemically or until it was removed by soil erosion.

Certainly DDT stays in the soil for a long time. Edwards (1966) has reviewed the extensive data on this subject, and an average figure for the so-called 'half life' in temperate soils is 2–3 years. There are many factors that can alter this figure, and in one extreme case, with a dry sandy loam soil, it took 17 years for 60% of the DDT to disappear (Nash & Woolson 1967).

The question then is, what happens to the DDT that disappears? Many factors play a part (Edwards 1970), but probably only a few are of major importance. At one time it was widely believed that, at least for agricultural soils, much of the DDT was transported in surface water, either in solution or absorbed on soil particles, into the rivers and thence to the seas and oceans. In some situations run-off can doubtless be a major cause of loss, but it seems unlikely to be of wide-spread importance. For example, the concentration of DDT in unfiltered river water in the U.S.A. during the mid-1960s appears to have ranged from below the limit of detection (about 10 pp 10^{12}) to 100 pp 10^{12} (Weaver *et al.* 1965; Breidenbach *et al.* 1967; Bailey *et al.* 1967). Higher concentrations are not often recorded. Amounts decreased sharply after 1966 (Lichtenberg *et al.* 1970). If the average concentration of DDT in water draining off the land into the rivers were 50 pp 10^{12}, this would imply that the annual rainfall in the U.S.A. would remove, by run-off, about 0·1% of the annual production of DDT in the U.S.A. (Woodwell *et al.* 1971).

Some analyses have been made for British rivers too. 'Clean' Scottish streams had up to 5 pp 10^{12} of DDE, with many samples containing less than a tenth of this concentration. DDT could not be detected (Holden & Marsden 1966). Similar results were obtained by Croll (1969) for rivers in Essex and Kent, where DDT and DDE could only be detected at times of heavy spraying in local orchards (limit of detection about 10 pp 10^{12}). One interesting result was that the concentrations of α and γ BHC, and of dieldrin, which could be detected most of the time, only fluctuated within a factor of two whilst the rate of flow in the rivers varied by 140-fold. Presumably the same situation applies to DDT, but concentrations in water were too low to test this. However, mud from one of these rivers, the Chelmer, contained 10,000–40,000 pp 10^{12} of DDT, including its metabolites DDD and DDE (Anon. 1971). So it may be difficult to assess, from concentrations in river water, the rate of loss from agricultural soils.

It is likely that rivers are important for the transport of DDT in industrial effluents. For example, rivers in Yorkshire and Lancashire had, in 1966, appreciably higher concentrations of DDT and DDE than occurred in the major river basins of the U.S.A. in 1965 (Lowden *et al.* 1969): many of the high concentrations, up to 250 pp 10^{12} of DDT, were associated with industrial effluents.

It is now becoming accepted that the major cause for the disappearance of DDT from the soil must be volatilization into the air. Much of the DDT that is, in theory, applied to the land volatilizes at once, and so never reaches the site of application, or volatilizes into the air again soon after application. A figure of 50% is commonly quoted for the fraction that never reaches the ground.

The rate at which DDT that has been deposited on the soil evaporates into the air is determined by its saturation vapour concentration and by its rate of diffusion through air close to the substance (Hartley 1969). Lloyd-Jones (1971) calculated, both from theory and by experiments on the evaporation of DDT from aluminium planchets, that about 2 lb of DDT could volatilize from each acre each year at summer temperatures. The figure drops to about 0·3 lb/acre for winter temperatures. Freed *et al.* (1972) argue that evaporation from the more absorbent structure that soils provide is likely, at the most, to be only a few tenths of a pound of DDT/acre/year. They suggest, from theory and experiment, that this rate is too slow to explain the loss of DDT from the soil, and that it is first converted to DDE and other related compounds, which volatilize more rapidly than does DDT. Many microorganisms can metabolize DDT (Spencer 1967), and the vapour pressure of DDE is almost an order of magnitude greater than that of DDT (Spencer & Cliath 1972).

However, it is not clear whether sufficient allowance has been made for the influence of soil moisture: water has two very important effects. First, many soil particles absorb DDT, and are also strongly hydrophilic. Such soils, when dry, absorb DDT to such an extent that the rate of evaporation is reduced (Spencer & Cliath 1972). When the soil is wetted again the DDT is displaced by water, for which the soil particles have a greater affinity.

Second, soil can be considered as a mass of capillaries. Water flows up these capillaries in moist soil to replenish the water as it evaporates from the soil surface. As the soil water moves up to the soil surface, so will dissolved DDT, and so the concentration of DDT at the surface will tend to increase. Diffusion of DDT down the concentration gradient into the soil again is too slow to counteract this mass flow, and so the DDT will tend to be lost from the soil surface, with the evaporating water, in the same ratio as in the undisturbed soil water (Hartley 1969). The depth from which the water migrates is critically important. For a depth of 1 cm the concentration at the surface increases twenty-fold over the original concentration. For a depth of 2 cm the increase is 400-fold. Presumably this mechanism explains why dry quartz sand retains all of its 10 ppm DDT after three weeks in an incubator at 37°C, whereas 17·1 ± 3·0% was lost from quartz sand maintained at or near field capacity for 56 days (Lichtenstein & Schulz 1961).

This wick evaporation is not to be confused with the idea of codistillation, propounded by Bowman *et al.* (1959), and by Acree *et al.* (1963). They suggested that the evaporation of water speeds up the diffusion of DDT from the soil, but their conclusion is based on incorrect values for the effective vapour pressure and on indirect measurements of the amount of DDT lost, and does appear to be wrong (Hartley 1969).

AMOUNTS AND TRANSPORT IN AIR

The evidence suggests then that most of the DDT in soil is lost by evaporation into the air, although it is uncertain how much is first metabolized to related compounds such as DDE. Several analyses have been made of air and rain-water, and in general they contain higher concentrations of DDT than of DDE.

Air is potentially a large reservoir for DDT. If DDT reached its equilibrium concentration as vapour in the air, an air pressure of 76 cm mercury would indicate 3×10^{-6} g/m^3 DDT. The saturation capacity of the atmosphere to the tropopause for DDT vapour is probably equivalent to the total amount of DDT produced so far (Woodwell *et al.* 1971). In practice, much lower concentrations occur. A sample of London air taken in August 1965, contained $3{\cdot}6 \times 10^{-9}$ g/m^3 (3 pp 10^{12} w/w) of DDT, of which an indeterminate amount was associated with particulate matter (Abbott *et al.* 1966). One of the highest observed concentrations must be the value of $1{\cdot}56 \times 10^{-6}$ g/m^3 over an agricultural area after spraying, but again most of the DDT was associated with particles (Stanley *et al.* 1971).

Atkins & Eggleton (1971) produced data to show that, over London, much of the DDT is probably associated with particles, whereas γ BHC and dieldrin occur mainly in the gaseous form. They did suggest, however, from the concentrations of DDT in rain falling in Wellesbourne, Warwickshire, and the concentration in air in London, that much of the DDT in the air is usually present as vapour. The argument is weak: it rests on the assumption that air in Wellesbourne and in London had similar concentrations of DDT vapour. Nevertheless, it is important that we check whether DDT is normally present as vapour or attached to particles in the air. The residence time of DDT vapour in the atmosphere may be quite different from that for DDT attached to particles.

There is good evidence that both gases and particles in the atmosphere can be transported great distances before they are deposited on the earth's surface again (Goldberg 1971). Data collected after a dust storm in the U.S.A. showed that organochlorine insecticides in the air could travel 1,000 miles or so (Cohen & Pinkerton 1966), but an even more striking example was described by Risebrough *et al.* (1968). They used nylon collecting screens to catch air-borne particles in Barbados. The screens had a collecting efficiency of about 50% for particles $> 1\ \mu$ diameter, and both the mineralogical and biological evidence suggested that most of the particles had come, in the equatorial easterlies, from Europe and Africa, which are over 6,000 km away. The average combined concentration of dieldrin, p,p′-DDT, DDD, p,p′-DDE and o,p′-DDT, attached to the dust, was $7{\cdot}8 \times 10^{-14}$ g/m^3 of air. The major part of the total was p,p′-DDT.

It has been calculated that dust settles on the tropical Atlantic at a rate of 0·6 μ/year. If the Barbados sample is representative of this dust, it indicates that 600 kg of pesticide are deposited each year into the Atlantic between the equator and 30°N. The authors point out that this is an underestimate—no allowance is made for DDT attached to small particles, or present as vapour. There is perhaps the risk too that some DDT is lost by volatilization from the particles after they have been trapped on the screens (Antommaria *et al.* 1965). It is also possible that the amount of DDT deposited decreases with distance from its origin, and Barbados is towards the end of the north-east trade winds' sweep across the Atlantic.

Sweden may provide the most striking example of the importance of aerial transport. There is believed to be more DDT in Swedish soils than has ever been used in Sweden (Odén 1972).

If we can assume that most of the DDT in the air is returned to the globe's surface by rain and snow (Abbott *et al.* 1965) then estimates of amounts of precipitation and of concentrations of DDT should enable us to assess the rate at which DDT is being transported in the air. So far we have very few data. Several analyses have been made in Great Britain (Wheatley & Hardman 1965; Abbott *et al.* 1965; Tarrant & Tatton 1968). The concentrations found were not obviously related to sampling sites, and in the last, most comprehensive, survey the range of mean concentrations was 18–66 pp 10^{12} of DDT, and 7–28 pp 10^{12} of DDE. Samples from three sites in the State of Ohio, U.S.A. contained 70–340 pp 10^{12} of DDT, and 5–30 pp 10^{12} of DDE (Cohen & Pinkerton 1966). By contrast, more recent samples of rainfall in Hawaii contained 3–4 pp 10^{12} of DDT (Bevenue *et al.* 1972). The Antarctic is presumably one of the least contaminated parts of the world, and five samples of melt-water from snow fallen within the previous two months (1966–7) contained from less than the limit of detection (5 pp 10^{12}) to more than 28 pp 10^{12} (Peterle 1969).

Woodwell *et al.* (1971) take 60 pp 10^{12} as an average global figure for the concentration of DDT in rainfall. This suggests, with assumptions about mean annual rainfall, that 3×10^{10}g DDT are removed annually from the air, which implies that about 30% of the world's annual production in the mid-1960s is deposited by rainfall each year. Obviously calculations of this sort involve large assumptions (Woodwell *et al.* 1972) and Stewart (1972) argues that relatively small amounts of DDT reach the oceans by aerial transport. Certainly the average figure for rainfall of 60 pp 10^{12} seems rather high. Comparison of the available data (Bevenue *et al.* 1972) suggests that it may be at least an order of magnitude too high for areas away from heavy applications. It does seem more likely that the concentration of DDT decreases away from major sites of application. Thus London air in 1965 contained nearly 10^5 times as high a concentration of DDT as the air reaching Barbados in 1965–6 (Abbott *et al.* 1966; Risebrough *et al.* 1968).

So we may suggest, tentatively, that most of the DDT in the air is associated with particles, that most of the DDT is returned to the globe's surface by rain and snow, that it can be transported enormous distances in the air, but that the major part is deposited relatively close to its point of origin. Obviously we need far more information before we can hope to produce a quantitative account. We cannot yet say how much of the World's annual production of DDT could be expected to volatilize into the air and be deposited again by rainfall each year. However, it is conceivable that a significant part of the total amount of DDT has not been accounted for. It is usually assumed that molecules of DDT are stable whilst in the atmosphere. Maugh (1973) has recently described some work by Moilanen & Crosby, first reported at the 165th meeting of the American Chemical Society, which shows that when DDT vapour is irradiated with ultra-violet light (290–310 nm) some of it is converted to DDE, which can then be transformed to various chlorinated biphenyls, which are quite inert. A little DDT was converted to DDD, which was also inert. It is at present an open question whether this is an important reaction outside of the laboratory, where conditions can be rather different (Rosen 1971). PCBs have of course been appreciated as pollutants in their own right for some years now (Jensen 1972).

Eventually much of the DDT that has been transported must end up in the oceans. It has not yet, so far as I know, been detected in ocean water. Much of it may well be associated with lipid layers (Seba & Corcoran 1969; Duce *et al.* 1972) or particles (Cox 1971) and may finally penetrate with organic sediments through the thermocline and be deposited at the bottom of the abyss.

Amounts in animals

So far, in this paper, any influence that living organisms may have on the distribution of DDT has been ignored. One might expect, since many organisms contain far higher concentrations of DDT than occur in the physical environment, that they would have an appreciable effect, but it has been calculated that the biota contain altogether about as much DDT as was produced in 3–4 days during the mid-1960s (Woodwell *et al.* 1971). To express it simply, DDT has important effects on organisms, but organisms do not appear to have significant effects on the distribution of DDT.

In contrast to the available information on distribution and movement in the physical environment, we have far more information about amounts in animals. Two reasons could be that detection is relatively easy, and that DDT is primarily of interest because of its biological effects. Several features about the results of residue analysis are noteworthy.

METABOLISM

Most of the p,p′-DDT that is retained is usually metabolized. The commonest metabolite to be found is usually p,p′-DDE: wildlife specimens commonly contain more DDE than DDT. There are many other metabolic pathways too (O'Brien 1967), and care has to be taken when reading the literature to note whether residue values refer specifically to p,p′-DDT, or to p,p′-DDT plus all of the commonly found metabolites and breakdown products.

Two controversies about the metabolism of DDT are relevant. The major impurity in technical grade DDT is the isomer o,p′-DDT, which may comprise 8–21% of the total product (Negherbon 1959). Klein *et al.* (1964) concluded, from analyses after feeding o,p′-DDT to rats, that the o,p′ isomer could be converted to the p,p′ isomer. This is most unlikely: o,p′-DDT is rapidly metabolized and then excreted. It is more likely that the p,p′-DDT was a minor impurity of the o,p′-DDT that was much more persistent in the rats' tissues than was the o,p′-DDT (Bitman *et al.* 1971).

The second controversy concerns p,p′-DDD (p,p′-TDE). Before the development of gas-liquid chromatography for the detection of organochlorine insecticides, these compounds were detected by the Schechter-Haller colorimetric test, which does not distinguish DDD from DDT. With the advent of gas-liquid chromatography in the 1960s, DDD was often found in specimens, and it was suggested that it was a common metabolite of DDT (Finley & Pillmore 1963). It is reasonably well established that p,p′-DDT can be converted to p,p′-DDD in the anaerobic conditions that can occur in muds and soils (Hassall & Forrest 1972). Similarly it has been suggested from experiments with mice carcasses and with livers of Bengalese Finches that the DDD found in wildlife specimens is formed from DDT after death (Barker & Morrison 1964; Jefferies & Walker 1966). However, Bailey *et al.* (1969) produced some contradictory data. They fed p,p′-DDT to feral pigeons for 24 days, and birds were killed for analysis 1–274 days later. The highest concentrations of DDD residues occurred in the livers, where, soon after the end of exposure, they were about 1 ppm. Other tissues contained less, and none could be detected in the omental fat. The authors concluded that this DDD was formed, as a metabolite of DDT, by the pigeons. But it is difficult, from their results, to exclude post-mortem changes as a sufficient explanation. They also found that when pigeons were fed with DDD the highest concentrations occurred in omental fat. Therefore, if DDD is formed during life, one would expect to detect some residues in the omental fat. The decrease in DDD concentrations with increase of time between exposure and death is not inconsistent with post-mortem conversion—the amount produced could depend on the amount of DDT present, which also decreased with time between exposure and death.

EXPOSURE AND RESIDUE

Different tissues within an organism have different concentrations of DDT and of its metabolites. Adipose tissue usually has by far the highest concentration, and differences between tissues can be as much as 1,000 fold. These differences have stimulated the use of compartmental models for quantitative explanations of the relationships between exposure and the residues found in different tissues (Fig. 2) (Moriarty in press). A compartment is defined as a quantity of DDT, or any other residue, that has uniform and distinctive kinetics of transport to and from that compartment and of metabolism.

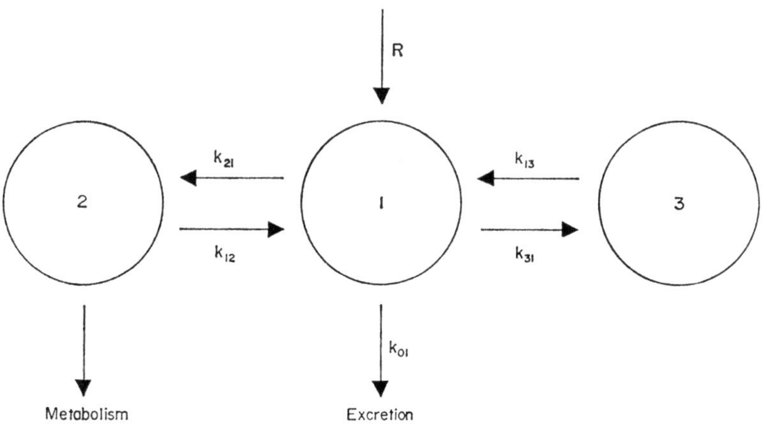

Figure 2. Model for the distribution of DDT within an organism that is considered to consist of three compartments. DDT is absorbed into the blood (compartment 1) at a steady rate R. Most metabolism occurs in the liver (compartment 2). Rates of transfer between compartments are indicated by the rate constants (k).

Residues enter the blood (Compartment 1) from the physical environment or in the food, and residues transfer between the other compartments within the body via the blood. Excretion occurs from the blood, whilst metabolism, in vertebrates, occurs predominantly in the liver (compartment 2). In practice, so far, no data have been fitted to anything more complicated than a two-compartment model, which must be an over-simplified picture of the real situation. Even so, several important conclusions have been reached: (1) Residues do not increase indefinitely when there is a constant, chronic, exposure. Eventually, a steady state is reached, when the amounts of residues remain constant.

Our own species provides some anomalies. American convicts were given one of four doses of DDT for 21·5 months: 0, 3·5 mg technical grade, 35 mg technical grade or 35 mg recrystallized DDT/man/day (Hayes *et al.* 1971). Fat samples were taken before exposure, and 12·2, 18·8 and 21·5 months after exposure began. The analytical results suggested that the concentration of DDT increased continuously in all four groups. The rise in the controls and in those receiving 3·5 mg DDT/day could be ascribed to receiving the wrong dose occasionally, but this cannot explain the continued rise for the two groups receiving 35 mg/day. Hayes *et al.* (1971) concluded that a steady state was probably reached within 18·8 months, because the values then were not significantly different from those taken 2·7 months later. This interpretation could be correct, but at first sight, without any preconceived ideas, the data appear to suggest a fairly steady rise with time (Fig. 3). This latter possibility is greatly strengthened by a similar experiment with dieldrin, another organochlorine insecticide. Adult male volunteers ingested 50 or 211 μg of HEOD (the active principle of dieldrin) per day for two years, and blood samples were taken at intervals for analysis (Hunter *et al.* 1969). Here too it seems, contrary to the original interpretation, that the concentration rises fairly steadily (see Fig. 4) during the whole two years' exposure (Moriarty 1974). This is not to imply that the size of these residues in man is steadily rising—the evidence suggests that, in the United Kingdom, residues have decreased since about 1965 (Abbott *et al.* 1972)—but the experimental results do deserve further investigation.

There is one other very long term experiment with DDT, in which rhesus monkeys were fed DDT for up to 7½ years (Durham *et al.* 1963). The concentration of DDT in body fat reached a steady state within 6 months, and there was little evidence of any subsequent major changes. In contrast, when beagle hounds were fed dieldrin a steady-state concentration was attained in the blood, but after 18 months' exposure the concentration of dieldrin in the blood started to rise again (Moriarty 1972). So it is an open question whether steady states persist during very long term chronic exposure.

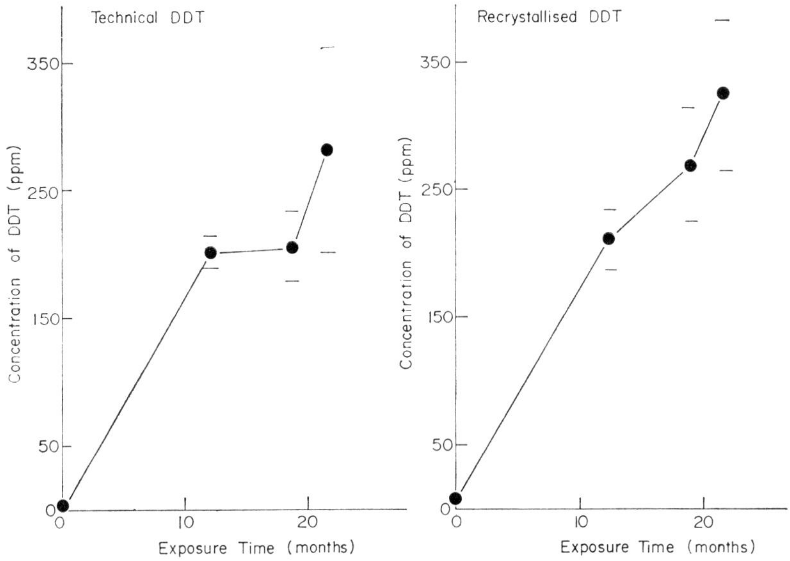

Figure 3. Amounts of DDT found in the body fat of adult men whilst ingesting 35 mg DDT (technical grade or recrystallised)/day. Standard errors are indicated by the horizontal bars (data from Hayes, Dale & Pirkle 1971).

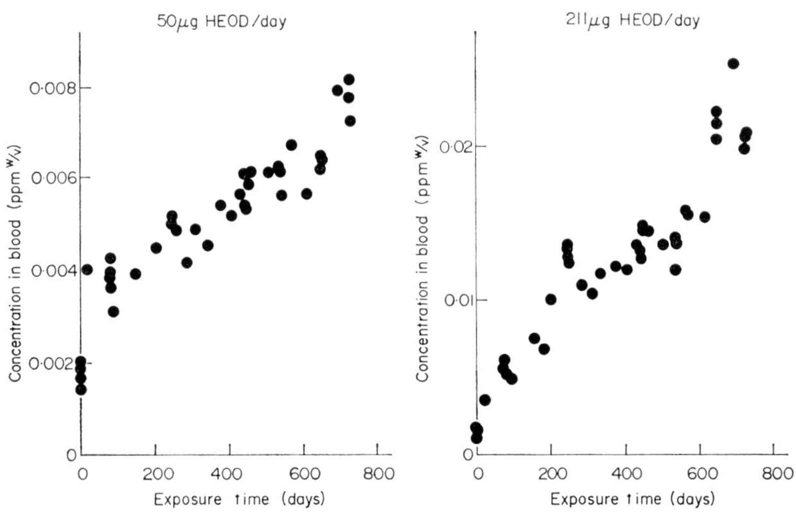

Figure 4. Amounts of HEOD found in the blood of adult men whilst ingesting 50 or 211 μg HEOD/day. Each point represents the mean value for samples taken from three men (data from Hunter, Robinson & Roberts 1969).

(2) It has become widely accepted that persistent pesticides such as DDT concentrate and accumulate along food chains (e.g. Carson 1962). However, the amount of residue is the balance between that absorbed and that lost by excretion and metabolism. We cannot therefore accept without question the idea that the amount of DDT, or of any other pesticide, depends on a species position in the food chain. In fact, the evidence suggests that aquatic species obtain their residues by direct uptake from the physical environment, and that any residue consumed with the food is of minor significance (Moriarty 1972). The only likely exception would seem to be a situation where the prey has a much greater direct exposure than the predator.

Terrestrial predators do presumably obtain their residues via food, and there is a considerable amount of field evidence to show that terrestrial predators do have higher residues than their prey species (e.g. Moore 1965). The compartmental model suggests that if the original explanation, of concentration along the food chain, be correct it should be restated: when species have reached steady-state concentrations of DDT, then comparable tissues contain higher concentration as one moves along a food chain.

Let us imagine two species, P_1 and P_2. Individuals of both species have reached steady state concentrations of DDT in their tissues, with mean body concentrations of C_1 and C_2 ppm. Individuals of species P_2 feed only on individuals of species P_1, their food weighs x% of their body weight per day, they lose y% of their DDT residues/day, and they have a body weight W_2. The daily intake of DDT by individuals of species P_2 must equal the daily loss, if the steady state is to be maintained.

Therefore

$$x.C_1W_2 = y.C_2W_2$$

or

$$\frac{C_2}{C_1} = \frac{x}{y}$$

This equation suggests that successive predators in a food chain will have higher residues if their food intake, expressed as a % of their body weight, is greater than the turnover of their residues, expressed as a % of the residues present. Conversely, if turnover exceeds intake, the predator will have lower residues than its prey. In practice of course animals consist of more than one compartment, but the same principle still applies, although the details become more complicated.

Estimates of food consumption in the field are difficult to obtain, but a figure of 10% of body weight/day is probably too low for many predators. A turnover for DDT residues of 10%/day implies a 'half life' of 6·9 days. A quicker turnover would imply an even shorter 'half life'. The available data suggest that the 'half life' for DDT is often longer than 10 days (Moriarty in press). These calculations do of course involve various assumptions, but it seems reasonable to conclude that, given a steady-state situation, residues will often increase as one passes along a terrestrial food chain, provided that exposure is chronic and that a steady state exists. Certainly this calculation should help to emphasize that the important determinants of the steady-state residues in any species are its exposure and its rate constants for excretion and metabolism.

I should like to make just three comments about the interpretation of field data:

(i) Residue size depends on the period of exposure. Eventually a steady state will be reached. Comparisons of residues between species are only meaningful if one knows how near the steady state each species is. One never does know with field specimens. In practice exposures are likely to be intermittent, so that an approximation to a steady state may take a long time. Successive predators in a food chain tend to be larger and longer lived, and may have larger residues because of this.

(ii) Predators may, as a group, be less efficient than herbivores at disposing of foreign compounds like DDT. Wit (1972) suggests that herbivorous birds may have evolved more effective detoxication mechanisms than predatory birds. Other things being equal, predators would then have higher residues than herbivores.

(iii) Predators may select the most heavily contaminated individuals of their prey species (Moriarty 1972).

(3) If a population of animals has attained steady-state concentrations of residues during chronic exposure, then one might expect analytical results to have a normal distribution. In practice, analyses of wildlife populations often have a positive skew (Moriarty 1972). Apart from any questions of bias, it is likely in fact that exposure is often intermittent, when residues may be declining most of the time, with occasional sudden increases. Loss of residues after exposure often appears to follow a simple exponential decline, and this alone would be sufficient to explain a skew distribution if exposures are intermittent. However, it is possible to show, with a compartmental model, that the loss of residues from any compartment after exposure can be expressed by the equation

$$C_t = X_1e^{-\lambda_1 t} + X_2e^{-\lambda_2 t} \dots X_ne^{-\lambda_n t}$$

where C_t is the concentration at time t after the end of exposure (Moriarty in press). The number of exponential terms depends on the number of compartments. The relevance of this equation is that, if the compartmental model be appropriate, then residues are lost inevitably in successively slower phases, and a small part of the original residue will be exceedingly persistent. This occurrence of phases could accentuate the tendency for a skew distribution of residues.

One of the outstanding features of DDT, and especially of its metabolite DDE, is its widespread occurrence in wildlife specimens. This has caused some surprise, despite DDT's persistence in the physical environment. The relatively great persistence of a small part of the residues may be part of the reason for the widespread occurrence of residues.

Discussion

It is obvious that more quantitative information is needed, about the occurrence and distribution of DDT in the physical environment, and about the relationships between exposure and residues, before we can answer some important practical questions about the possible risks of using DDT on a large scale. One fruitful approach is to consider both organisms and the physical environment as a series of compartments, and to estimate the sizes of these compartments and the rates at which DDT enters and leaves them.

Summary

(1) Data on the amounts manufactured and used are very inadequate.

(2) Rivers can contain large residues of DDT from industrial effluents.

(3) Most of the DDT that is applied to soil is probably lost by volatilization into the air.

(4) Most DDT in the atmosphere is probably associated with particles, is returned to the earth's surface by rain and snow, and can travel enormous distances in the air.

(5) Some of this DDT vapour may be converted by ultra-violet light to DDE and PCBs.

(6) Conversion (metabolism) of DDT is a very important process in animals.

(7) Residues within animals are the balance between uptake and loss, and can usefully be described with compartmental models.

(8) Chronic exposure leads to a steady-state situation. However, there is some doubt whether these steady states will persist indefinitely.

(9) Predators will tend to have higher residues than their prey provided that their food intake, as a percentage of the body weight, exceeds the percentage turnover of their residues. Many factors can affect the truth of this statement, including the length of exposure, rates of detoxication and selection of prey.

(10) Residues are lost in successively slower phases after exposure, so that a small part of the total residues will be more persistent than the rest.

References

ABBOTT D.C., HARRISON R.B., TATTON J.O'G. & THOMSON J. (1965) Organochlorine pesticides in the atmospheric environment. *Nature, Lond.* 208, 1317–8.

ABBOTT D.C., HARRISON R.B., TATTON J.O'G. & THOMSON J. (1966) Organochlorine pesticides in the atmosphere. *Nature, Lond.* **211**, 259–61.

ABBOTT D.C., COLLINS G.B. & GOULDING R. (1972) Organochlorine pesticide residues in human fat in the United Kingdom 1969–1971. *Br. med. J.* 1972, 553–6.

ACREE F., BEROZA M. & BOWMAN M.C. (1963) Codistillation of DDT with water. *J. agric. Fd Chem.* 11, 278–80.

ANON. (1971) *Wat. Pollut. Res.* 1970.

ANTOMMARIA P., CORN M. & DEMAIO L. (1965) Airborne particulates in Pittsburgh: association with p,p'-DDT. *Science, N.Y.* **150**, 1476–7.

ATKINS D.H.F. & EGGLETON A.E.J. (1971) Studies of atmospheric wash-out and deposition of γ-BHC, dieldrin and p-p DDT using radio-labelled pesticides. *Nuclear Techniques in Environmental Pollution*, Proc. Symposium IAEA, Salzburg.

BAILEY S., BUNYAN P.J., RENNISON B.D. & TAYLOR A. (1969) The metabolism of 1,1-Di (*p*-chlorophenyl)-2,2,2-trichloroethane and 1,1-Di(*p*-chlorophenyl)-2,2-dichloroethane in the Pigeon. *Toxic. appl. Pharmac.* **14**, 13–22.

BAILEY T.E., ASCE A.M. & HANNUM J.R. (1967) Distribution of pesticides in California. *J. Sanit. Eng. Div.* 93, 27–43.

BALSON E.W. (1947) Studies in vapour pressure measurement. III. An effusion manometer sensitive to 5×10^{-6} millimeters of mercury: vapour pressure of DDT and other slightly volatile substances. *Trans. Faraday Soc.* 43, 54–60.

BARKER P.S. & MORRISON F.O. (1964) Breakdown of DDT to DDD in mouse tissue. *Can. J. Zool.* 42, 324–5.

BEVENUE A., OGATA J.N. & HYLIN J.W. (1972) Organochlorine pesticides in rainwater, Oahu, Hawaii, 1971–1972. *Bull. environ. Contam. Toxicol.* 8, 238–41.

BITMAN, J., CECIL H.C. & FRIES G.F. (1971) Nonconversion of o,p'-DDT to p,p'-DDT in rats, sheep, chickens and quail. *Science, N.Y.* 174, 64–6.

BOWMAN M.C., ACREE F., SCHMIDT C.H. & BEROZA M. (1959) Fate of DDT in larvicide suspensions. *J. econ. Ent.* 52, 1038–42.

BOWMAN M.C., ACREE F. & CORBETT M.K. (1960) Solubility of carbon-14 DDT in water. *J. agric. Fd Chem.* 8, 406–8.

BREIDENBACH A.W., GUNNERSON A.B., KAWAHARA F.K., LICHTENBERG J.J. & GREEN R.S. (1967) Chlorinated hydrocarbon pesticides in major river basins, 1957–1965. *Public Health Reports* 82, 139–56.

CARSON R. (1962) *Silent Spring*. Houghton Mifflin, Boston.

COHEN J.M. & PINKERTON C. (1966) Widespread translocation of pesticides by air transport and rain-out. In *Organic Pesticides in the Environment* (Ed. by A.A. Rosen & H.F. Kraybill), *Adv. Chem. Series* 60, *Am. chem. Soc.*

COX, J.L. (1971) DDT residues in seawater and particulate matter in the California current system. *Fish. Bull. U.S.D.C.* 69, 443–50.

CROLL B.T. (1969) Organochlorine insecticides in water. I. *J. Soc. Wat. Treat. Exam.* 18, 255–74.

DUCE R.A., QUINN J.G., OLNEY C.E., PIOTROWICZ S.R., RAY B.J. & WADE T.L. (1972) Enrichment of heavy metals and organic compounds in the surface microlayer of Narrogansett Bay, Rhode Island. *Science, N.Y.* 176, 161–3.

DURHAM W.F., ORTEGA P. & HAYES W.J. (1963) The effect of various dietary levels of

DDT on liver function, cell morphology, and DDT storage in the rhesus monkey. *Archs. int. Pharmacodyn. Thér.* **141**, 11–129.

EDWARDS C.A. (1966) Insecticide residues in soils. *Residue Rev.* **13**, 83–132.

EDWARDS C.A. (1970) Persistent pesticides in the environment. *Crit. Rev. environ. Contr.* **1**, 7–67.

FINLEY R.B. & PILLMORE R.E. (1963) Conversion of DDT to DDD in animal tissue. *BSCS Bull.* **13**, 41–2.

FREED, V.H., HAQUE R. & SCHMEDDING D. (1972) Vaporization and environmental contamination by DDT. *Chemosphere* **1**, 61–6.

GOLDBERG E.D. (1971) Atmospheric transport. In *Impingement of Man on the Oceans* (Ed. by D.W. Hood), Wiley, New York.

HARTLEY G.S. (1969) Evaporation of pesticides. In *Pesticidal Formulations Research—Physical and Colloidal Chemical Aspects*, *Adv. Chem. Ser.* 86.

HASSALL K.A. & FORREST T.J. (1972) Reductive dechlorination of DDT by heated liver. *Nature, Lond.* **236**, 214–16.

HAYES W.J., DALE W.E. & PIRKLE C.I. (1971) Evidence of safety of long-term, high, oral doses of DDT for man. *Archs environ. Hlth* **22**, 119–35.

HOLDEN A.V. & MARSDEN K. (1966) The examination of surface waters and sewage effluents for organo-chlorine pesticides. *J. Proc. Inst. Sew. Purif.* 4, 295–9.

HUNTER C.G., ROBINSON J. & ROBERTS M. (1969) Pharmacodynamics of dieldrin (HEOD). Ingestion by human subjects for 18 to 24 months, and postexposure for eight months. *Archs environ. Hlth* 18, 12–21.

JEFFERIES D.J. & WALKER C.H. (1966) Uptake of pp'-DDT and its post-mortem breakdown in the avian liver. *Nature, Lond.* **212**, 533–4.

JENSEN S. (1972) The PCB story. *Ambio* **1**, 123–31.

KLEIN A.K., LAUG E.P., DATTA P.R., WATTS J.O. & CHEN J.T. (1964) Metabolites: reductive dechlorination of DDT to DDD and isomeric transformation of o,p'-DDT to p,p'-DDT *in vivo*. *J. Ass. off. agric. Chem.* **47**, 1129–45.

LICHTENBERG J.L., EICHELBERGER J.W., DRESSMAN R.C. & LONGBOTTOM J.E. (1970) Pesticides in surface waters of the United States—a 5-year summary, 1964–68. *Pestic. Monit. J.* 4, 71–86.

LICHTENSTEIN E.P. & SCHULZ K.R. (1961) Effect of soil cultivation, soil surface and water on the persistence of insecticidal residues in soils. *J. econ. Ent.* **54**, 517–22.

LLOYD-JONES C.P. (1971) Evaporation of DDT. *Nature, Lond.* **229**, 65–6.

LOWDEN G.F., SAUNDERS C.L. & EDWARDS R.W. (1969) Organo-chlorine insecticides in water. II. *J. Soc. Wat. Treat. Exam.* 18, 275–87.

MAUGH T.H. (1973) DDT: an unrecognised source of polychlorinated biphenyls (PCBs). *Science, N.Y.* **180**, 578–9.

MOORE N.W. (1965) Pesticides and birds—a review of the situation in Great Britain in 1965. *Bird Study* **12**, 222–52.

MORIARTY F. (1972) The effects of pesticides on wildlife: exposure and residues. *Sci. Ttl Environ.* **1**, 267–88.

MORIARTY F. (in press) Exposures and residues. In *Organochlorine Insecticides: Persistent Organic Pollutants* (Ed. by F. Moriarty), Academic Press, London.

MORIARTY F. (1973) Residues in animals during chronic exposure to dieldrin. *Environmental Quality and Safety* 3, 104–12.

NASH R.G. & WOOLSON E.A. (1967) Persistence of chlorinated hydrocarbon insecticides in soils. *Science, N.Y.* **157**, 924–7.

NEGHERBON W.O. (1959) Handbook of Toxicology. Vol. III: Insecticides. Saunders, Philadelphia.

O'BRIEN R.D. (1967) *Insecticides: Action and Metabolism.* Academic Press, New York.

ODEN S. (1972) The extent and effects of atmospheric pollution on soils. *FAO Soils Bull.* 16.

PETERLE T.J. (1969) DDT in antarctic snow. *Nature, Lond.* **224**, 620.

RISEBROUGH R.W., HUGGETT R.J., GRIFFIN J.J. & GOLDBERG E.D. (1968) Pesticides: transatlantic movements in the north east trades. *Science, N.Y.* **159**, 1233–6.

ROSEN J.D. (1971) Photodecomposition of organic pesticides. In *Organic Compounds in Aquatic Environments* (Ed. by S.D. Faust & J.V. Hunter), Dekker, New York.

SEBA D.B. & CORCORAN E.F. (1969) Surface slicks as concentrators of pesticides in the marine environment. *Pestic. Monit. J.* **3**, 190–3.

SPENCER D.A. (1967) Problems in monitoring DDT and its metabolites in the environment. *Pestic. Monit. J.* **1**, 54–7.

SPENCER W.F. & CLIATH M.M. (1972) Volatility of DDT and related compounds. *J. agr. Fd Chem.* **20**, 645–9.

STANLEY C.W., BARNEY J.E., HELTON M.R. & YOBS A.R. (1971) Measurement of atmospheric levels of pesticides. *Environ. Sci. Technol.* **5**, 430–5.

STEWART C.A. (1972) Atmospheric circulation of DDT. *Science, N.Y.* **177**, 724–5.

STRICKLAND A.H. (1965) Amounts of organochlorine insecticides used annually on agricultural, and some horticultural, crops in England and Wales. *Ann. appl. Biol.* **55**, 319–25.

TARRANT K.R. & TATTON J.O'G (1968) Organochlorine pesticides in rainwater in the British Isles. *Nature, Lond.* **219**, 725–7.

WEAVER L., GUNNERSON C.G., BREIDENBACH A.W. & LICHTENBERG J.J. (1965) Chlorinated hydrocarbon pesticides in major U.S. river basins. *Public Health Reports* **80**, 481–93.

WHEATLEY G.A. & HARDMAN J.A. (1965) Indications of the presence of organochlorine insecticides in rainwater in central England. *Nature, Lond.* **207**, 486–7.

WIT J.G. (1972) Metabolism of foreign compounds by different classes of birds. *Proc. 15th Int. orn. Congr.*, 466–74.

WOODWELL G.M., CRAIG P.P. & JOHNSON H.A. (1971) DDT in the biosphere: where does it go? *Science, N.Y.* **174**, 1101–7.

WOODWELL G.M., CRAIG P.P. & JOHNSON H.A. (1972) Atmospheric circulation of DDT. *Science, N.Y.* **177**, 725.

Cycling of mercury in the environment

ARNE JERNELÖV *Institutet för Vatten- och Luftvardsforskning, Stockholm, Sweden*

As far as we know today mercury undergoes two different cycling processes in the environment. The first and smallest involves only local aquatic ecosystems while the larger cycle is a more or less global one involving the earth's crust, the atmosphere and oceans.

The mercury turnover in a local aquatic ecosystem is comparatively well known and includes the following principal steps (Langley 1971; Gavis & Ferguson 1972; Jernelöv & Lann 1973):

1. Release of mercury into the watercourse.

2. Accumulation of mercury in organic particles in the sediment or in suspension.

3.a. Gradual biological (chemical) conversion to mono- or dimethyl mercury. The former will be released to the water and the latter may evaporate into the atmosphere, *alternatively*

3.b. Conversion to mercuric sulphide under anaerobic conditions and/or binding to ferrioxides under aerobic conditions. In these forms mercury will remain inactive as long as the anaerobic or the aerobic conditions remain unchanged. These forms can be regarded as temporary sinks for mercury.

4. Mercury converted into dimethyl mercury may be degradated into monomethyl mercury in contact with acid conditions (pH approx 5·6) or converted into elementary mercury in contact with UV-light.

5. Methyl mercury released to the watermass will be accumulated in water-living organisms.

6. A further accumulation will take place along the food chain. However, the food intake will be of larger or similar importance to the direct uptake only for the top predators.

7. (Methyl) mercury will be returned to the sediment with dead organisms or removed from the water body with the organisms through catch or

migration. (Naturally, mercury will also be removed when in solution and when attached to suspended particles through water and mass transport.)

Most physical, chemical and biological factors in the water system will affect the process of biological conversion and accumulation of mercury.

As a brief summary and general rule, it can be said that the rate of biological formation of methyl mercury is directly related to microbiological activity in the sediment or in the suspension where the inorganic (divalent) mercury is present.

A few special cases should, however, be noted:

1. When mercury is in the form of mercuric sulphide, the methylation rate (also under aerobic conditions) will be significantly reduced.
2. When mercury containing organic matter is exposed to air and water, e.g. drying land, deposed dredged material and tidal areas, the biological methylation may be very much accelerated.
3. Differences in the net result (mono- or dimethyl mercury) of the biological methylation process (where pH is a determining factor) may be of great importance to the mercury contamination of local waterliving organisms.

Another complicating factor in transferring laboratory results to the ecosystem level is the de-methylating capacity that has been described, as in cultures of *Pseudomonas*. Most experiments on rates of methylation have been performed in such a way that they do not allow for a discrimination between the two competing processes, methylation and demethylation. Accordingly, it is possible that experimental results that have been interpreted as measures of gross methylation rates in fact have been net methylation rates. The kinetics of the response of such a competitive system in relation to external stimuli such as temperature is of course more complicated; thereby making any data interpretation more uncertain than if only methylation were presupposed to occur.

The importance of de-methylating microorganisms have been stressed from experiments with sediments from the St. Clair area in the Great Lake System (Sprangler *et al.* 1973). The de-methylation is likely to be specially important when alkyl mercury is formed not only through biological processes but also through chemical ones like transalkylation from ethyl and methyl lead compounds discharged to the same water system as inorganic mercury (Jernelöv *et al.* 1972).

One important and troublesome aspect of mercury contamination of lakes and rivers is that unless an effective flushing system exists and a high rate of sediment transport occurs, the mercury deposits in the sediments are likely to continue to release methylated mercury and maintain an elevated mercury level in aquatic organisms for a very long time. Estimates of 'ecological half-life' for mercury levels in contaminated Swedish lakes after a hypothetical cessation of discharge from all sources has given figures of 10 to 100 years.

The recovery process will be much faster in eutrophic lakes with high production, high sedimentation rates, high pH and anaerobic conditions in the sediments than in oligotrophic acid areas with well oxygenated sediments.

Restoration methods for contaminated water bodies

During the last five years, a research programme has been carried out in Sweden with the purpose of finding methods for restoration of mercury-contaminated water bodies. The following methods have been suggested from field and/or laboratory tests.

1. Removal of mercury.
2. Conversion of mercury into mercuric sulphide with a low availability for biological methylation.
3. Binding of mercury to inorganic material—like silica minerals—where availability for methylation is low. Under aerobic conditions, ferric ions and manganese ions bind heavy metals including mercury when forming oxides and crystallizing.
4. Covering mercury deposits with mercury binding or inert material that decreases the release of methylated mercury to water.
5. Increase of pH so that the biological methylation process will give volatile dimethyl mercury rather than monomethyl mercury and thereby lend to a lower accumulation rate of methyl mercury in fish in the primary recipient.

No *direct* restoration attempts have been carried out in Sweden so far. The laboratory tests, pilot field studies and observation of effects of mercury turnover on dredging performed for other reasons indicate that *technically* at least, the methods (1), (2) and (4) are feasible.

However, from an *economic* point of view, the costs for restoration measurements according to any of these principles, will outweigh the value of the fishing that may be restored with, generally, orders of magnitude.

From an *ecological* point of view, all the methods, specially (2), (3), (4) and (5) will have adverse effects on the ecosystems concerned. Naturally, before any attempt is made, it has to be evaluated, from the local conditions, how severe these adverse effects may be in relation to the advantages of reducing mercury contamination of fish.

In view of these facts, direct restoration measurements, according to any of the methods presented, seem unlikely to be carried out in Sweden in the near future, except perhaps in very localized areas with very high recreational values.

During the last two years, restoration techniques have been studied intensively, also in the U.S.A. and Canada. The principles of the suggested methods, however, have to a large extent been those originally suggested and tried out in Sweden. (Larsson *pers. comm.*). The technical experience in North America is not very different from that reported above from Sweden. However, in connection with certain court cases, very divergent opinions exist on the cost-benefit relation of large scale dredging and covering operations.

During the last few years, a few mathematical models have been presented that describe the turnover of mercury in global and local ecosystems. Naturally, mathematical models do not in themselves create any new knowledge, they only help to organize existing information. From a restoration point of view, those describing local ecosystems are interesting as they do so through sensitivity analyses—where different factors can be tested as to their effect on the end result in the form of mercury levels in fish, and thus can screen possibilities of new principles and techniques.

One alternative method of controlling mercury levels in fish populations, that has recently resulted from sensitivity tests of a mathematical model for methyl mercury accumulation in limnic food chains (Fagerström & Åsell 1973), and that may deserve more attention, is the possibility of using fish population management as a means of controlling maximum and average mercury concentrations in fish.

In areas where the present level is approximately 100% above the acceptable level, fish population management may provide a means of bringing the mercury levels below the critical level at a moderate cost and with the social advantage that those benefiting from the money spent on the measurement would be those suffering most from the loss of the fishing opportunities—the fishermen.

THE OCEANS

For a study of the global mercury turnover, the oceans represent a natural focus and starting point.

In their classical work, Stock & Cucuel (1934) reported mercury concentration in sea surface water of 0·03–0·04 ppb. Several later reports have indicated similar figures although others and, in the author's personal opinion, more reliable investigations, have found surface water to contain approximately 0·13 ppb and deep ocean water 0·30 ppb.

Dependant on which group of data is preferred for testing, with a total volume of $1{\cdot}4 \times 10^{21}$ l, the oceans would thus contain between 4·5 and 21×10^{13} g of mercury.

Estimates by Weiss *et al.* (1971) of mercury transports to the sea indicates that rivers may contribute approximately 10^{10} g/year including natural as well as man-released mercury, while the washout with precipitation from the atmosphere could be $(2 \cdot 5\text{–}15)10^{10}$ g/year.

The $1 \cdot 5 \times 10^{11}$ g/year of mercury in the rain is based on an estimated atmospheric load of 4×10^9 g, a close to complete washout in connection with precipitation and an average time between rainfalls of 10 days. It has been argued that the estimate of atmospheric mercury burden is an overestimate of 10 times or more. The alternative figure presented ($2 \cdot 5 \times 10^{10}$ g) is derived by multiplying the amount of rain falling over the oceans with a low estimate of average mercury concentration in precipitation of 0·06 ppb derived from mercury content in apparent uncontaminated glacial ice (Weiss *et al.* 1971). Accepting the idea that $1 \cdot 5 \times 10^{11}$ g/year may represent an overestimate and $2 \cdot 5 \times 10^{10}$ g/year an underestimate, a compromise of 5×10^{10} g/year will be used by the author for further calculations.

The world production of mercury is in the order of 1×10^{10} g. To this should be added direct release of mercury in connection with burning of fossil fuels, heating of mercury containing minerals etc. At the very most, the sources could account for a release of mercury equivalent to that from 'intentional' production and use. Evidently, as large parts of the man-used mercury will not be released into the environment immediately, but only, if at all, after a considerate lapse of time.

Analyses of the Greenland ice-sheet have demonstrated a tendency to an increase in mercury concentrations during the last decades (Weiss *et al.* 1971). When the estimated atmospheric flux of mercury is related to the human production, it is apparent that the observed increase, with roughly a factor of two, that has occurred during the last decades cannot be accounted for by the mercury released from mercury handling industries, fossil fuel burning etc.

In the 'natural' flux of mercury 'degassing from the earth crust' is of dominating importance. Weiss *et al.* (1971) have estimated it to be in the order of $(2 \cdot 4\text{–}44) \times 10^{10}$ g/year.

It is possible that man, through his activities has affected this degassing during the last decades. One way in which such an enhancement could have occurred is through the changed agricultural practice involving more efficient ploughing, the use of fertilizers, etc. that have accelerated the biological processes in the soil thereby perhaps increasing the formation rate of volatile mercury compounds like elementary and dimethyl mercury.

The average residue time for mercury in the upper 200 m layers of the oceans has been calculated to be 50–150 years (Jernelöv 1974), while for the oceans as a whole the average residue time has been estimated to be 6×10^4 years (Gavis & Ferguson 1972).

Inorganic mercury, as well as methyl mercury, has a high affinity for organic substances and will therefore in the sea as well as in fresh water tend to accumulate in particulate matter of organic origin and settle with these in the general sedimentation process.

Bioaccumulation and excretion with fecopellets or sedimentation with the dead organisms also contribute to this process. This provides the mechanism for elimination of mercury from sea water and causes the higher mercury concentration in the bottom water compared with surface water.

In areas with up-welling water and extremely high biological activity, the combination of the higher mercury content in the water and the higher methylation rate could be a natural cause for higher than usual methyl mercury concentrations in marine organisms.

Conclusions

The conclusions of this review of the global mercury circulation is that man, through his activities, may have affected the atmospheric flux of mercury. The main part of this effect may, however, be due to increased degassing from earth crust perhaps through the new agricultural practices and not to his industrial use and release of mercury.

Out of the total amount of mercury present in the sea, man's impact to date is negligible.

Summary

Mercury cycling is a function of biological and chemical transformations which mercury may undergo in the environment. The local aquatic mercury cycle is affected by mercury discharge, eutrophication and water-level manipulations. Man's impingement on the global cycle involves industrial emission to the atmosphere but may also include changes in evaporation from soil due to changes in agricultural practices.

References

FAGERSTROM T. & ASELL B. (1973) Methyl mercury accumulation in the aquatic food chain. A model and some implications for research planning. *Ambio.* 2, 164–71.

GAVIS J. & FERGUSON J.F. (1972) The cycling of mercury through the environment. *Wat. Res.* 6, 989–1008.

JERNELOV A. (1974) Heavy metals, metalloids and synthetic organics. *The Sea V* (ed. by E.D. Goldberg), Wiley Interscience, New York.

JERNELOV A. & LANN H. (1973) Studies in Sweden on feasibility of some methods for restoration of mercury-contaminated bodies of water. *Environ. Sci. Technol.* 7, 712–18.

JERNELOV A., LANN H. & WENNERGREN G. (1972) Rate of biological methylation in the St. Clair System. Report from IVL.

LANGLEY D.G. (1971) Proc. 162nd Am. chem Soc. Meeting.

SPRANGLER W.J., SPIGARELLI J.L., ROSE J.M. & MILLER H.M. (1973) Methyl mercury: bacterial degradation in lake sediments. *Science, N.Y.* **180**, 192–3.

STOCK A. & CUCUEL F. (1934) Die Verbreitung des Quicksilvers. *Naturwissenschaften* **22**, 390–3.

WEISS H.V., KOIDE M. & GOLDBERG E.D. (1971) Mercury in a Greenland ice sheet: evidence of recent input by man. *Science, N.Y.* **174**, 692–4.

Eutrophication and algal growths in Scottish fresh water lochs

W. D. P. STEWART, S. B. TUCKWELL and E. MAY
Department of Biological Sciences, University of Dundee, Scotland, U.K.

Introduction

Within recent years a massive literature has accumulated on factors which govern the eutrophication of fresh water ecosystems and on the effects which eutrophication may have on the physical, chemical and biological parameters of such aquatic ecosystems. The approaches used vary from the purely chemical to the purely biological, the quality of the various studies is diverse, and in some instances (fortunately few) there is difficulty in disentangling facts from emotions. Nevertheless several points emerge rather clearly from a review of the literature.

First, it is agreed generally that the main nutrients responsible for increased eutrophication of fresh water systems are nitrogen and phosphorus (Vollenweider 1968; Lund 1972; Moss 1972; Lee 1973) although some workers implicate carbon as the main nutrient limiting primary production in fresh waters (Kuenztel 1969; but see Goldman *et al.* 1972). Even when it is accepted that the key nutrients are nitrogen and phosphorus, there is no general agreement as to which of the two is the more important.

Second, urban effluent and agricultural run-off are considered as the main sources of nitrogen and phosphorus. Third, there is a general belief that increased eutrophication results in increased growths of primary producers, particularly of planktonic blue-green algae (Fogg *et al.* 1973).

Despite the data which are available, there is a need for detailed information on nutrient input and primary productivity of the type available for Lake Washington (Edmondson 1970, 1971), Lake Windermere (Lund 1972) and Lake Mendota (Lee 1966). In this paper we should like to present preliminary comparative data on two bodies of water in Scotland which we are studying at present in some detail—Forfar Loch and Loch Rescobie. Both lochs are surrounded by rich agricultural land, but in addition Forfar Loch receives discharge from a sewage works which treats the sewage of 11,600 persons, mainly from the town of Forfar. This has provided us with a unique opportunity to compare the input from agriculture and urbanization of those nutrients generally accepted as being mainly responsible for the eutrophication of fresh waters, i.e. nitrate-nitrogen, ammonium-nitrogen and phosphorus, and to assess their effect on the standing crops of planktonic algae in the lochs.

The lochs

In Scotland there are several hundred fresh water lochs which show differing degrees of eutrophication. The physical characteristics of many of these are reviewed by Murray & Pullar (1910), and Brook (1964) and Spence (1964) have studied the phytoplankton and aquatic macrophytes respectively of some of these. In general the lochs of the north and west are oligotrophic or mesotrophic, but in certain other areas of Scotland, particularly in the midland belt and in areas of rich agricultural land, increased nutrient enrichment from agriculture and urbanization has resulted in certain of the lochs becoming eutrophic. There have, however, been few detailed studies of the relationship between nutrient level and primary production in Scottish waters apart from a recent investigation of Loch Leven by an International Biological Programme team (Morgan 1972; Bindloss 1974; Holden & Caines 1974).

Rescobie Loch and Forfar Loch lie approximately 5 miles apart near the town of Forfar in the county of Angus, Scotland. Despite their close proximity and the fact that they are surrounded by almost identical agricultural land, the two systems have quite distinct catchment areas because they are separated by a watershed which runs in a north to south direction. The area is shown in Fig. 1, and in this study we have considered Balgavies Loch, a small loch into which Rescobie Loch flows and from which it is separated by several hundred metres of marshy ground, as part of Loch Rescobie because, as can be seen from Fig. 1, the catchment areas of both are virtually the same.

The physical characteristics of the lochs are summarized in Table 1. Data on loch dimensions were obtained from Murray & Pullar (1910) and there appears to have been little change in physical detail since then. Retention times were calculated from loch volume and inflow measurements.

The dominant soils of the catchments are iron podsols of two major types: the Forfar series which is an imperfectly drained loam and the Vinny series which is a freely drained soil, low in nutrients. The parent material is old red sandstone. These catchments contain excellent arable soils, which are well

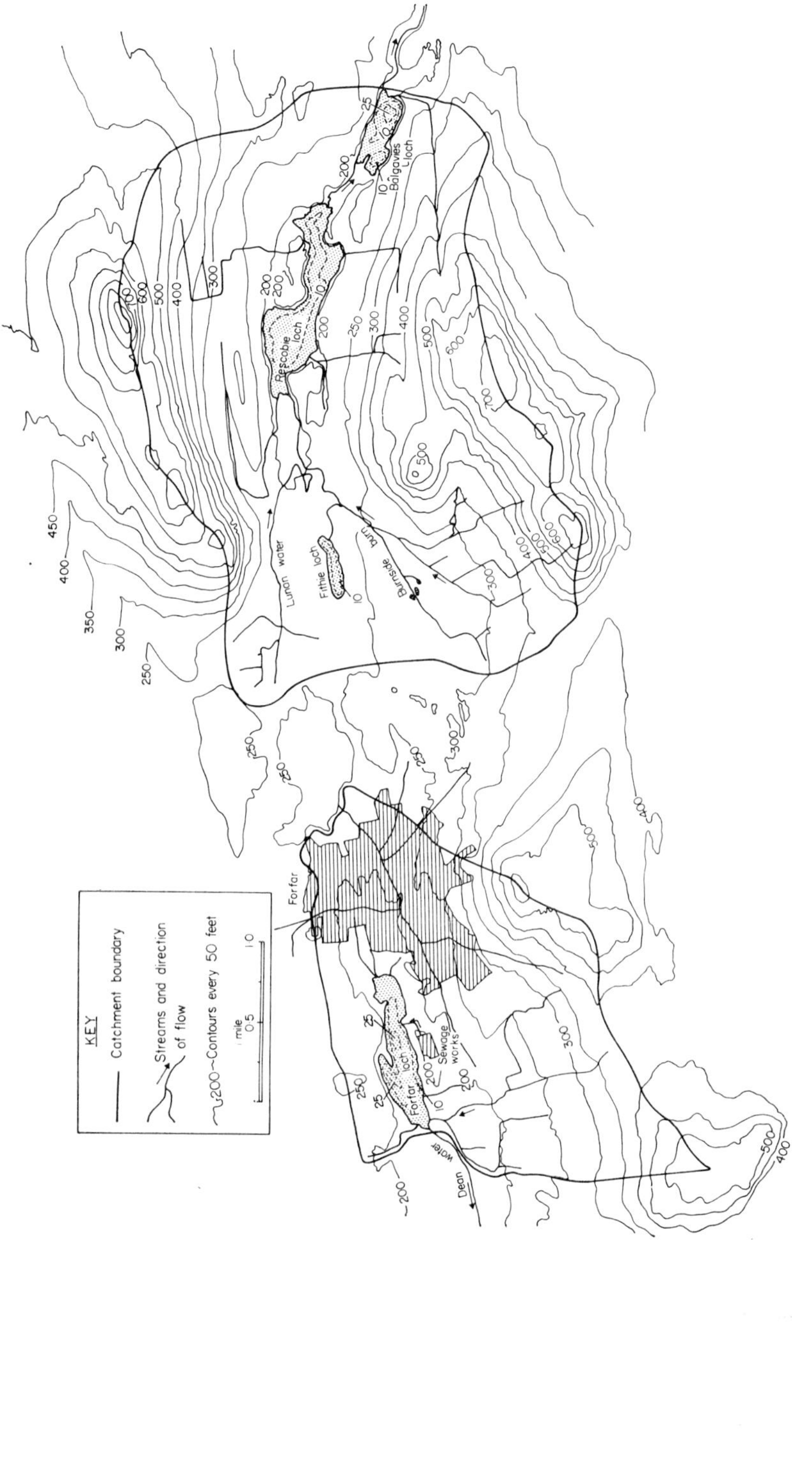

Figure 1. Diagram showing the location of the lochs under study and their catchments.

managed by farmers who make extensive use of available advisory services. The soils are cultivated intensely, particularly for soft fruit (raspberries, strawberries and rhubarb comprise 36% of the arable acreage), cereals (27%), turnips (5%) and potatoes (3%). Approximately 29% of the catchment is grassland.

Table 1. Physical characteristics of Loch Rescobie and Forfar Loch.

Loch	Area (ha.)	Volume (1×10^{-6})	Mean depth (m)	Retention time (y.)	Catchment (ha.)
Rescobie*	85·5	2,577	3·00	0·45	1,874
Forfar	41·4	1,444	3·48	0·39	690

* Balgavies Loch comprises 24% of the total volume listed here (see text).

The sewage works which discharges into Forfar Loch (see Fig. 1) is of the trickling filter type and has an outflow of $4{\cdot}5\times10^6$ l day^{-1}. There is no discharge of sewage into Loch Rescobie.

Levels of nutrients in the lochs

The lochs under study are generally shallow (see Table 1) and during our study period (Dec. 1971 to 1972) remained mixed, except in one or two small pockets in both lochs which became thermally stratified in summer. Thus, there was little change in nutrient level with depth throughout the year in more than 90% of the areas of the lochs. This is exemplified by the data for nitrate-nitrogen levels with depth in Loch Rescobie (Fig. 2) and rather similar data have been obtained with other nutrients.

Data on seasonal variations in the levels of soluble nitrate-nitrogen (all nitrate-nitrogen values include nitrite-nitrogen in this paper), ammonium-nitrogen and total phosphorus in the surface waters of both lochs are presented in Fig. 3. These data show that the lochs contain high levels of all these nutrients with the levels in Forfar Loch usually exceeding those in Rescobie. This was particularly so in the case of phosphorus where the minimum levels in Forfar Loch exceeded the maximum levels recorded for Rescobie Loch. In the case of nitrate-nitrogen, peaks occurred in both lochs in January to early February and again in late June when there was a sharp maximum, which followed very heavy rain. In the autumn the levels were fairly steady but tended to increase again towards the end of the year. Peaks of nitrate-nitrogen were recorded in Forfar Loch on 31 January (10,800 μg l^{-1}), 28 June (20,000 μg l^{-1}) and 12 December (9,200 μg l^{-1}). In Rescobie the maximum nitrate-nitrogen level recorded was 8,032 μg l^{-1}

in mid-January. The annual mean nitrate-nitrogen values were much lower, being approximately 8,000 μg l^{-1} in Forfar Loch and 2,500 μg l^{-1} in Loch Rescobie. The total phosphorus levels in both lochs showed distinct variations throughout the year, particularly in Forfar Loch where they fluctuated markedly with peaks of 1,080 μg l^{-1} on 10 January and 1,427 μg l^{-1} on 4 May. The Rescobie levels, on the other hand, remained remarkably steady throughout the year and although there was a general increase in autumn and early winter they did not exceed 139 μg l^{-1}. The annual mean total phosphorus levels in Forfar Loch and Loch Rescobie were 968 μg l^{-1} and 59 μg l^{-1} respectively.

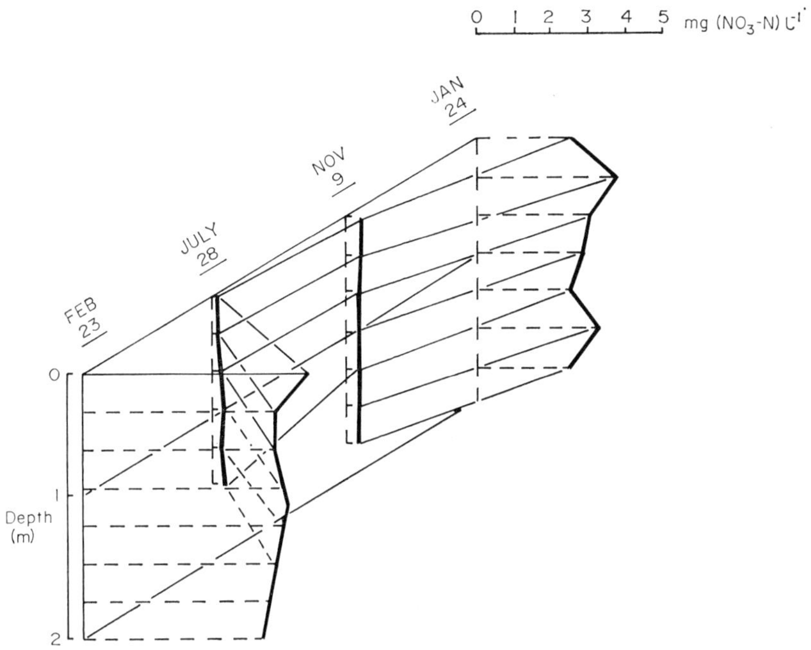

Figure 2. Variation with depth in mean level of nitrate-nitrogen in Loch Rescobie at various times.

From the nitrate and phosphate levels recorded, both lochs can be considered, on the basis of Vollenweider's (1968) classification, as 'polytrophic' (i.e. very highly eutrophic). Furthermore, when these levels are compared with maximum values reported for other British fresh water lakes (Table 2) it is seen that Forfar Loch, in particular, has nitrate and phosphorus levels which are higher than any of these. It may be noted, in particular, that the nitrate-nitrogen maximum, even for Loch Rescobie, is several times higher than that for Loch Leven in Scotland (Holden & Caines 1974). The reasons for this are at least two-fold: first, the ratios of catchment area to loch area

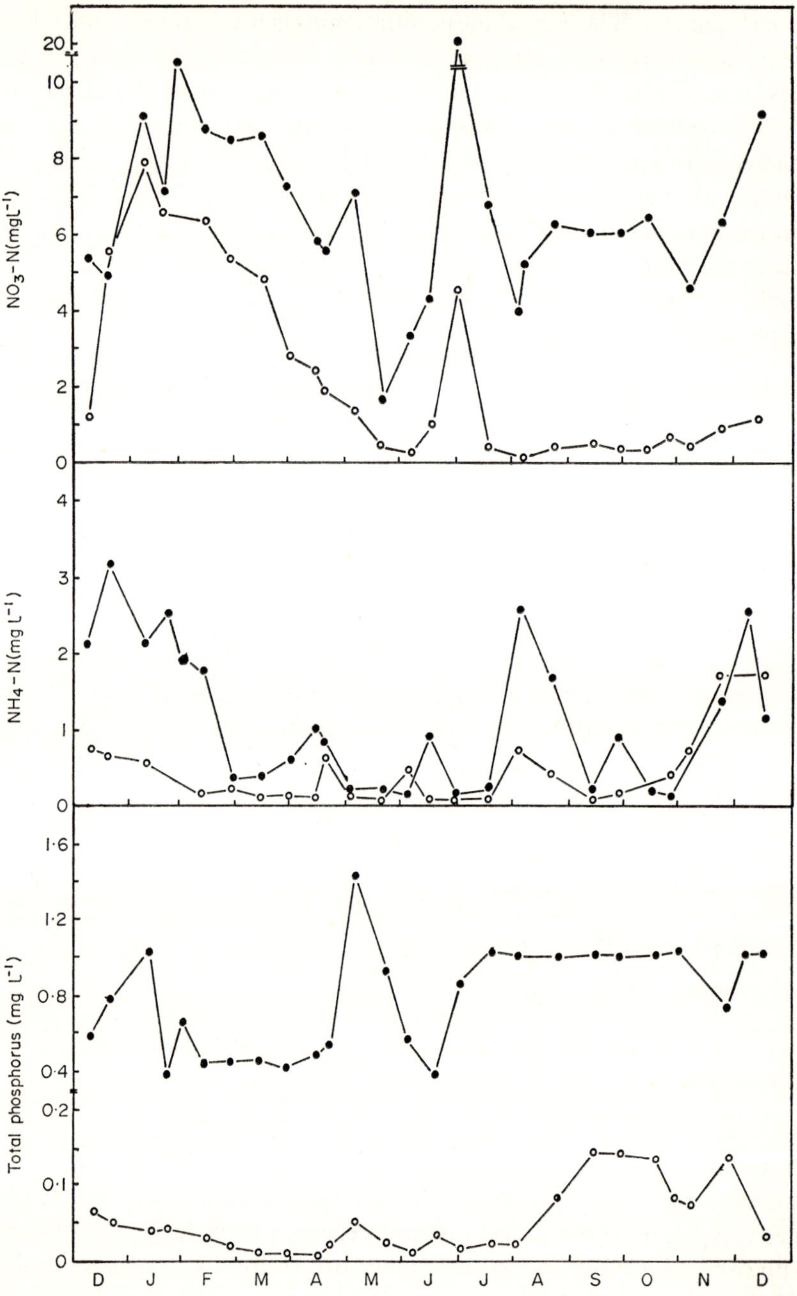

Figure 3. Seasonal variation in the mean levels of soluble nitrate-nitrogen, ammonium-nitrogen and total phosphorus in the surface waters of Loch Rescobie (○—○) and Forfar Loch (●—●) from December 1971 to December 1972.

for Forfar Loch (16·7:1) and Rescobie (21·9:1) are higher than that for Loch Leven (9:1), and second, the land in our catchments is very intensely cultivated and heavily fertilized and this may supply much of the nitrate. The exceptionally high phosphorus levels in Forfar Loch are due largely to a high input from the sewage works (see below).

Table 2. Comparison of maximum levels of nitrogen and phosphorus in Loch Rescobie, Forfar Loch and several other eutrophic lochs in the United Kingdom.

	NO_3-N (μg l^{-1})	PO_4-P (μg l^{-1})	References
Forfar Loch	20,000	1,427	Present data
Loch Rescobie	8,032	222	Present data
Lough Neagh (Kinnego Bay)	1,760	190	Wood & Gibson (1973)
Lough Leven	2,070	200	Holden & Caines (1974)
Lough Neagh (open water)	900	60	Wood & Gibson (1973)
Esthwaite Water	590	8	J.F. Talling, in Wood & Gibson (1973)

The input of nutrients to Rescobie and Forfar Lochs

The measurement of nutrient input to the lochs from the catchments involved setting up water-gauging stations at certain inflows. This enabled the inflow of water to be calculated, and nutrient levels in the waters were analysed every two weeks over a twelve-month period. This procedure was preferred to sampling more frequently and pooling the samples on site, because experiments indicated the presence of enzymes and bacteria in the waters which could have affected the nutrient levels in the pooled samples on standing.

Table 3 shows the contribution of nutrients to the loch from agricultural run-off, from direct precipitation, and from sewage in the case of Forfar Loch. The values for nutrient input from the catchment should be considered as minimum values because there was no way of measuring sub-surface seepage and this could be appreciable (Lee 1966). In addition, particulate material retained by a Millipore filter (0·45 μ porosity) was not analysed. The input of nutrients to the lochs is very high and comes mainly from run-off and from sewage and not from direct precipitation. There is a particularly high input of nitrate-nitrogen and a comparatively low input of phosphorus from the catchment area. Compared with nitrate-nitrogen, input of soluble ammonium-nitrogen is low, representing only 4% and 10% of the total soluble nitrogen (nitrate- plus ammonium-nitrogen) entering Rescobie Loch and Forfar Loch respectively. The Forfar Loch data show, however, that more ammonium-nitrogen enters from sewage than from agriculture. The total input of phosphorus into Forfar Loch, over 96% of which comes from sewage, is approximately 34 times higher than the total input from the agricultural catchment. When input of phosphorus from sewage is calculated on a population discharge basis the input *per capita* is 1·6 g day^{-1}. This value compares with a value of 1·8 calculated for the Great Ouse catchment area by Owens (1970). Wood & Gibson (1973) hypothesize that the *per capita* input in Northern Ireland is nearer 4·0 g day^{-1}.

Table 3. Input per annum of soluble phosphorus and inorganic nitrogen to Loch Rescobie and Forfar Loch.

Loch	Nutrient (kg)	Catchment	Source Direct Precipitation	Sewage Effluent	Total
	Ortho-phosphate phosphorus	489	140	—	629
	Total phosphorus	552	140	—	692
Rescobie	Nitrate-nitrogen	96,248	1,106	—	97,354
	Ammonium-nitrogen	3,710	248	—	3,958
	Total nitrogen	99,958	1,354	—	101,312
	Ortho-phosphate phosphorus	180	66	5,586	5,832
	Total phosphorus	203	66	6,525	6,794
Forfar	Nitrate-nitrogen	35,438	535	22,033	58,006
	Ammonium-nitrogen	1,366	120	5,004	6,490
	Total nitrogen	36,804	656	27,036	64,496

The data in Table 4 compare the levels of nitrogen and phosphorus being added as fertilizer per unit area of Loch Rescobie catchment, with the levels of nutrients being added per unit area of loch surface. It is seen that the inorganic-nitrogen being added per unit area of loch surface is approximately 10 times higher than the addition of fertilizer nitrogen per unit area to the catchment and that nitrogen equivalent to 43% of the fertilizer nitrogen being applied to the catchment is entering the loch each year in surface run-off. In the case of phosphorus the levels being added per unit area of loch surface from the catchment are only 27% of those being added as fertilizer per unit area of catchment. Furthermore the nitrogen: phosphorus ratio of the fertilizer being added to the catchment is 4·2:1 while the mean nitrogen:phosphorus ratio of the run-off water from the catchment is 181:1. Thus, compared with nitrogen input, input of phosphorus from agriculture is very low, due presumably to adsorption of inorganic phosphate on to the the soil particles, as well as to the lower levels of phosphorus being added to the soils.

Table 4. Total nitrogen and phosphorus applied as fertiliser to the catchment of Loch Rescobie per annum and total soluble nitrogen and phosphorus being added per unit area of loch surface.*

	Surface area (ha.)	Total inorganic N added (kg.)	Inorganic N added (kg. ha.$^{-1}$)	Total soluble P added (kg.)	Total soluble P added (kg. ha.$^{-1}$)
Catchment	1,874	233,875	125	55,845	29·8
Loch	85·5	101,312	1,185	692	8·1

* 1972 data.

A further source of nitrogen input into both lochs, but Rescobie Loch in particular, is input from biological nitrogen fixation. This process is associated with the presence of heterocystous blue-green algae in the surface of fresh water lakes (see Horne & Goldman 1972; Fogg *et al.* 1973; Stewart 1973) and nitrogenase activity in the surface waters of the nearby less eutrophic Long Loch has been reported on recently (Stewart 1972). Nitrogenase activity was measured over a twelve-month period, both in Forfar Loch and in Rescobie Loch, using the acetylene reduction technique (Stewart *et al.* 1967) but although low rates were obtained in the surface waters it was calculated, after taking into account the fact that the gas vacuolated nitrogen-fixing planktonic algae were dominant in the surface or near surface waters (see Fogg & Walsby 1971), that the total nitrogen input into the lochs from biological nitrogen fixation by the plankton amounted to less than 0·1% of the input of total nitrogen from other sources. The reasons for the relative unimportance of nitrogen fixation are at least two-fold. First, in these nitrogen-rich lochs nitrogen-fixing algae represented only a small proportion of the total standing crop, which itself was low (see Fig. 9). Second, there is evidence from a variety of studies (Stewart *et al.* 1968; Bone 1971; Ohmori & Hattori 1972) that high levels of combined nitrogen may partially, or completely, inhibit the synthesis of nitrogenase in blue-green algae, so that even in potential nitrogen-fixing species there may be some repression of nitrogen fixation by the levels of combined nitrogen present.

The accumulation of nutrients in the phytoplankton

The high levels of nitrogen and phosphorus entering, and in, these lochs suggested that there was probably an excess of these nutrients in the waters for algal productivity. Direct evidence that this was so comes from several lines of investigation. When the levels of nitrogen and phosphorus in the algae per unit volume of loch water are compared (Table 5) it is seen that over the year the nitrogen and phosphorus in the algae represent only 0·5% and 2·0% of the nutrients present in the surface waters of Loch Rescobie, while the Forfar Loch values are considerably lower. Even at the times when nitrogen and phosphorus levels are at a minimum in the waters, the quantities in the algae as a percentage of the total available in the water are 10·7% and 30·5% for nitrogen and phosphorus in Rescobie and 0·2% and 0·5% in Forfar Loch. We conclude, therefore, that in both lochs the levels of nitrogen and phosphorus did not limit productivity during our year of study. It is possible, however, that if conditions especially conducive to bloom development occurred in Rescobie at other times there then might arise a limitation of algal growth due to an insufficiency of phosphorus. However, two additional lines of evidence suggest that this is seldom likely to be the case.

Table 5. Average and minimum quantities of total nitrogen and total phosphorus found in algae and in surface waters of Loch Rescobie and Forfar Loch.

Nutrient values		Total P in algae (μg l^{-1})	Total P in water (μg l^{-1})	P in algae as % of total P available	Total N in algae (μg l^{-1})	Total N in water (μg l^{-1})	N in algae as % of total N available
Mean	Loch Rescobie	1·19	59	2·0	12·5	2,520	0·5
	Forfar Loch	0·49	968	0·05	5·2	8,010	0·06
Minimum	Loch Rescobie	4·6	15·2	30·5	41·7	390	10·7
	Forfar Loch	3·0	555	0·5	4·0	1,840	0·2

First, we checked for the presence of surplus nitrogen and phosphorus in phytoplankton collected at different times during 1971–2. This was done by examining for the presence of storage phosphorus bodies (polyphosphate bodies, Jensen 1968, 1969) and storage nitrogen bodies (structured granules, see Simon 1971; Lang *et al.* 1972) within the cells. Phytoplankton samples were collected, fixed immediately and examined for storage bodies under the electron microscope. The findings, which will be detailed elsewhere, showed that blue-green algae in these two bodies of water always contained nitrogen and phosphorus storage bodies even when the dissolved levels of nitrogen and phosphorus in the lochs were at a minimum.

Second, the presence of surplus phosphorus within the algae of Loch Rescobie was demonstrated using the acetylene reduction technique as a bioassay for available phosphorus (Stewart *et al.* 1971). The basis of the method is that when phosphorus-deficient nitrogen-fixing blue-green algae are provided with available phosphorus they respond by showing a rapid increase in nitrogenase activity (the reduction of acetylene to ethylene), while algae with surplus phosphorus show no such response. Typical data are shown in Fig. 4. It is seen that while phosphorus-starved algae respond to the addition of phosphorus there was no such response by algae from Rescobie Loch. This is not due to some other nutrient being limiting (see later) but apparently to a sufficiency of available phosphorus in the algae. Other experiments of this type have been detailed elsewhere (Stewart *et al.* 1970; Stewart & Alexander 1971).

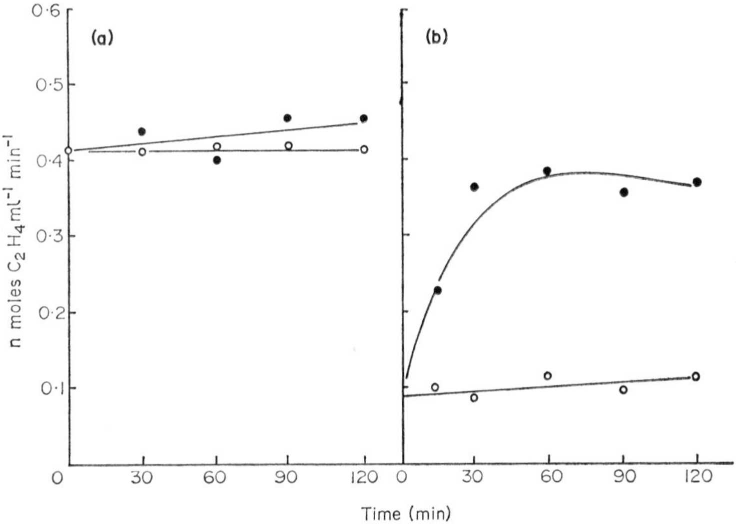

Figure 4. Acetylene reduction by (a) a natural population of *Anabaena* and (b) a phosphorus starved laboratory culture of *Anabaena flos-aquae* supplied at 0 time with orthophosphate phosphorus (50 μg l^{-1}). The natural population was collected from Loch Rescobie in August 1970 and transferred into fresh water collected from that loch for 60 min. prior to the addition of phosphorus. The laboratory population was grown in the medium of Allen & Arnon (1955) without added phosphorus or combined nitrogen for 60 h prior to the re-addition of phosphorus. ●—●, denotes samples to which phosphorus was added; ○—○, denotes samples to which no phosphorus was added. The temperature during the experiment was 25°C and the light intensity was continuous at 3,000 lumens/ft^2. For full details of the method see Stewart *et al.* (1970).

Nutrient losses from the lochs

Despite the high levels of nitrogen and phosphorus entering the lochs, calculations of nitrogen inflow and outflow from Loch Rescobie showed that over 30% of the nitrogen entering the loch was not leaving via the outflow.

This could be due to various factors such as nitrogen assimilation into primary and secondary producers which remain in the loch, nitrogen accumulation in the sediments, or denitrification (i.e. the reduction of nitrate and nitrite to elemental nitrogen). It seemed evident from standing crop measurements that these were unlikely to tie up all this nitrogen, and the accumulation of high levels of nitrogen in the sediments was apparently not an important factor either, because the levels of nitrate nitrogen there were not high (15 to 20 $\mu g\ l^{-1}$). The high carbon:nitrogen ratios of the sediments (10·4–14·0:1) also suggested that there was no large scale accumulation of nitrogen there.

Evidence has been obtained, however, that denitrification is occurring in the sediments of these lochs. Various workers have hypothesized that denitrification may be a major route of nitrogen loss from freshwater ecosystems although there have been few detailed studies (see however Goering & Dugdale 1966; Owens 1970; Chen *et al.* 1972). This may be due in part to methodological difficulties. In our opinion the most convincing way of demonstrating denitrification is to carry out tracer experiments in which ^{15}N-labelled nitrate is added to the sample and the production of $^{15}N_2$ monitored. This technique, first used in denitrification studies in lakes by Goering & Dugdale (1966), provides a direct and fairly specific method of measuring denitrification and has been used in the present study.

Numbers of denitrifying bacteria in the water column and in the sediments of both lochs are shown in Table 6. It is seen that denitrifying bacteria are much more abundant in surface sediments than in the water column and that they are more abundant in Forfar Loch than in Loch Rescobie. The numbers of denitrifiers present do vary however during the year as evidenced by the following counts for numbers of denitrifying bacteria in the sediments of Loch Rescobie during the summer of 1973. Figures in brackets are values for denitrifying bacteria as a percentage of the total bacteria isolated: March, 17 ml^{-1} (0·2%); April, 21 ml^{-1} (0·01%); May, 1,700 ml^{-1} (0.18%); June, $1{\cdot}7\times10^5$ ml^{-1} (13%); July, $5{\cdot}4\times10^7$ ml^{-1} (100%); August, $0{\cdot}79\times10^7$ ml^{-1} (17·2%). There are few other reports of numbers of denitrifiers in lake sediments and the numbers obtained will, of course, vary depending on the conditions used to isolate the bacteria. Nevertheless Fischer (1972) working on lake sediments in Poland observed seasonal fluctuations in the numbers of denitrifying organisms ranging from 0–10^7 ml^{-1} with maximum numbers in the summer, while Niewolak (1970) found numbers of denitrifiers to be fairly constant at about 10^3 ml^{-1} in Ilawa lakes in Poland. Our data for Rescobie and Forfar Lochs (Table 6) are similar to those found by Fischer (1972) and correlate well with the high denitrifying activity of our sediments (see below).

These numbers indicate the potential for denitrification in the lochs and actual measurements of denitrification using $^{15}NO_3$ have confirmed that the process occurs. Such studies have involved taking sediment cores, placing sections from them in specially constructed flasks, adding $^{15}NO_3$ and measuring $^{15}N_2$ evolution by mass spectrometry. Tests were carried out in the field and in the laboratory. In the laboratory, test cores taken with a mud sampler (Freshwater Biological Association type), were maintained at 4°C until tested for denitrification (the delay in testing after collection was never more than 24 hrs). To date we have analysed cores not only from Forfar Loch and Loch Rescobie, but also from the nearby Long Loch,

Table 6. Numbers of denitrifying bacteria ($NO_3 \rightarrow N_2$) in the water column and sediment of Loch Rescobie and Forfar Loch.

Sample	Depth (cm)	Denitrifiers ml^{-1} Loch Rescobie	Forfar Loch
Water column	0	3	4
	200	20	n.d.
	400	120	n.d.
Sediment	0–5	$5{\cdot}4 \times 10^3$	$5{\cdot}4 \times 10^6$
	5–10	$1{\cdot}4 \times 10^3$	n.d.
	10–15	$0{\cdot}8 \times 10^3$	n.d.

Sampling date: Rescobie, 9.5.73; Forfar, 24.5.73; n.d., not determined.
Each value is the most probable number of organisms obtained by dilution counts. All counts performed within 4 hours of sampling.

Clunie Loch and Loch of the Lowes and we have obtained evidence of denitrification of $^{15}NO_3$ to $^{15}N_2$ in all of these lochs. The typical pattern of $^{15}N_2$ production is shown in Fig. 5 and Fig. 6. Figure 5 shows that a linear response in $^{15}N_2$ evolution can be obtained for at least 16 hrs although we have routinely used an incubation period of 4 hrs in the presence of $^{15}NO_3$. Figure 6 shows that the release of $^{15}N_2$ is accompanied by a decrease in the levels of nitrate in the sediment and the pattern of disappearance is rather similar to that observed by Chen *et al.* (1972), being rapid first, and then proceeding at a slower rate thereafter. A proportion of the $^{15}NO_3$ is presumably removed also by assimilatory nitrate-reduction (see Keeney *et al.* 1971). Small amounts of nitrite appear in the sediments for a short period after the addition of nitrate but these subsequently disappear. The temporary accumulation of nitrite is due presumably to some bacteria denitrifying only to nitrite, or to the release of nitrite during nitrate reduction to N_2.

Important factors governing *in situ* rates of denitrification, given that a population of denitrifying bacteria is present, are: temperature, the reducing capacity of the environment in which the bacteria occur, the presence of oxidizable substrate and the availability of nitrate. The latter may not

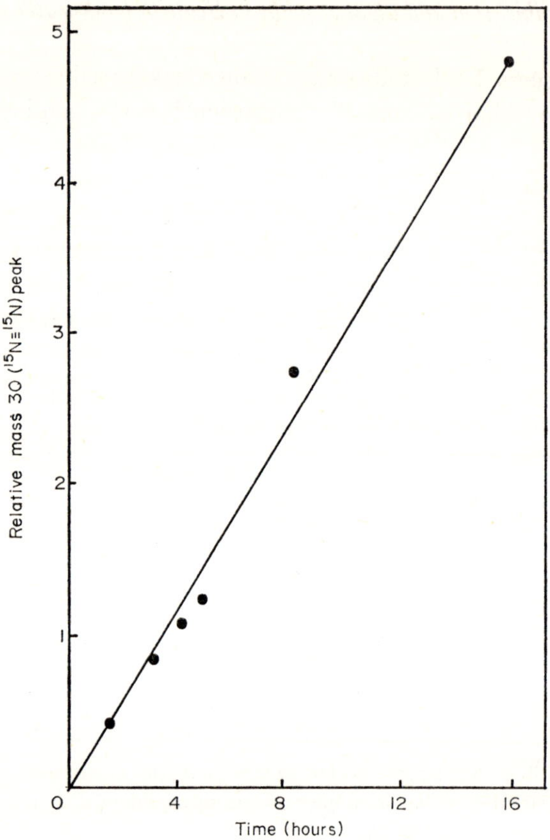

Figure 5. Time course of $^{15}N_2$ release from $Na^{15}NO_3$ (95 atom % excess ^{15}N) by surface sediment cores from Loch Rescobie, incubated at 18°C. Each point is the mean of duplicate samples.

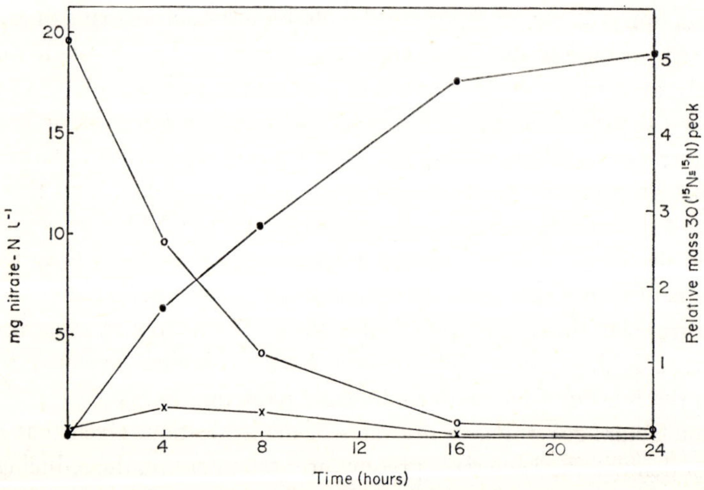

Figure 6. Time course of disappearance of ^{15}N-labelled (95 atom % excess ^{15}N) $NaNO_3$ added to surface sediments from Loch Rescobie at 25°C and the concomitant appearance of $^{15}N_2$. Each point is the mean of duplicate samples. ●—●, $^{15}N_2$; ○—○, NO_3^-; ×—×, NO_2^-.

present a problem in our lochs, and we have also found, in short-term studies (up to 16 hrs), that there is no shortage of oxidizable substrates in the sediments. The most important factor regulating denitrification rates in our sediments is probably temperature. As Fig. 7 shows, the minimum *in situ* winter temperature and maximum summer temperature of the sediments (4°C and 15°C respectively) both allow denitrification to occur, but at lower rates than at the optimum temperature range for denitrification of 24°C to 30°C. As Fig. 8 shows, the levels of dissolved O_2 in the water above the sediment have little effect on N_2 production from nitrate in the sediment, indicating that the sediments maintain reducing conditions irrespective of the O_2 levels in the water column and that deoxygenation of the water column is not a prerequisite for denitrification to occur. Similar conclusions have been arrived at by other workers (Brezonik & Lee 1966; Owens 1970; Chen *et al.* 1972).

We are currently accumulating data on rates of denitrification in Loch Rescobie over a twelve-month period and the results will be presented in detail elsewhere in due course. Our preliminary results, based on experiments carried out on over 50 cores, suggest that in this loch system most of the nitrogen entering in the surface waters and not being lost from it via the outflow is being lost by denitrification.

Standing crop measurements

The above data on nutrient chemistry indicate that both lochs are highly eutrophic and in view of the often reported correlations between nutrient input and standing crop we followed the levels of the latter with special interest.

Loch Rescobie has a rich flora of blue-green algae as is typical of many mesotrophic and eutrophic waters. The most common species in 1972 were *Microcystis aeruginosa*, *M. flos-aquae*, *Aphanizomenon flos-aquae* and *Anabaena* species together with smaller quantities of *Oscillatoria limnetica*, and of the diatom *Asterionella formosa* which occurred in abundance mainly in spring (April–May). In Forfar Loch blue-green algae are rare except for *Microcystis* sp.; diatoms, particularly *Asterionella* sp., *Scenedesmus* sp. and unicellular green algae are usually dominant. Figure 9 shows the levels of standing crop on a seasonal basis. The pattern is rather similar in both lochs with levels increasing to a maximum in the summer months. In general there is a direct correlation between levels of standing crop and both the temperatures of the surface waters and the available quantities of light energy. This pattern is characteristic of many well mixed eutrophic lakes. The most striking feature, however, is that the levels of standing crop are not high

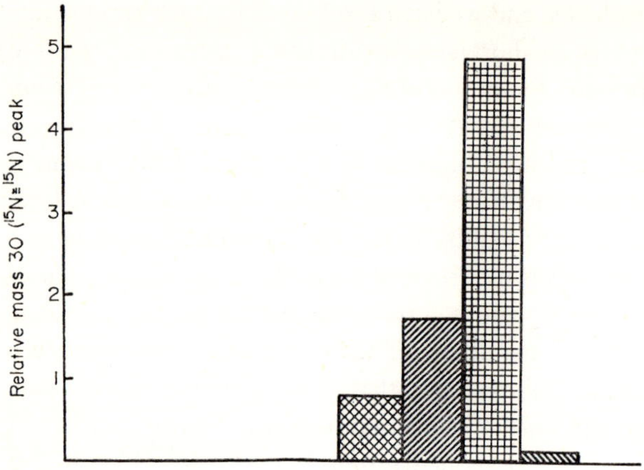

Figure 7. Relative release of $^{15}N_2$ from $Na^{15}NO_3$ (95 atom % excess ^{15}N) at different temperatures after a 24 h incubation period. Temperatures from left to right, 4°C, 15°C, 25°C and 40°C. Each value is the mean of duplicate samples.

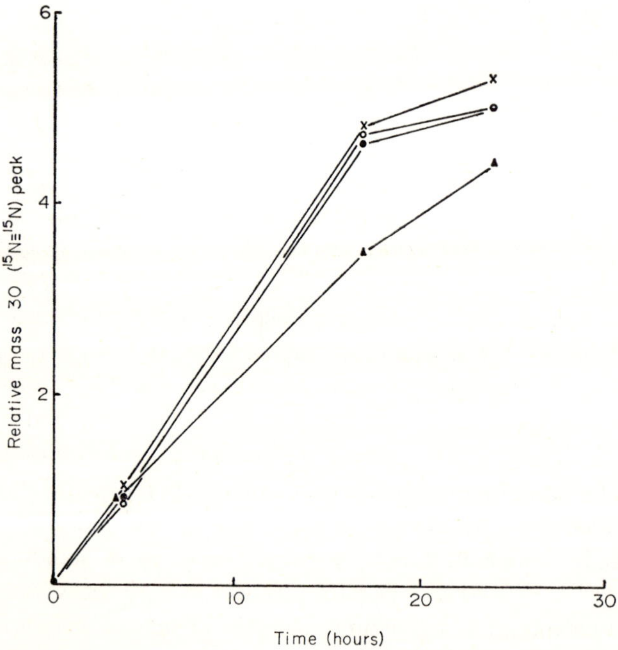

Figure 8. $^{15}N_2$ production from $Na^{15}NO_3$ (95 atom % excess ^{15}N) by sediment cores from Loch Rescobie at 25°C in the presence of various O_2 tensions in the water above the sediment. Each value given is the mean of duplicate samples. ×—×, 0% O_2; ○—○, 5% O_2; ●—●, 10% O_2; ▲—▲, 20% O_2.

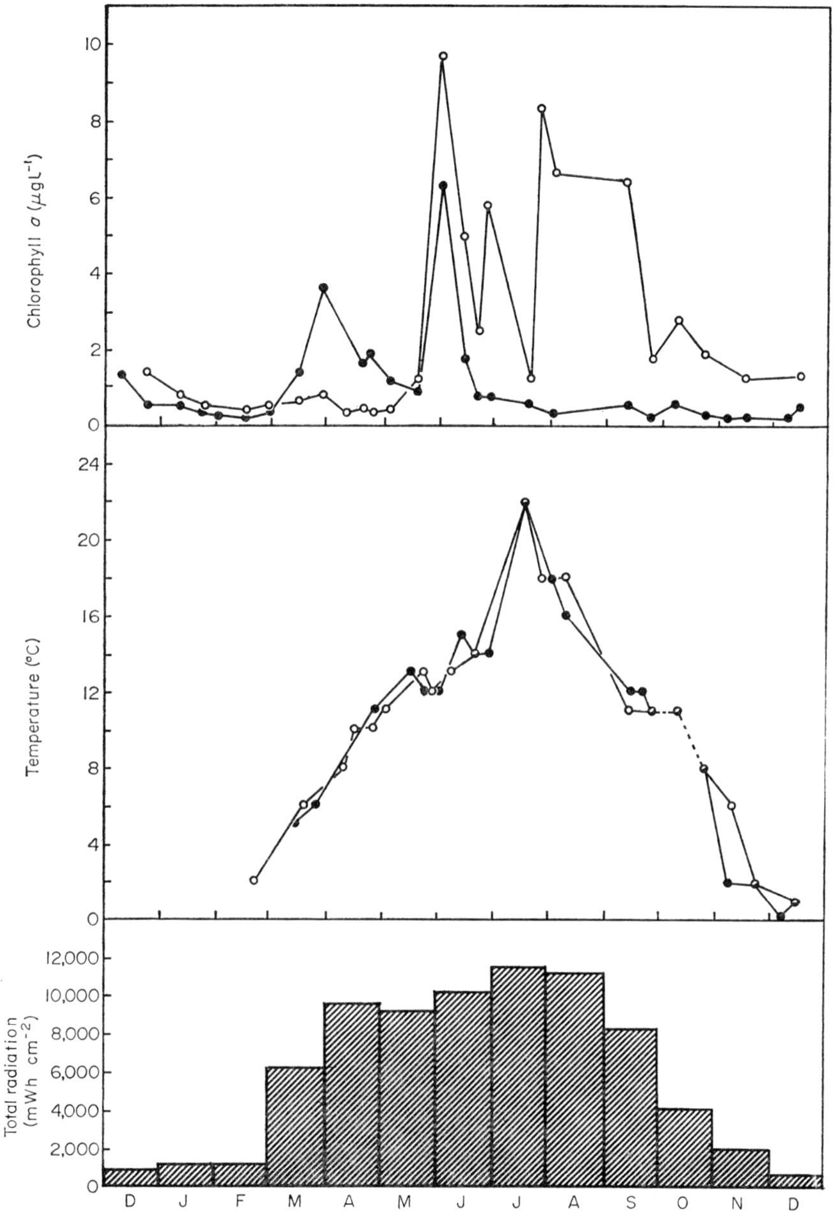

Figure 9. Seasonal variation in the mean chlorophyll *a* levels in the surface waters of Loch Rescobie (○—○) and Forfar Loch (●—●) together with mean temperatures of the surface waters in Loch Rescobie (○—○) and Forfar Loch (●—●) and total solar radiation recorded at the nearby Scottish Horticultural Research Institute, Dundee.

and not only are they lower in Forfar Loch than in Loch Rescobie, but they are much lower than those recorded for various other lakes in the United Kingdom, for example in Loch Leven (Bailey-Watts 1974; Bindloss 1974), Lough Neagh (Wood & Gibson 1973) and Lake Windermere (Lund 1972), none of which is any more eutrophic than the two waters studied here. Possible reasons for these findings are considered in the Discussion.

Discussion

The present study provides comparative data on the relative inputs of nitrogen and phosphorus from agricultural land and from domestic discharge into two freshwater bodies of water in Scotland only. One must be careful therefore before applying the data obtained to other freshwater ecosystems where physical characteristics of the lochs, soil type, agricultural practices and catchment/loch surface areas etc. are probably different. Nevertheless there are some points of general relevance to the eutrophication problem as well as being of particular interest in our study area.

First, there is the question of phosphorus. As expected, the major input of phosphorus into Forfar Loch is from the sewage works and although input *per capita* into this loch (1·6 g *capita*$^{-1}$ day^{-1}) is no higher than values obtained elsewhere by other workers it results in very high levels in the loch which is small relative to the discharge it receives. It goes without saying that such a phosphorus loading should be reduced if at all possible, but it also needs to be pointed out that about a thousand-fold reduction in the maximum level of phosphorus recorded would be necessary to provide a generally acceptable phosphorus value. This could be achieved only by the installation of a much more efficient sewage treatment plant and/or a major reduction in the use of phosphorus-based domestic products. It is questionable whether this is economically feasible, and as will be discussed later, one must question the necessity of such drastic action in our particular ecosystem. Input of phosphorus from agriculture, although much less than from sewage discharge, is still appreciable, as can be gauged from the fact that there is excess phosphorus available for algal growth in Loch Rescobie even at the time of minimum soluble phosphorus in the water.

Second, our data show that the bulk of the nitrogen input comes from agricultural run-off as nitrate, but that the input from urbanization, much of it as ammonium-nitrogen, is also too high for complacency, even although input from point sources such as sewage treatment plants can be minimized by the installation of more efficient units if need be. Input from agriculture is not only higher but is more difficult to control. At present nitrogen equivalent to over 43% of the total fertilizer nitrogen applied reaches the loch waters and this from land which is well managed. Such an input of nitrogen, which appears to result from current agricultural fertilizer usage, is unlikely to be reduced unless the use of slow release fertilizers becomes established practice.

In the present study the concentration of total soluble inorganic nitrogen and total soluble phosphorus in the run-off waters entering Loch Rescobie ranged from 750–23,300 μg N l^{-1} and 22–199 μg P l^{-1} respectively, with annual mean values of 8,474 μg N l^{-1} and 47 μg P l^{-1} respectively. Other published values for nutrient levels in agricultural drainage in the United Kingdom are 500 to 91,000 μg N l^{-1} in water from boulder clays in East Anglia and 1,200 to 26,000 μg N l^{-1} for drift deposits over Oxford clay in Bedfordshire, with mean values of 12,200 μg N l^{-1} and 14,800 μg N l^{-1} respectively. Corresponding values for phosphorus are 0–700 μg P l^{-1} (mean 60 μg P l^{-1}) and 0–750 μg P l^{-1} (mean 120 μg P l^{-1}) respectively (Williams 1971). These values, and those of Owens & Wood (1968) for land drainage entering the Great Ouse in Bedfordshire thus show ranges in concentration which are rather similar to the values obtained in the present study. Our values however are much higher than those reported for Loch Leven inflows in Scotland by Holden & Caines (1974) (annual means of eight years' data range from 2,000 to 2,030 μg N l^{-1} and 18 to 31 μg P l^{-1}).

The addition of nitrogen to our lochs is partially balanced by nitrogen losses which result from biological denitrification. Various workers have shown an unexplained deficit between nitrogen input to lakes and nitrogen output which they usually attribute to denitrification, often without any direct experimental evidence to back this up. In our study we have consistently demonstrated denitrification using ^{15}N tracer techniques, and our preliminary data for Loch Rescobie suggest that almost all the nitrogen lost from the lake, other than via the outflow, may be due to sediment denitrification. Rather similar data have been obtained for Lake Mendota by Chen *et al.* (1972) and for Lake Kinneret by B. Cavari (*pers. comm.*). It seems possible that the low levels of nitrate in Loch Rescobie during the summer are due mainly to denitrification (even the peak of nitrate which followed heavy rain rapidly disappeared, see Fig. 3). The high nitrate levels in winter no doubt result from the high input of nitrogen into the lochs at that time, but possibly also because optimum rates of denitrification may not be occurring in these lochs during the winter because the temperature of the sediment is low. Other workers have already demonstrated the importance of temperature in regulating denitrification rates. For example, Bailey & Beauchamp (1973) noted that a temperature of 5°C inhibited denitrification in saturated soil systems and Hardy & Holsten (1973) state that denitrification rates increase rapidly over the temperature range 2°–25°C. It may be noted that we have observed denitrification at 4°C and our temperature data correlate well with the studies of Dawson & Murphy (1972) on denitrification rates in activated sludge. Seasonal data on denitrification rates are required.

Temperature exerts its effect on other biological processes as well, including primary production. However it is very unlikely that this alone could explain the paradox that although the nutrient levels in our lochs are very high, the standing crops of algae are low, with the chlorophyll *a* levels in the surface waters in 1972 never exceeding 10 μg l^{-1} and similar results were obtained in 1971. Indeed it could be argued that if these data are typical (and one cannot be sure of this until several years' data have been obtained) then increasing the levels of nitrogen and phosphorus in our lochs even further would not result in increased algal growths, if other regulatory factors remain unchanged.

The factors responsible for the low standing crops are uncertain but several may be of importance. First, physical features such as temperature and light may limit productivity. Laboratory data show that field populations incubated in the laboratory show increased metabolism and growth as the temperature is increased from 12°C to 28°C at 400 lumens/ft^2. This agrees with laboratory data for axenic cultures of blue-green algae which have temperature optima of 30°–35°C (Allen & Stanier 1968). Similarly increased productivity resulted when field populations were exposed to increasing light intensities from 100 lumens/ft^2 to 500 lumens/ft^2 at constant temperature (18°C) in the laboratory. Such findings are backed up by field observations which show that standing crop is related more to temperature and light intensity than to the levels of nitrogen and phosphorus (see Fig. 3 and Fig. 9) and it is likely that in both these lochs shortage of light may seriously limit productivity. (Secchi disc values in summer may on occasion be less than 0·7 m.) Temperature is probably also an important limiting factor although greater standing crops occur in other lochs with rather similar temperatures, e.g. in Loch Leven (Bailey-Watts 1974; Bindloss 1974). The fact that increased growths of the algae occur when the temperature and/or light intensity is increased also indicates indirectly that the lake waters do not contain a chemical toxin, or that if they do, it can be detoxified rapidly.

Second, grazing of the phytoplankton, or attack by pathogenic fungi (Canter 1972), bacteria and viruses (see Fogg *et al.* 1973) may be important. Indeed, in Forfar Loch, some of the most common planktonic invertebrates, for example, *Daphnia magna*, *D. hyalina* var. *lacustris* and *Diaptomus gracilis* are herbivores and we have isolated algal lysing viruses and bacteria from these and nearby lochs (Daft *et al.* 1970; Daft & Stewart 1971, 1973, unpublished). We are currently obtaining data on the role of grazers and pathogens on the levels of standing crops in these lochs, as well as on primary production rates as distinct from standing crop measurements.

Irrespective of what the factors which regulate the standing crops in our lochs eventually turn out to be, it may be noted that collectively they appear, at least in 1972, to provide a well balanced ecosystem. Despite this, the levels of nitrogen and phosphorus in the lochs are potentially capable of supporting extensive algal blooms should the existing balance be upset, for example if exceptionally high temperatures persisted in particular years, or if the level of grazers is altered, for example by the accumulation of pesticides in the lochs. Indeed the latter could have a more disastrous effect on the levels of primary, as well as secondary producers, than the addition of more nitrogen and phosphorus. It would thus be prudent to reduce the nutrient levels in the lochs if at all possible. This is particularly so in the case of phosphorus which in our opinion appears to be the key nutrient in the ecosystem, because even if it were possible to deplete the waters completely of combined nitrogen, blooms of nitrogen-fixing algae could still occur if other nutrients, including phosphorus were available. If, on the other hand, the input of nutrients into these lochs cannot be reduced because of economic pressures it is essential that the lochs remain well managed so that the low levels of primary producers which occur at present can be maintained.

Acknowledgments

This work was made possible through grants to Professor W. D. P. Stewart from N.E.R.C., S.R.C., and the Royal Society. We are grateful to Mr D. Harper of this department for useful discussion of many of the aspects of this work.

Summary

1. Two Scottish lochs which we have studied contain very high levels of nitrogen and phosphorus and on the basis of their nutrient status can be classified as polytrophic.
2. Run-off from rich agricultural land contributes most of the nitrate, and nitrogen equivalent to over 43% of the nitrogen added as fertilizer is leached out of the soil during the year.
3. Input of phosphorus from agricultural run-off is much lower than input of nitrate but there is a substantial contribution of phosphorus to one loch from sewage discharge.
4. The algae in the lochs have levels of nitrogen and phosphorus which are

much in excess of their immediate requirements for growth, even when the nutrient levels in the lochs are at a minimum.

5. Nitrogen input from biological nitrogen fixation by planktonic algae is negligible.

6. Studies using ^{15}N as tracer show that there is a substantial loss of nitrate from the system by denitrification. Preliminary data for one loch indicate that this could account for most of the total nitrogen added which is not lost from the loch via the outflow.

7. The standing crops of algae in the lochs are not high and extensive algal blooms did not occur during 1971–2. The reasons for this are uncertain but the relatively low temperatures in the lochs, grazing, possibly the effects of algal pathogens, and the low retention times of the water in the lochs may all contribute to the low levels of standing crop observed.

8. Although existing nitrogen and phosphorus levels are high they do not cause a problem at present, but they provide a potential danger if the ecosystems which exist at present become modified.

References

ALLEN M.B. & ARNON D.I. (1955) Studies on nitrogen-fixing blue-green algae. I. Growth and nitrogen fixation by *Anabaena cylindrica* Lemm. *Pl. Physiol., Lancaster* 30, 366–72.

ALLEN M.M. & STANIER R.Y. (1968) Selective isolation of blue-green algae from water and soil. *J. gen. Microbiol.* 51, 203–9.

BAILEY L.D. & BEAUCHAMP E.G. (1973) Effects of temperature on NO_3 and NO_2 reduction, nitrogenous gas production, and redox potential in a saturated soil. *Can. J. Soil Sci.* 53, 213–18.

BAILEY-WATTS A.E. (1974) Algal plankton of Loch Leven, Kinross. *Proc. R. Soc. Edinb.* B. 74, 135–56.

BINDLOSS M.E. (1974) Primary productivity of phytoplankton in Loch Leven, Kinross. *Proc. R. Soc. Edinb.* B. 74, 157–82.

BONE, D.H. (1971) Nitrogenase activity and nitrogen assimilation in *Anabaena flos-aquae* growing in continuous culture. *Arch. Mikrobiol.* 80, 234–41.

BREZONIK P.L. & LEE, G.F. (1966) Sources of elemental nitrogen in fermentation gases. *Air & Wat. Pollut. Int. J.*, 10, 145–60.

BROOK A.J. (1964) The phytoplankton of the Scottish freshwater lochs. In *The Vegetation of Scotland* (Ed. by J.H. Burnett), Oliver & Boyd, Edinburgh and London.

CANTER H.M. (1972) A guide to the fungi occurring on planktonic blue-green algae. In *Taxonomy and Biology of Blue-Green Algae* (Ed. by T.V. Desikachary), University of Madras, Madras.

CHEN R.L., KEENEY D.R., GRAETZ D.A. & HOLDING A.J. (1972) Denitrification and nitrate reduction in Wisconsin lake sediments. *J. environ. Quality* 1, 158–62.

DAFT M.J., BEGG J. & STEWART, W.D.P. (1970) A virus of blue-green algae from fresh water habitats in Scotland. *New Phytol.* 69, 953–61.

DAFT M.J. & STEWART, W.D.P. (1971) Bacterial pathogens of freshwater blue-green algae. *New Phytol.* 70, 812–29.

DAFT M.J. & STEWART, W.D.P. (1973) Light and electron microscope observations on algal lysis by bacterium CP-1. *New Phytol.* 72, 799–808.

DAWSON R.N. & MURPHY, K.L. (1972) The temperature dependency of biological denitrification. *Wat. Res.* 6, 71–83.

EDMONDSON W.T. (1970) Phosphorus, nitrogen and algae in Lake Washington after diversion of sewage. *Science, N.Y.* 169, 690–1.

EDMONDSON W.T. (1971) Nutrients and phytoplankton in Lake Washington. In *Nutrients and Eutrophication* (Ed. by G.E. Likens), Am. Soc. Limnol. Ocean.

FISCHER E. (1972) Seasonal changes of the number of nitrogen cycle bacteria in bottom sediments of a pool. *Polskie Arch. Hydrobiol.* 19, 37–52.

FOGG G.E. & WALSBY A.E. (1971) Buoyancy regulation and the growth of planktonic blue-green algae. *Mitt. Internat. Verein. Limnol.* 19, 182–8.

FOGG G.E., STEWART W.D.P., FAY P. & WALSBY A.E. (1973) *The Blue-green Algae.* Academic Press, London and New York.

GOERING J.J. & DUGDALE V.A. (1966) Estimate of rates of denitrification in a subarctic lake. *Limnol. Oceanogr.* 9, 448–51.

GOLDMAN J.C., PORCELLA D.B., MIDDLEBROOKS E.J. & TORIEN D.F. (1972) The effect of carbon on algal growth—its relationship to eutrophication. *Wat. Res.* 6, 637–79.

HARDY R.W.F. & HOLSTEN R.D. (1973) Global nitrogen cycling: pools, evolution,transformations, transfers, quantitation and research needs. In *The Aquatic Environment: Microbial Transformations and Water Quality Management Implications* (Ed. by R.K. Ballantine & L.J. Guaraia), (in press).

HOLDEN A.V. & CAINES L.A. (1974) Nutrient chemistry of Loch Leven. *Proc. R. Soc. Edinb.* B. 74, 101–22.

HORNE A.J. & GOLDMAN C.R. (1972) Nitrogen fixation in Clear Lake, California, I. Seasonal variation and the role of heterocysts. *Limnol. & Oceanogr.* 17, 678–92.

JENSEN T.E. (1968) Electron microscopy of polyphosphate bodies in a blue-green alga, *Nostoc pruniforme. Arch. Mikrobiol.* 62, 144–52.

JENSEN T.E. (1969) Fine structure of developing polyphosphate bodies in a blue-green alga, *Plectonema boryanum. Arch. Mikrobiol.* 67, 328–38.

KEENEY D.R., HERBERT R.A. & HOLDING A.J. (1971) Microbiological aspects of the pollution of fresh water with inorganic nutrients. In *Microbial Aspects of Pollution* (Ed. by G. Sykes & F.A. Skinner), Academic Press, London and New York.

KUENTZEL L.E. (1969) Bacteria, carbon dioxide and algal blooms. *J. wat. Pollut. Control Fed.*, 41, 1737–47.

LANG N.J., SIMON R.D. & WOLK C.P. (1972) Correspondence of cyanophycin granules with structured granules in *Anabaena cylindrica. Arch. Mikrobiol.* 83, 313–20.

LEE G.F. (1966) Report on nutrient sources of Lake Mendota. *Tech. Rep. by Nutrient Sources Subcommittee of Lake Mendota Problems Committee* 37 pp. (mimeo).

LEE G.F. (1973) Role of phosphorus in eutrophication and diffuse source control. *Water Res.* 7, 111–28.

LUND J.W.G. (1972) Eutrophication. *Proc. R. Soc. Lond.* B. 180, 371–82.

MORGAN N.C. (1972) Productivity studies at Loch Leven. *Proc. IBP/UNESCO Symp. on Productivity Problems of Freshwaters.* Poland.

MOSS B. (1972) Studies on Gull Lake, Michigan II. Eutrophication—evidence and prognosis. *Freshwat. Biol.* 2, 309–20.

MURRAY J. & PULLAR L. (1910) *Bathymetric Survey of The Scottish Fresh-water Lochs.* Vol. II. Challenger Office, Edinburgh.

NIEWOLAK S. (1970) Seasonal changes of nitrogen-fixing and nitrifying and denitrifying bacteria in the bottom deposits of the Ilawa lakes. *Polskie Archiv. Hydrobiol.* **17**, 509–23.

OHMORI M. & HATTORI A. (1972) Effect of nitrate on nitrogen fixation by the blue-green alga *Anabaena cylindrica. Pl. & Cell Physiol.* **13**, 589–99.

OWENS M. (1970) Nutrient balances in rivers. *Wat. Treat. Exam.* **19**, 239–47.

OWENS M. & WOOD G. (1968) Some aspects of the eutrophication of water. *Wat. Res.* **2**, 151–9.

SIMON R.D. (1971) Cyanophycin granules from the blue-green alga *Anabaena cylindrica*: a reserve material consisting of copolymers of aspartic acid and arginine. *Proc. natn. Acad. Sci. U.S.A.* 68, 265–7.

SPENCE D.H.N. (1964) The macrophytic vegetation of freshwater lochs, swamps and associated fens. In *The Vegetation of Scotland* (Ed. by J.H. Burnett), Oliver & Boyd, Edinburgh and London.

STEWART W.D.P. (1972) Algal metabolism and water pollution in the Tay region. *Proc. R. Soc. Edinb.* B **71**, 209–24.

STEWART W.D.P. (1973) Nitrogen fixation by photosynthetic micro-organisms. *Ann. Rev. Microbiol.* 27, 283–316.

STEWART W.D.P. & Alexander G. (1971) Phosphorus availability and nitrogenase activity in aquatic blue-green algae *Freshwat. Biol.* **1**, 389–404.

STEWART W.D.P., FITZGERALD G.P. & BURRIS R.H. (1967) *In situ* studies on N_2 fixation using the acetylene reduction technique. *Proc. natn. Acad. Sci. U.S.A.* 58, 2071–8.

STEWART W.D.P., FITZGERALD G.P. & BURRIS R.H. (1968) Acetylene reduction by nitrogen-fixing blue-green algae. *Arch. Mikrobiol.* 62, 336–48.

STEWART W.D.P., FITZGERALD G.P. & BURRIS R.H. (1970) Acetylene reduction assay for determination of phosphorus availability in Wisconsin lakes. *Proc. natn. Acad. Sci. U.S.A.* 66, 1104–11.

STEWART W.D.P., MAGUE T., FITZGERALD G.P. & BURRIS R.H. (1971) Nitrogenase activity in Wisconsin lakes of differing degrees of eutrophication. *New Phytol.* **70**, 497–509.

VOLLENWEIDER R.A. (1968) Scientific fundamentals of the eutrophication of lakes and flowing waters, with particular reference to nitrogen and phosphorus as factors in eutrophication. *Organisation for Economic Co-operation and Development Report, Paris.*

WILLIAMS R.J.B. (1971) The chemical composition of water from land drains at Saxmundham and Woburn and the influence of rainfall upon nutrient losses. *Rep. Rothamsted exp. Stn. for 1970*, **2**, 36–67.

WOOD R.B. & GIBSON C.E. (1973) Eutrophication and Lough Neagh. *Wat. Res.* 7, 173–87.

The passage of nuclear weapon debris through the atmosphere

D. H. PEIRSON *Atomic Energy Research Establishment, Harwell, U.K.*

Insertion into the atmosphere

Nuclear weapons have been exploded in the atmosphere for 25 years. Each nuclear explosion spreads traceable radioactive substances through the atmosphere and into the human environment. The radioactive debris from a nuclear weapon consists of fission products, unexpended fissile material and activation products—the result of the capture of excess neutrons in the weapon material or in the neighbouring air and land. This vaporized mixture is forced upwards into the atmosphere by the tremendous release of energy in the explosion. As the cloud rises and cools, the radioactive products condense to form particles of solid debris.

At first the explosions were comparatively small so that the nuclear debris was confined zonally, within the troposphere and roughly to the latitude of the explosion site. The stratosphere was penetrated for the first time by a thermonuclear device exploded by the United States in 1952. Since then there have been many large explosions (Anon. 1964) so that the subsequent dispersal of the debris, controlled by the meteorology of the stratosphere, has been worldwide. A convenient general indication of the incidence and effect of nuclear fallout is given in Fig. 1 which describes the rate of deposition of a long-lived fission product strontium-90 (Peirson 1971). The shapes of the curves of deposition from high yield explosions during the three phases of weapon testing—up to 1959, 1961–2 and after 1966—show the peaks of deposition rate immediately after the first and second phases and the levelling during the third phase when the rate of injection into the stratosphere has been in fortuitous balance with the rate of depletion of the reservoir.

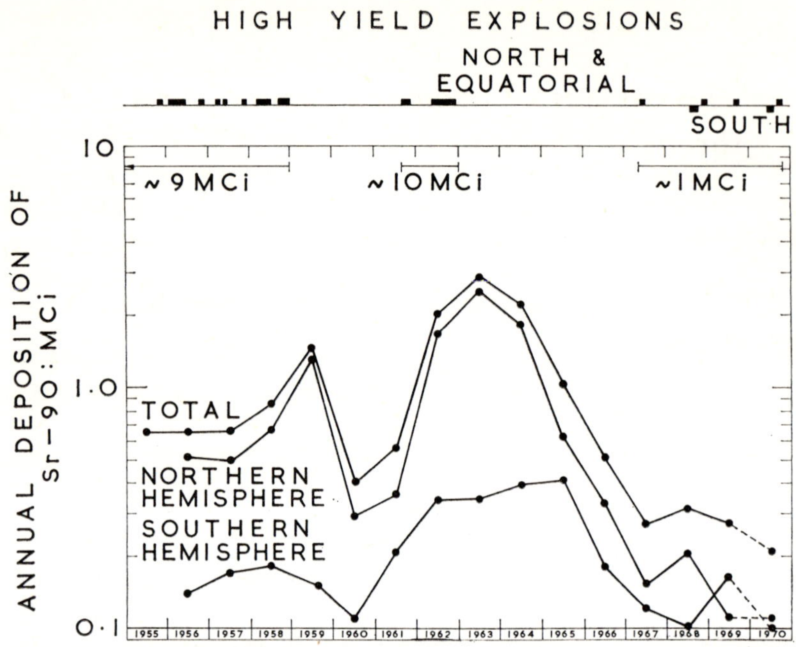

Figure 1. Annual deposition of strontium-90.

Stratosphere and troposphere

The stratosphere and troposphere are the two parts of the atmosphere separated by the tropopause, which varies in altitude from about 11 km at the poles to about 18 km at the equator, the stratosphere being above and the troposphere below. The tropopause is determined by the change of temperature gradient from negative in the troposphere to zero (or small positive) in the stratosphere. It follows that the stratosphere is a region where convection is inhibited and there is small or slow vertical mixing. On the other hand in the troposphere there is considerable vertical mixing, cloud, precipitation and 'weather'.

Observation of stratospheric debris

The radioactive debris injected into the stratosphere has been sampled from aircraft and balloon so that a direct measurement can be made of the stratospheric inventory and its variation with time (Krey & Krajewski 1972). However of more immediate concern to the human environment is the behaviour of this debris when it has returned through the troposphere to the earth's surface.

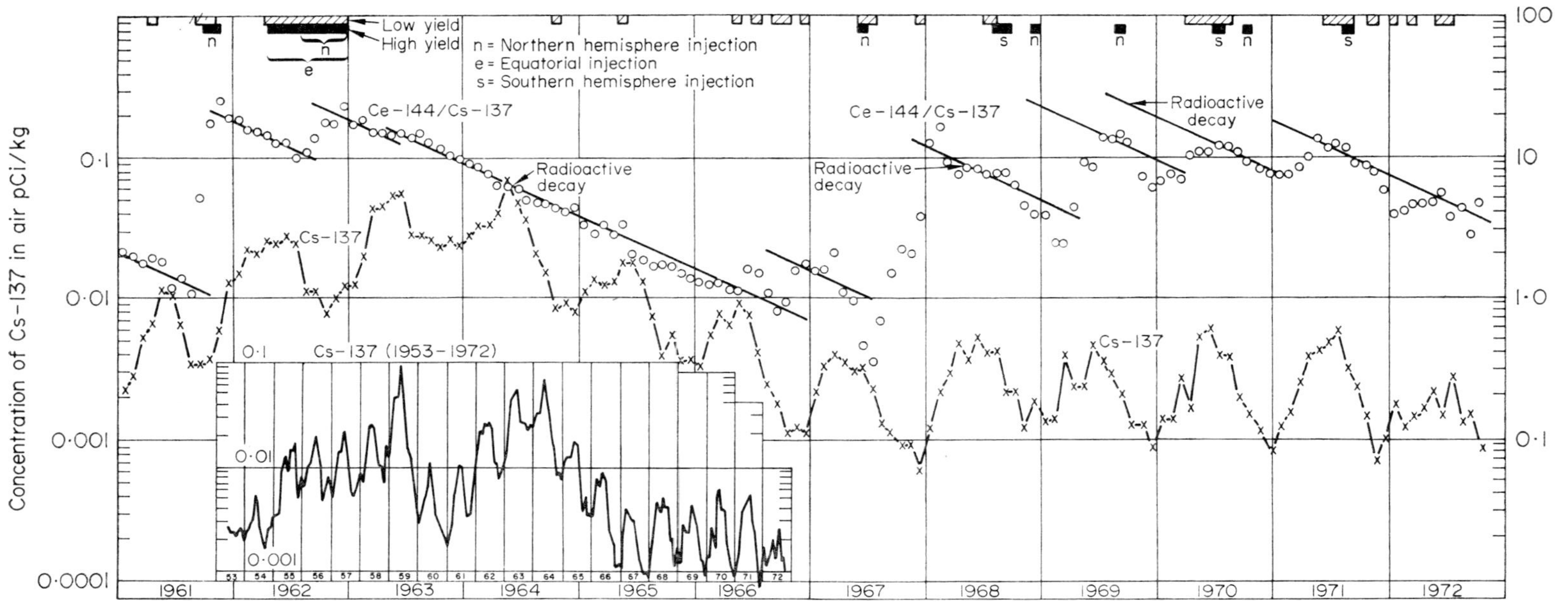

Figure 2. Concentration of caesium-137 in air near ground level at Chilton (Berkshire).

Figure 2 shows the concentration of caesium-137 in surface air at Chilton, Berkshire, since 1952 (Cambray *et al.* 1972). The general level of caesium-137 follows that of the worldwide deposition of strontium-90 in Fig. 1. The remarkable feature of this more detailed record of concentration in air is the regular seasonable variation such that a peak in concentration occurs in the spring or early summer of each year, superimposed upon the variation due to the incidence of nuclear explosions.

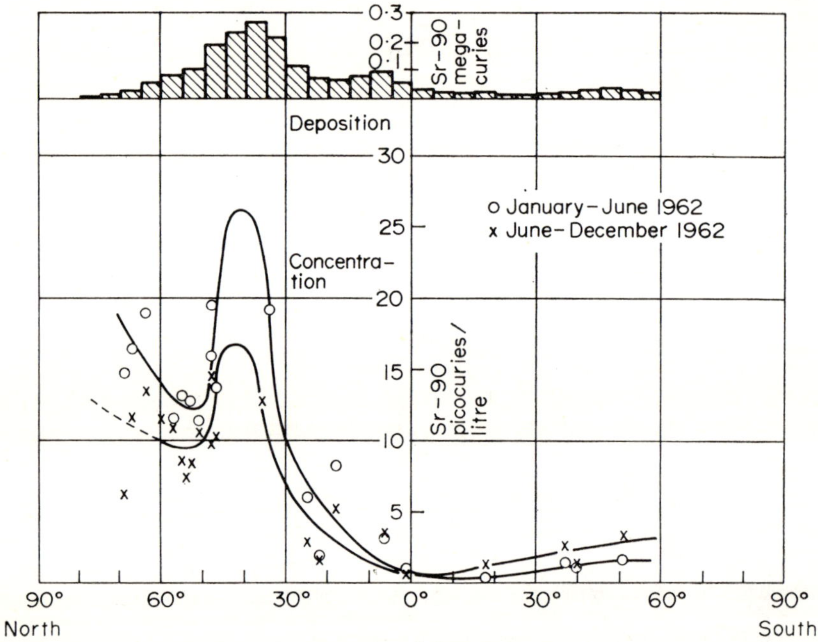

Figure 3. Strontium-90 in rain versus latitude, 1962.

Another feature of stratospheric behaviour is demonstrated in Fig. 3. This shows the variation with latitude of strontium-90 in rain, both as concentration and deposition (Peirson & Cambray 1965). There are well-marked peaks at about 40°N and a low minimum at the equator. Levels in the southern hemisphere were much lower at this time (1962) because of the preponderance of nuclear explosions in the northern hemisphere.

These observations of seasonable and latitude effects may be explained by the following mechanisms (Peirson & Cambray 1965; Stewart *et al.* 1958; Peirson 1961). There appears to be a subsidence of air over the pole in mid-winter that brings the debris into the lower stratosphere. From here it will settle through the tropopause and, more markedly, transfer through the mid-latitude gap in the tropopause. This could explain the latitudinal variation of deposited debris: the combination of winter subsidence in the stratosphere, inferred from meteorological observation, and of transfer activity across the tropopause—less well-defined in time—is responsible for the observed seasonal peak in surface air concentration.

The stratosphere has another characteristic that has been determined by observation of the fission products. The mean residence time is about 16 months, corresponding to a half-residence time of about one year (Peirson & Cambray 1967). This must apply to the lower stratosphere for there is evidence, based on the injection of special radioactive tracers into the upper stratosphere and mesosphere, that the residence times in these upper regions may range from 5 to 10 years. It has been possible to estimate the exchange between the northern and southern hemispheres essentially through the stratosphere (Peirson & Cambray 1967). Thus the mean residence time against interhemispheric transfer is from 3 to 5 years. This phenomena accounts for the increase in level in the southern hemisphere during 1962–5 (Fig. 1) before substantial testing had started in the south.

Observation of tropospheric debris

Apart from the radioactivity in the troposphere, that is passing from the stratosphere to the ground, there is the debris injected directly into the troposphere from small explosions or as a residue at low altitude from the larger explosions. This material, dominated by short-lived fission products, may be traced in its passage through the troposphere (Peirson 1967; Peirson & Cambray 1965). The trajectory shown in Fig. 4 describes the path of debris from the Russian Arctic site, under the influence of westerly winds (at 5·5 km), to the British Isles arriving some 11 days later. The trajectory was constructed from a knowledge of the explosion site and date, the date of arrival of the radioactive material in southern England and by making use of the routine meteorological data (winds and pressure). In Fig. 5 are shown two trajectories, a short path around the north pole and another at 2·5 km occurring in the opposite direction because the debris at this lower altitude has sheared from the main cloud (moving eastwards) under the influence of the then prevailing anti-cyclone over northern Scandinavia.

Other features of tropospheric behaviour have been elucidated by observation of these radioactive tracers. It has been shown that the mean residence time against removal by rain is about one month (half-residence time 3 weeks) (Burton & Stewart 1960). This value must apply to the troposphere as a whole, since the residence time below cloud level is known to be much shorter. The 'washout factor' for radioactive debris is about 700 (Cambray *et al.* 1970), expressed as a ratio of radioactivities for equal masses of rain and air. Except in dry climates rain is the main agent of removal of dust from the air. However, the mechanism of dry deposition is not insignificant. The 'dry deposition velocity' is typically 2 mm sec^{-1}, given by the surface deposition rate divided by the volumetric concentration in air.

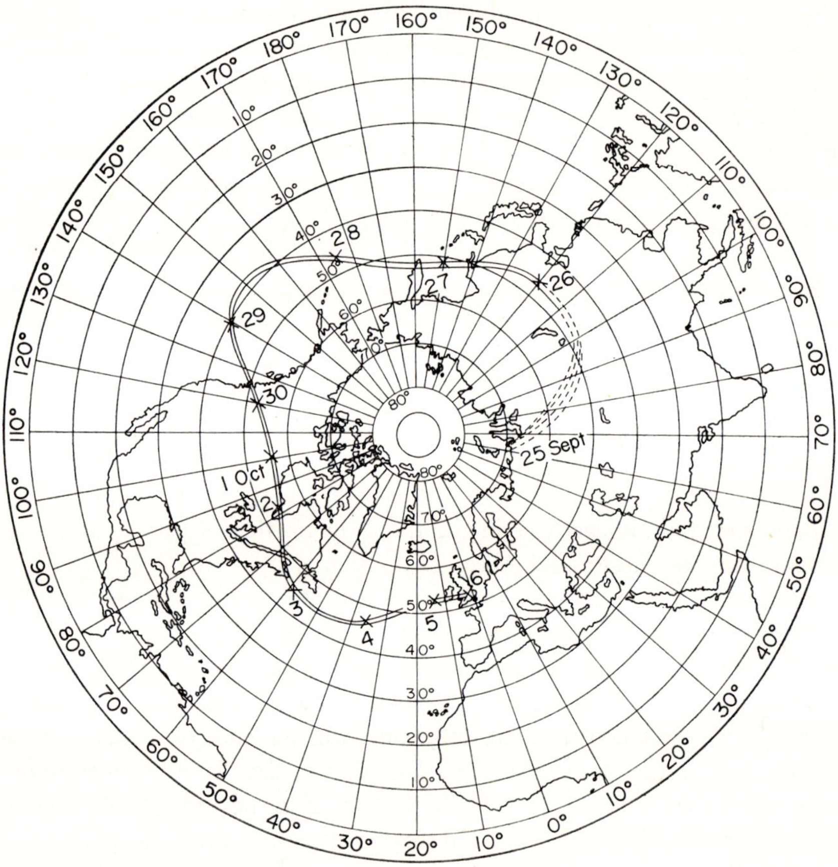

Figure 4. Trajectory at 500 mb (5·5 km) September–October 1962.

Conclusion

It has been demonstrated that the passage of nuclear weapon debris from the site of the explosion can be followed through the atmosphere and back to the ground. By observation of this radioactive tracer it has been possible to elucidate atmospheric behaviour and mechanisms that play important roles in the distribution and dispersal of any material, natural or artificial, that is injected into the atmosphere.

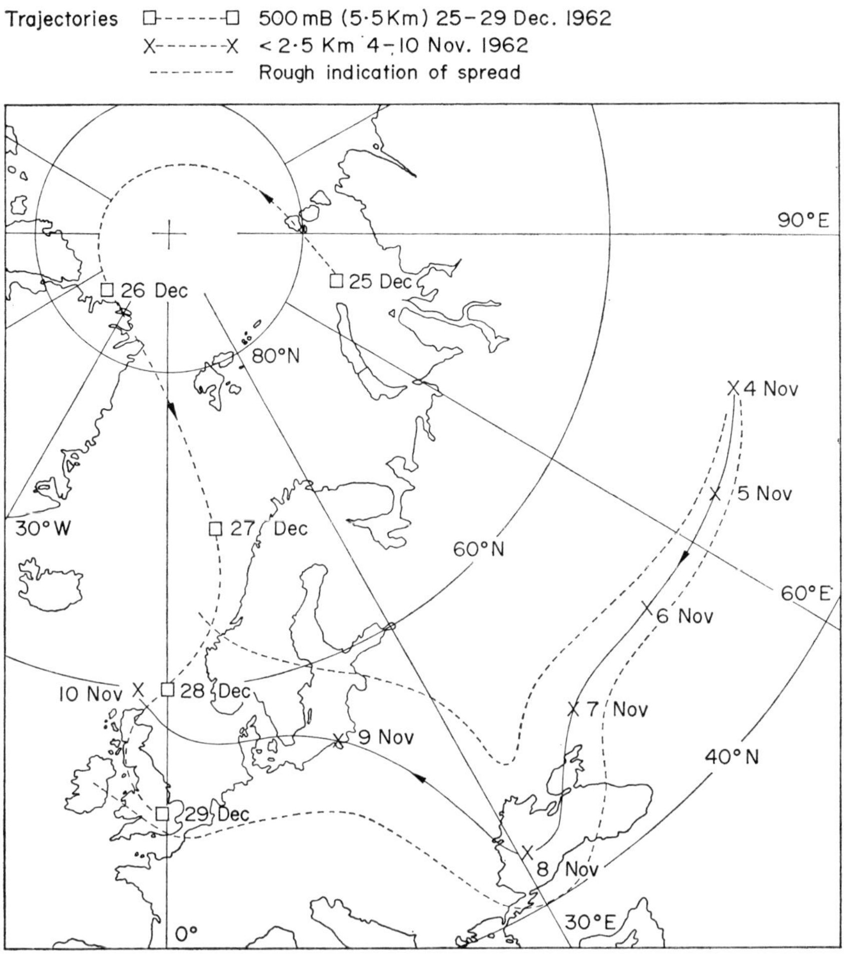

Figure 5. Trajectories November–December 1962.

Acknowledgments

Permission was kindly given by *Nature* to reproduce Figs 1, 3, 4 and5.

Summary

Radioactive debris from nuclear explosions have been inserted into the atmosphere at various times, places and to various altitudes. The passage of material can be traced from the troposphere, or from the stratosphere through the troposphere down to the ground. Fallout is by wet or dry deposition.

References

ANON. (1964) *Federal Radiation Council Report No. 4.* U.S. Gov. Printing Office, Washington, D.C.

BURTON W.M. & STEWART N.G. (1960) Use of long-lived natural radioactivity as an atmospheric tracer. *Nature, Lond.* **186**, 584–9.

CAMBRAY R.S., FISHER E.M.R., BROOKS W.L. & PEIRSON D.H. (1970) *Radioactive Fallout in Air and Rain: Results to the Middle of 1970.* AERE—R6556, H.M.S.O., London.

CAMBRAY R.S., FISHER E.M.R., PEIRSON D.H. & PARKER A. (1972) *Radioactive Fallout in Air and Rain: Results to the Middle of 1972.* AERE—R7245, H.M.S.O., London.

KREY P.W. & KRAJEWSKI B.T. (1972) *Stratospheric Inventories.* HASL—257, U.S.A.E.C., New York.

PEIRSON D.H. (1961) Transfer of stratospheric fission products in the troposphere. *Nature, Lond.* **192**, 497–500.

PEIRSON D.H. (1969) Interhemispheric transfer of radioactive pollution from nuclear explosions. *Phil. Trans. Roy. Soc. Lond. A* **265**, 295–300.

PEIRSON D.H. (1971) World-wide deposition of long-lived fission products from nuclear explosions. *Nature, Lond.* **234**, 79–80.

PEIRSON D.H. & CAMBRAY R.S. (1965) Fission product fallout from the nuclear explosions of 1961 and 1962. *Nature, Lond.* **205**, 433–40.

PEIRSON D.H. & CAMBRAY R.S. (1967) Interhemispheric transfer of debris from nuclear explosions using a simple atmospheric model. *Nature, Lond.* **216**, 755–8.

STEWART N.G., OSMOND R.D.G., CROOKS R.N. & FISHER E.M.R. (1958) *The Worldwide Deposition of Long-lived Fission Products From Nuclear Test Explosions.* AERE—HP/R2354, H.M.S.O., London.

SECTION TWO

Air and water degradation

Residence times in the stratosphere

A. D. CHRISTIE *Environment Canada, Toronto, Canada*

Introduction

Over the past decade there has been a growing awareness of the many ways by which man is modifying his environment. Many of the harmful changes become quickly apparent by their deleterious effects on the quality of life, ranging from adverse effects on human health, aesthetically offensive changes or mere environmental nuisances. The aerial effects of industrialization with its need for power, transportation, chemical control of insects and many other polluting practices are restricted to the lower atmosphere. Some environmental changes are brought about in a much more cryptic manner and become apparent on a longer time scale either through modification of global or regional climate.

The intention of this paper is to clarify our understanding of the impact that the stratosphere (that 15% or so of the atmosphere above the tropopause) may have in both the obvious and the more subtle type of degradation outlined above. The stratospheric processes will be described with a view to critically examining the various models for interpreting the effects of different trace contaminants, particular emphasis being given to the residence time concept in the light of its widespread use.

Stratospheric residence time and atmospheric transfer processes

The lifetime of a constituent in a reservoir is simply defined when the rate of depletion of the constituent is proportional to its amount. The solution of such a depletion equation: $dn/dt = -\lambda_D n$ is $n = n_0 \exp -\lambda_D t$, and permits us to define the lifetime as that time necessary for the constituent, n, to fall to $^1/e$ of its initial amount, n_0. i.e. $n/n_0 = e^{-1}$ when $\tau_D = 1/\lambda_D$.

In a reservoir with more than one process acting simultaneously to deplete the constituent each governed by a linear coefficient the solution is of the form $n = n_0 \exp -(\lambda_1+\lambda_2+ \ldots \lambda_n)t$, where λ_i, denotes the depletion coefficient for the i'th process, the *removal time* $\tau = 1/\lambda_1+\lambda_2+ \ldots \lambda_n$, and the net lifetime, τ, is expressible in terms of the lifetimes of each individual process by the expression: $1/\tau = 1/\tau_1+1/\tau_2+ \ldots 1/\tau_n$.

A simple example of the application of this rule to two processes is the depletion of tropospheric radioactivity by radioactive decay and washout processes (Engelmann 1968). It is assumed that the washout is equally effective throughout the entire reservoir which is not strictly true as the washout is dependent on the initial distribution of radioisotopes and on the nucleation and droplet growth processes.

The definition of a *residence time* depends on our being able to specify exchange between two reservoirs by a linear transfer coefficient. For such transfer to have any general application it must act impartially on all constituents and should be independent of any other reservoir that may simultaneously be transferring some of the constituent in the reverse direction. (In fact such a return transport is necessary if we are not to evacuate all mass from one reservoir to another.)

The fundamental difference between the *residence time* and the *removal time*, two parameters frequently mistakenly considered identical, can be simply examined by means of the equations. The atmosphere is represented in Fig. 1 by two reservoirs with the subscript S, denoting the stratosphere and T, the troposphere. Transfer from one reservoir to another is effected by the coefficient a with the subscripts defining the direction of transfer (a_{ST} is transfer *from* stratosphere *to* troposphere) and removal processes by λ. The S's represent the net source of the constituent and will be neglected in this part of the discussion. The equation governing depletion of constituent, n in the stratosphere is:

$$\frac{dn_S}{dt} = -a_{ST}n_S+a_{TS}n_T \tag{1}$$

Considering only the efflux $(-a_{ST}n_S)$, $n_S(t)/n_S(o) = \exp -a_{ST}t$ where t is the time from the initial time $t = 0$. The *residence time* for the stratosphere $\tau_S = 1/a_{ST}$ is independent of any other reservoir.

Under this linear depletion constraint the average lifetime, L, of a molecule in reservoir S is given by the expression:

$$L = \int_{n_S(o)}^{o} t dn_S \Big/ \int_{n_S(o)}^{o} dn_S = 1/a_{ST} = \tau_S \tag{2}$$

The *removal time* for depletion by more than one process in a single reservoir was shown to be simply expressible in terms of the lifetime of each process but the solution becomes less simple with consideration of an increasing number of reservoirs as will be illustrated by consideration of the stratosphere-troposphere case.

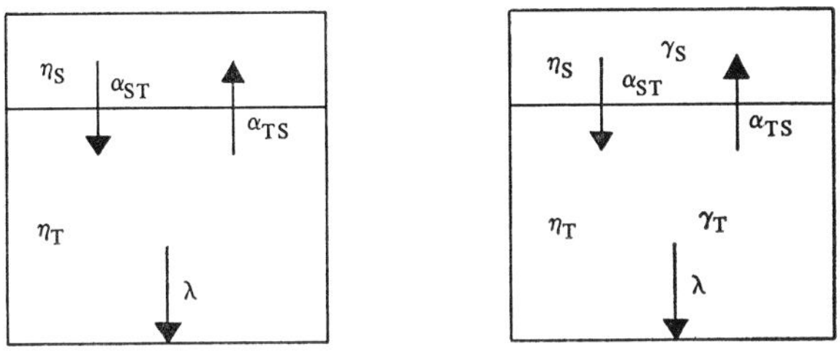

(a) (b)

$$\frac{d\eta_S}{dt} = -\alpha_{ST}\,\eta_S + \alpha_{TS}\,\eta_T + S_S(t) \qquad \frac{d\eta_S}{dt} = -(\alpha_{ST} + \gamma_S)\,\eta_S, + \alpha_{TS} + S_S(t)$$

$$\frac{d\eta_T}{dt} = \alpha_{ST}\,\eta_S - (\alpha_{TS} + \lambda)\,\eta_T + S_T(t) \qquad \frac{d\eta_T}{dt} = \alpha_{ST}\,\eta_S - (\alpha_{TS} + \gamma_T + \lambda)\,\eta_T + S_T(t)$$

Figure 1. Graphical representation of source, transfer and removal in a two reservoir system such as stratosphere and troposphere. (a) Source and transfer considered i.e. conservative constituent. (b) Depletion by chemical or other sink processes also considered.

The general equations corresponding to Fig. 1(a) where the differential operation is represented by $D = d/dt$ are:

$$(D+a_{ST})n_S - a_{TS}n_T = S_S(t) \tag{3}$$

$$-a_{ST}n_S + (D+a_{TS}+\lambda)n_T = S_T(t) \tag{4}$$

Separating variables

$$[(D+a_{ST})(D+a_{ST}+\lambda) - a_{ST}a_{TS}]n_S = (D+a_{ST}+\lambda)S_S(t) + a_{TS}S_T(t)$$
$$[(D+a_{ST})(D+a_{ST}+\lambda) - a_{ST}a_{ST}]n_T = (D+a_{ST})S_T(t) + a_{ST}S_S(t)$$

where a_{ST}, a_{TS} and λ are constants the solutions are of form:

$$n_S = C_{1S}\exp -\delta_1 t + C_{2S}\exp -\delta_2 t + \pi_S$$
$$n_T = C_{1T}\exp -\delta_1 t + C_{2T}\exp -\delta_2 t + \pi_T$$

δ_1 and δ_2 are solutions of the algebraic quadratic

$$\delta^2 + (\lambda + a_{ST} + a_{TS})\delta + a_{TS}a_{ST} = 0$$

π_S and π_T are particular integrals depending on $S_S(t)$ and $S_T(t)$.

It is apparent then that there is no longer a simple factor for the *removal time* and the situation becomes progressively more complicated as further reservoirs are added in 'box' models. To the extent that we can obtain analytical solutions from the linear models we must derive residence time values for use in the box models.

Information from radioisotopes injected into the stratosphere by atmospheric testing of nuclear bombs provides the best source of data for deriving the stratospheric residence time, and the appropriate equation is:

$$\frac{dn_S}{dt} = -a_{ST}n_S + a_{TS}n_T$$

Since a_{ST} is uniform for all molecules $dm_S/dt = -a_{ST}m_S + a_{TS}m_T$ where m_i represents all molecules in reservoir i. If a_{ST} and a_{TS} are both constants we must assume a steady state condition or the mass in one reservoir will be transferred into the other so $a_{TS} = a_{ST}\ (m_S/m_T)$. Since $n_T < n_S$ and $m_S \simeq 14\%$ of m_T, $a_{TS}n_T \ll a_{ST}n_S$. To a good approximation the depletion in the stratospheric inventory of a constituent $\triangle n_S = -a_{ST}\,n_S\,\triangle t$. n_S has been obtained from knowledge of bomb yields and from global assessments made from extensive aircraft and balloon sampling programmes. $\triangle n_S$ may also be inferred from the latter, or with less accuracy from global deposition measurements since the tropospheric removal time is short relative to that of the stratosphere as discussed by Sheppard (1963) Engelmann (1968) and others.

The radioactivity associated with a particular series of tests may be isolated either by a specific identifying isotope or by isotope ratio techniques. I propose to simply comment briefly on the results rather than review the evidence.

The early estimates of 5 to 10 years reviewed by Martell (1959) have been discredited in favour of values ranging from a few months to a few years depending on the location and season of injection. Studies carried out by Feely (1960), Staley (1960), Stebbins (1961), Gustafson *et al.* (1962) and others have given values of 2 to 4 months for the lower polar stratosphere, 6 to 12 months for the tropical lower stratosphere depending on the season of injection and 12 to 48 months at middle stratospheric levels.

The large variations in the experimentally determined residence time strongly suggest that the assumption of a linear transfer process is unrepresentative of the atmospheric exchange. Moreover, the global monitoring of surface deposition showed clearly that there is distinctive meridional and seasonal variation in fallout from which we must conclude that a_{ST} is not constant with latitude or season, further corroborating that conclusion.

To permit more accurate prediction of fallout patterns Staley (1962) and Peirson & Cambray (1965) modified the transfer coefficient to incorporate factors involving both seasonal and meridional variations. There will not, in general, be a simple residence time associated with such an exchange coefficient unless the analytic solution represents an exponentially damped periodic function. It is clear that the value of a_{TS} must be simultaneously modified and verified to be consistent with net mass continuity considerations.

The total stratospheric mass is only 13·1% or 15·7% of the net hemispheric mass in summer and winter respectively. This 2·6% variability is neglected in the steady state assumption implicit in the constant transfer coefficient but should be simulated by the varying transfer coefficients.

In 'box' models where the equation for net content with constant coefficients are solved, the seasonal and meridional variations are obtained using statistically derived empirical adjustments (Krajewski & Krey 1971). These answers have utility in special cases but the solution is unrepresentative of the atmospheric processes.

Many pollutants entering the atmosphere interact chemically with a variety of constituents in the ambient air and we now consider to what extent the photochemical processes may be represented by a further linear depletion coefficient. A complete reaction scheme comprises S reactions of types:

$n_l + n_m + n_o \rightarrow n_p + n_q$ with rate k_{lmo}
$n_l + n_m \rightarrow n_p + n_q$ with rate k_{lm}
and $n_l + h\nu \rightarrow n_p + n_q$ with photodissociation J_l between Z chemical species ($Z < S$), where n is the number density of the constituent identified by an integral subscript l, m, o, p, q, between l and Z. J_l varies with solar zenith angle (location, time of day and date) and k_{lmo} and k_{lm} may be temperature dependent.

The net rate of change of constituent n_p attributable to chemical reactions is

$$\frac{\partial n_p}{\partial t} = P_p - Q_p n_p - R_p n_p{}^2 = \triangle_p \qquad (5)$$

where P_p, Q_p and R_p are functions of type:

$$\begin{aligned} P_p &= \sum J_l n_l + \sum k_{lmo} n_l n_m n_o + \sum k_{lm} n_l n_m, \; p \neq l, m, o. \\ Q_p &= \sum J_l + \sum k_{lmp} n_l n_m + \sum k_{lp} n_l \\ R_p &= \sum k_{lpp} n_l + \sum k_{pp} \end{aligned}$$

Only if $R_p \equiv O$ and Q_p is constant can the photochemical lifetime defined in a manner analogous to the residence time (Hesstvedt 1955) be considered meaningful.

In general Q_p varies with height and season for most trace gases as illustrated in the studies of ozone (Hesstvedt 1968; Crutzen 1969, 1971) so that representation of the removal process by a linear coefficient for the entire stratosphere must be viewed with considerable scepticism.

In the special case where Q_p may be considered uniform throughout the reservoirs the relative influence on the distribution of removal by chemical and transfer processes may be seen from comparison of Q_p and the transfer coefficient α. If $Q_p \gg \alpha$ the transports may be neglected whereas $Q_p \ll \alpha$, the constituent may be treated as chemically inert (i.e. conservative).

For the many cases where $Q_p \simeq a$, residence time 'box' models will be ineffective. There will, moreover, be no basic improvement if the number of reservoirs is increased or the time constants modified, the flaw is fundamental and lies in the representation of the transfer and removal processes by linear functions.

To improve our models requires solution of the basic non-linear continuity equations consistent with the atmospheric general circulation and these have no general analytical solutions.

Atmospheric thermal structure and convection in the general circulation

The general properties of the circulation affecting transfer or trace constituents will be reviewed briefly to aid us in understanding the processes that must be incorporated in a model to assess the impact of pollution on the environment.

The *mean* thermal lapse rate in the troposphere is much greater than that in the stratosphere, but is still statically stable (Appendix 1). As shown in Fig. 2 the boundary between these two regions, the tropopause, in general

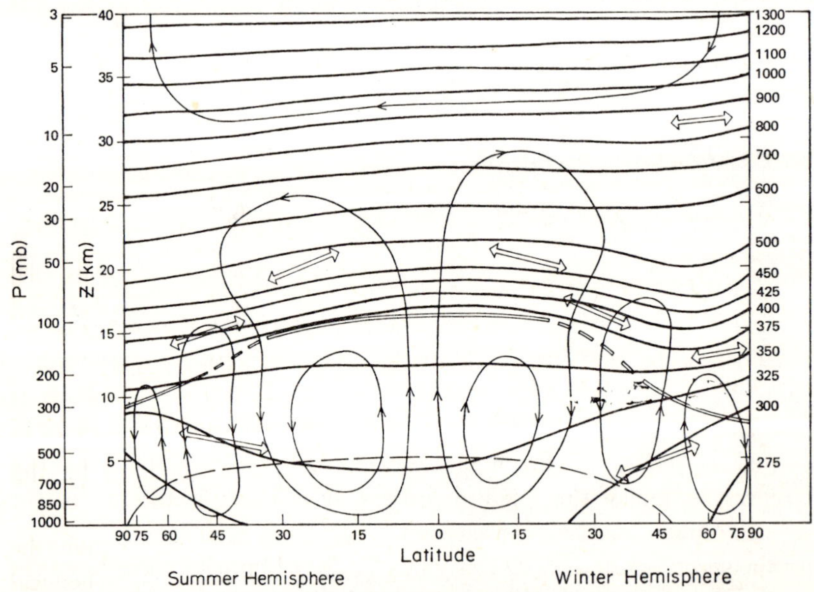

Figure 2. Meridional section of potential temperature (continuous thick lines) and tropopause (double line) with schematic presentation of the transport processes. Mean meridional cells are shown by continuous thin lines and maers eddy mixing by the arrows.

slopes downward from high, cold values in the tropics to lower warmer levels over the poles throughout the year.

The mean isentropic surfaces (surfaces of constant potential temperature, see Appendix 1) dip equatorward from the lower stratosphere into the troposphere clearly illustrating that, to the extent that diabatic heating may be neglected, the tropopause does not constitute a barrier to flow between the two regions.

On individual cross sections it is often difficult to identify a unique physically meaningful tropopause from a vertical thermal profile on the basis of stability criteria such as proposed by WMO (Appendix 2). There exists an almost infinite variety of profiles in the transition region separating the troposphere and stratosphere particularly in middle latitudes where thermal inversions of variable thickness are often observed. Several authors (Reed 1955; McClain & Danielsen 1955; Reed & Danielsen 1959; Danielsen 1959) have shown by detailed analysis that thermal inversions noted on individual temperature profiles have quasi-horizontal continuity and represent undulating layers of large static stability separated by less stable layers. These migrating stable laminae whose boundaries are almost isentropic surfaces may be included in the stratosphere at some locations and the troposphere at others. Danielsen (1959), moreover, demonstrated the temporal continuity of these stable layers over short periods of about a day and showed that the tropopause did not constitute a barrier to quasi-horizontal exchange of air between stratosphere and troposphere. Reference to Fig. 3 clearly demonstrates that heating rates are less than a degree Kelvin a day which implies that potential temperature is indeed conserved over periods quite long enough for exchange to take place.

The thermal field is only one of the constraints on the movement of air in the atmosphere. The velocity components are also limited by momentum balance considerations and any meaningful model must incorporate the variety of scales and modes of motion that occur. The average fields comprise zonal currents and mean meridional cells upon which are superimposed a variety of oscillatory systems.

The mean zonal wind field consistent with the thermal field is illustrated in Fig. 4. In the troposphere and lower stratosphere the mean winds, as derived from several years of data, are fairly symmetrical about the equator with low latitude easterlies giving way to westerlies in the sub tropics, these (Ferrel) westerlies reaching a maximum value that varies between 17 and 37 m sec^{-1}, depending on season and hemisphere. Seasonal variations result in a poleward or equatorward movement of the wind maximum in summer and winter respectively as well as in intensity changes.

Above the tropopause the horizontal thermal gradient reverses as shown in Fig. 2 and a consequent weakening of the winds consistent with the

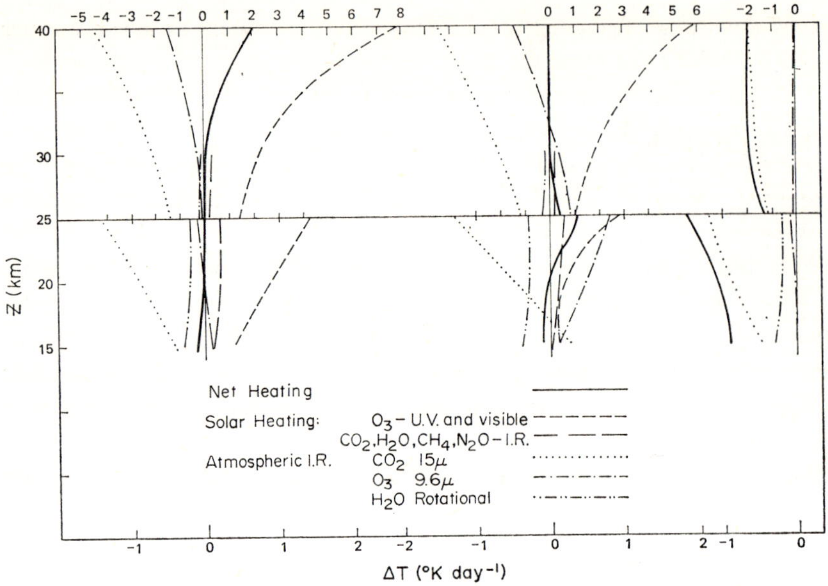

Figure 3. Profiles of heating rates in the atmosphere, net and components.

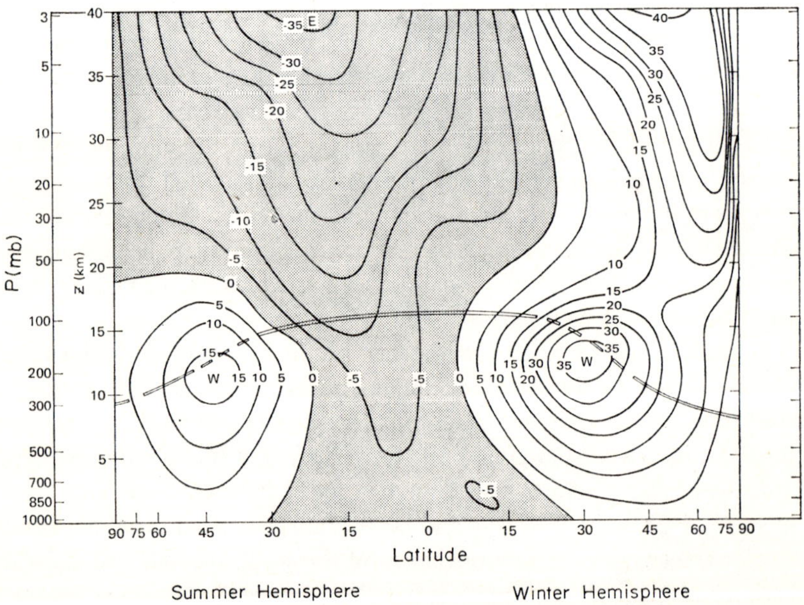

Figure 4. Meridional section of winter and summer seasonal average zonal wind component in m sec^{-1} in troposphere and stratosphere.

thermal wind relationship takes place.* The compensation is complete by the 50 mb level in summer when the stratospheric circulation is zonally symmetric and anticyclonic (Fig. 5).

In the winter the radiative cooling in middle stratosphere at high latitudes results in the development of a strong, cold cyclonic vortex with winds rising to values of 40 to 50 m sec^{-1} around 60° latitude in northern and southern hemispheres respectively as illustrated by Newell *et al.* (1970).

The tropical stratospheric zonal winds, moreover, exhibit a quasi-biennial fluctuation with period of approximately 26 months which has a maximum amplitude between 25 and 30 km where the wind fluctuates between values of about 15 m sec^{-1} westerly to 25 m sec^{-1} easterly (Reed 1964, 1965; Murgatroyd 1970).

Superimposed upon the mean zonal circulation there exists a variety of large scale fluctuations.

In the tropical stratosphere two types of waves have been detected in the equatorial westerlies: The Rossby-type waves called Yanai Waves (Yanai & Maruyama 1966; Maruyama & Yanai 1967; Maruyama 1969) which have a wavelength of about 10,000 km, have phase lines that tilt westward with height and propagate westward at about 2,000 km day. The Kelvin type wave (Wallace & Kousky 1968; Holton & Lindzen 1968; Maruyama 1969) which show fluctuations in the zonal wind component and temperature with periods of about 15 days.

In the Ferrel westerlies that predominate from the sub-tropics to high latitudes in troposphere and lower stratosphere there exist baroclinic Rossby waves (Bjerknes & Solberg 1922; Rossby 1939, 1949; Namias & Clapp 1944; Palmer 1948). These waves have trough and ridge lines that slope from NNE to SSW, tilt westward with height, and their thermal field lags the contour field by up to a quarter the wavelength. The instability of the

* The thermal wind relationship is derived from an approximate form of the Eulerian momentum equations known as the geostrophic approximation which implies a balance between the Coriolis and horizontal pressure-gradient forces.

Expressed mathematically

$$\frac{dV_G}{dt} = (R/f)k\wedge\nabla pT$$

where V_G: geostrophic wind

f: Coriolis parameter 2Ω Sinϕ

Ω: angular speed of the earth's rotation

ϕ: latitude

k: unit vector in the vertical

$\wedge$: vector multiplication

∇p: gradient on an isobaric surface

zonal vortex giving rise to these waves has been widely studied (Charney 1947; Eady 1949; Kuo 1952; Green 1960; Pedlosky 1965, 1970; Miles 1964a, 1964b; Brown 1969a, 1969b).

In the middle and upper stratosphere of the northern hemisphere the winter polar vortex which develops and intensifies from autumn to early winter as a symmetrical cyclonic current, becomes asymmetrical with a persistent warm ridge over the Aleutians dominating the circulation (Boville *et al.* 1961) after which time the circulation is dominated by developments of two main types (Wilson & Godson 1963). In the wave number one type major warming a warm cell from mid-latitudes displaces the cold vortex from the pole and the displaced vortex tends to become less intense or, in the case of a final warming, disappear. In the case of a 'bi-polar' type, two warm cells converge on the pole from the Siberian and Atlantic regions, ultimately splitting the vortex and displacing both cold lows southwards. In either case the net effect is to warm the polar region and reverse the wind field to easterly. If either process takes place early in the winter the radiative cooling will tend to re-establish the cold vortex once more, and the warmings in the regions invaded by the warm ridges are termed 'major' warmings, when the winter regime is terminated by the polar warming it is classified as a final warming event. In the Southern Hemisphere, though the vortex becomes more intense, the circulation there does not show nearly as marked a tendency to wave number two development in late winter nor the occurrence of such major warmings prior to the return of solar heating.

The mean wind components will control the general drift of air, however, the wave patterns noted above will result in large scale mixing or dispersion which will now be briefly discussed.

The *horizontal* dispersion characteristics of diffusion from a continuous fixed source have been studied by means of constant level balloons (Neiburger & Angell 1956; Angell 1969, 1961) and geostrophic trajectories for both tropospheric (Durst *et al.* 1959; Kao & Bullock 1964; Kao 1965) and stratospheric levels (Murgatroyd 1969). The dispersion in summer and winter seasons is shown graphically in Figs 5 to 10 by the ellipses with semi axes represented by the standard deviation of the zonal and meridional displacement vectors. In the case of dispersion of particles from an instantaneous source the *relative* motion between particles must be studied and this has been done for a variety of levels and latitudes (Kao 1968; Kao & Gain 1968; Kao & Hill 1970) and in a variety of locations in synoptic trough/ridge patterns (Kao & Henderson 1970). In general the relative dispersion of a cluster will be less than that of a continuing series of trajectories from a point source as the latter will be strongly affected by the changing position of the long wave troughs and ridges, consequently we would expect that the mixing from point sources such as nuclear bombs would proceed more slowly than implied by Figs 5 to 10. We see clearly, however, that large scale horizontal eddy mixing is greater in the upper troposphere than in the stratosphere and that it is considerably less effective in the summer, middle stratospheric anticyclonic circulation. The application of this kind of information to atmospheric modelling will be discussed later.

Figures 5–10. Contour charts on various isobaric surfaces in summer and winter seasons representing the flow fields. The diversion over four days is shown by the dotted envelopes.

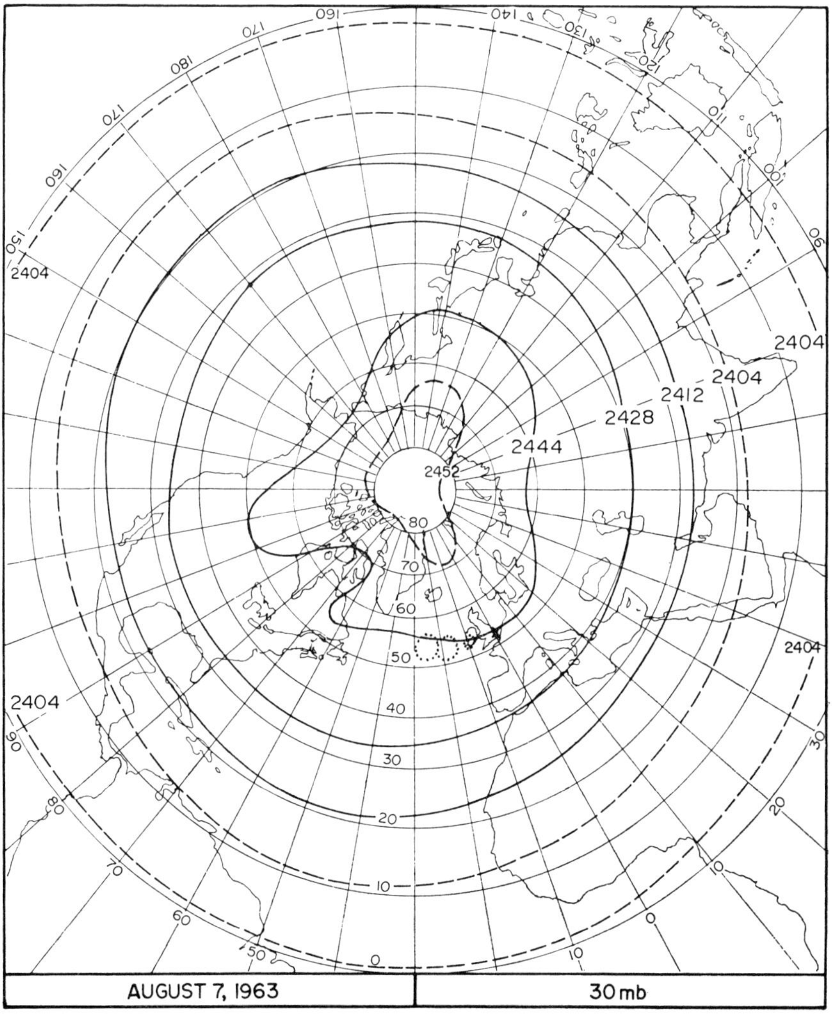

5: 30 mb level, summer

No direct studies of the dispersion in the vertical have been attempted though in a recent critical examination of trajectory methods (Danielsen 1973) has shown that the use of isentropic trajectories would permit assessment of the vertical dispersion and give more realistic estimates of the horizontal dispersion.

Much has been inferred about the relative importance of mean and large scale eddy transports from studies of a variety of atmospheric tracers.

Seeking to interpret the observed meridional gradient of total ozone and the absence of diffusive separation in the stratosphere Dobson *et al.* (1929)

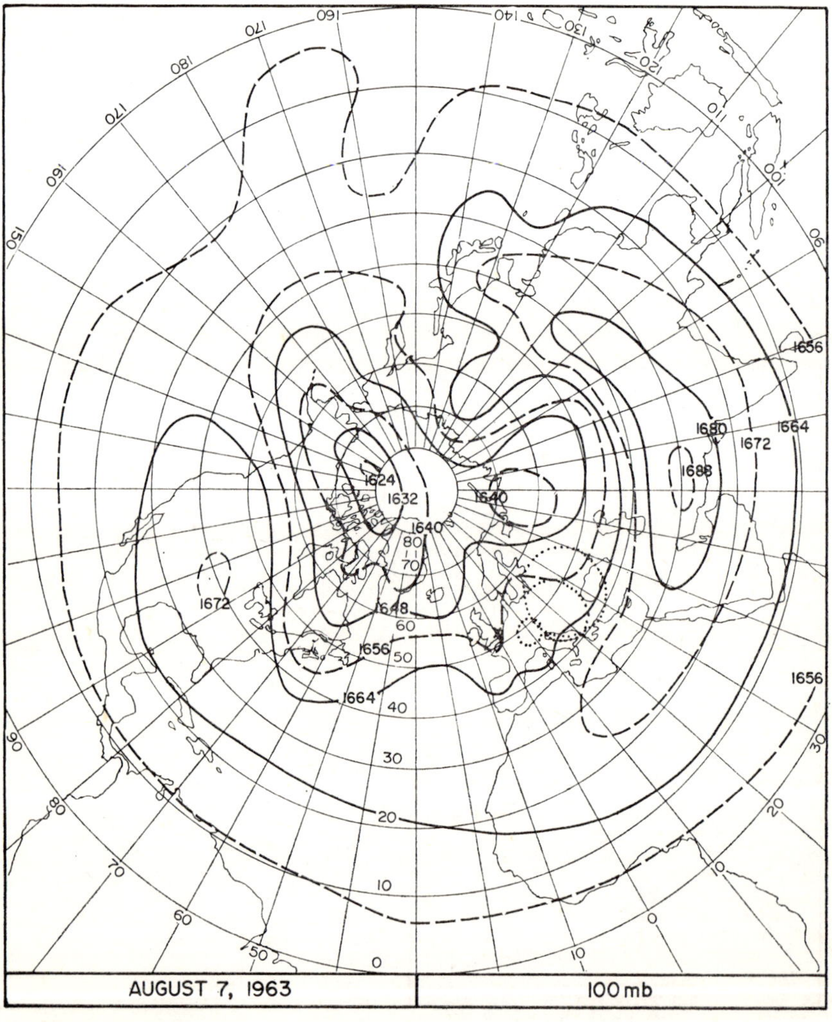

6: 100 mb level, summer

postulated the simplistic model of transfer incorporating slow large scale meridional overturning and small scale vertical eddy diffusion. It was proposed that air rose through the equatorial tropopause, spread polewards and downwards through the middle and high latitude tropopause and that vertical eddy mixing took place through the tropopause and lower stratosphere. Brewer (1949) estimated mean velocities for this circulation from profiles of water vapour in the descending current in middle latitudes. Further support for the existence of the mean transtropopause ascending current in the tropics may be inferred from the relative paucity of ozone

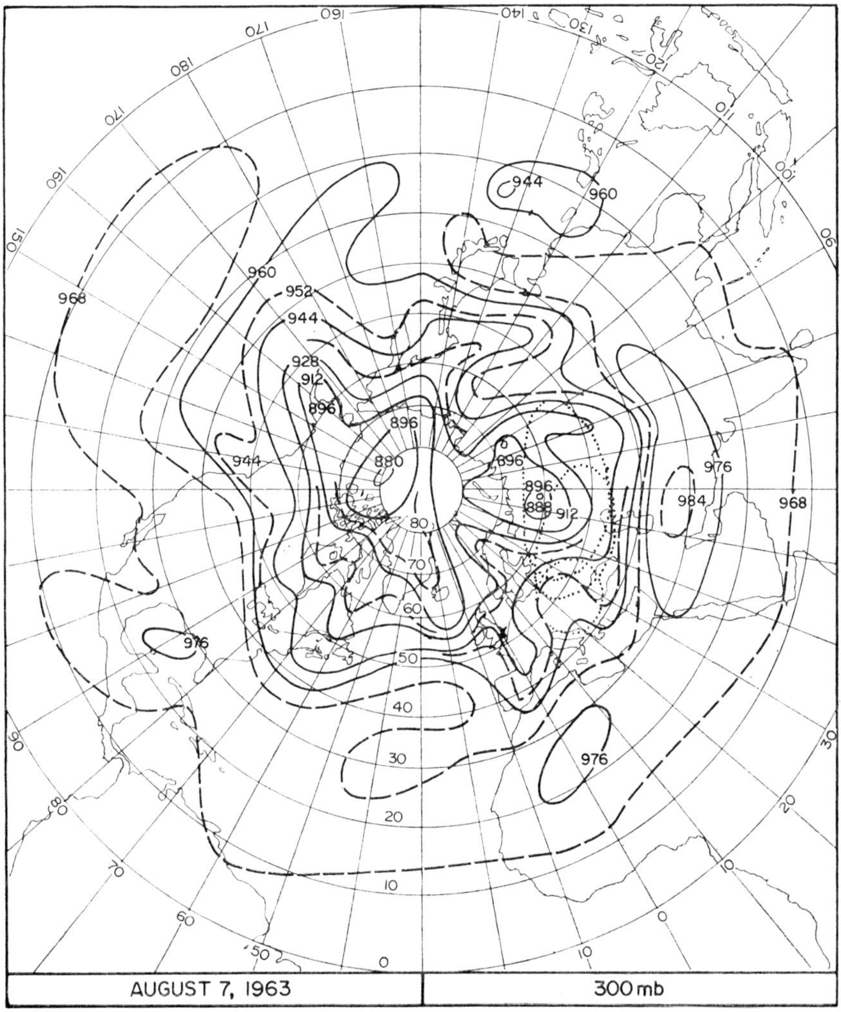

7: 300 mb level, summer

(Dutsch 1956; Ramanathan & Kulkarni 1960) and radioisotopes from stratospheric sources including tungsten-185 (Feely 1960; Feely & Spar 1960; Newell 1963; Martell 1968), strontium 90 (Machta & List 1959; Telegadas & List 1969), strontium 89 (Feely *et al.* 1965), excess carbon 14 (Hagemann *et al.* 1959, 1965) and from rhodium 102, cadmium 109 and platonium 238 (List *et al.* 1966; List & Telegadas 1969). Telegadas & List (1969), moreover, suggest on the basis of an analysis of C14/Sr 90 ratios, that air of tropospheric origin rises through the tropopause at latitudes higher than 70°N. The high values of the ratio found in the high latitude

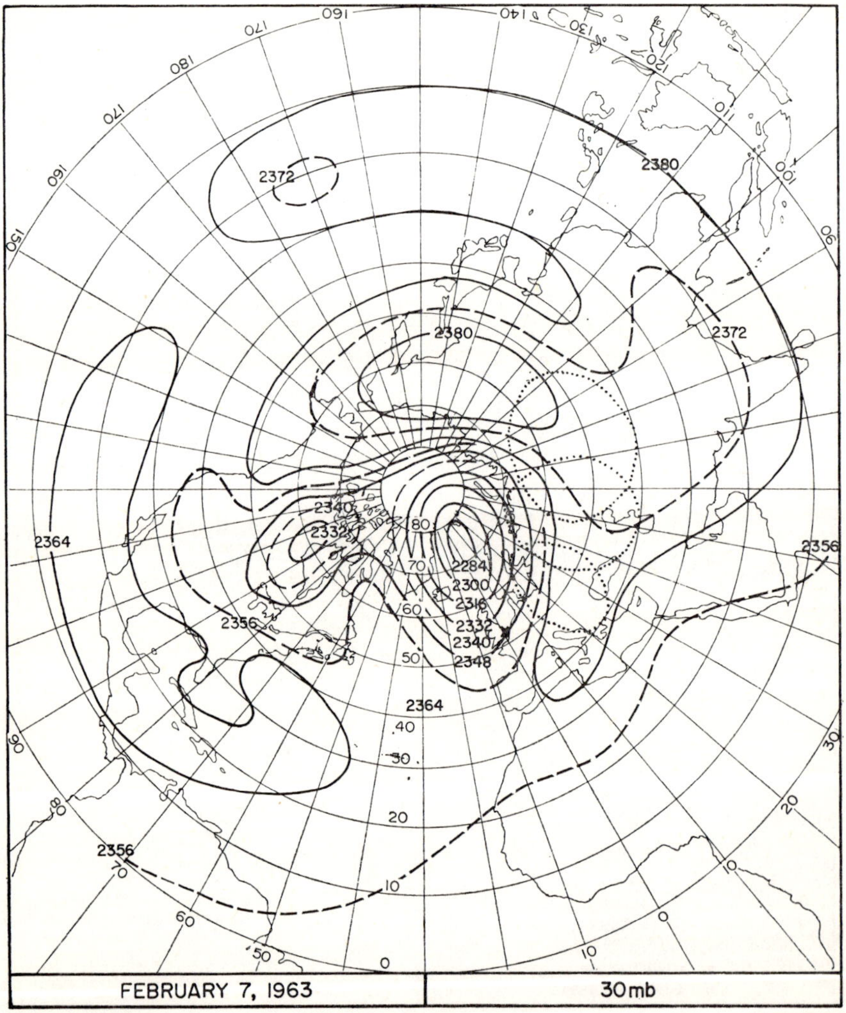

8: 30 mb level, winter

lower stratosphere are attributable to the more efficient scavenging of particulates, to which Sr 90 adheres, than carbon dioxide gas by rain in the troposphere. High relative values of the C14/Sr 90 ratio implies air of recent tropospheric history.

These inferences from tracer data are consistent with analyses of the observed velocities in the mean meridional cells as reported by Newell *et al.* (1970).

The necessity of large scale convective eddy transports to balance the global budgets of momentum and heat have been reviewed recently by

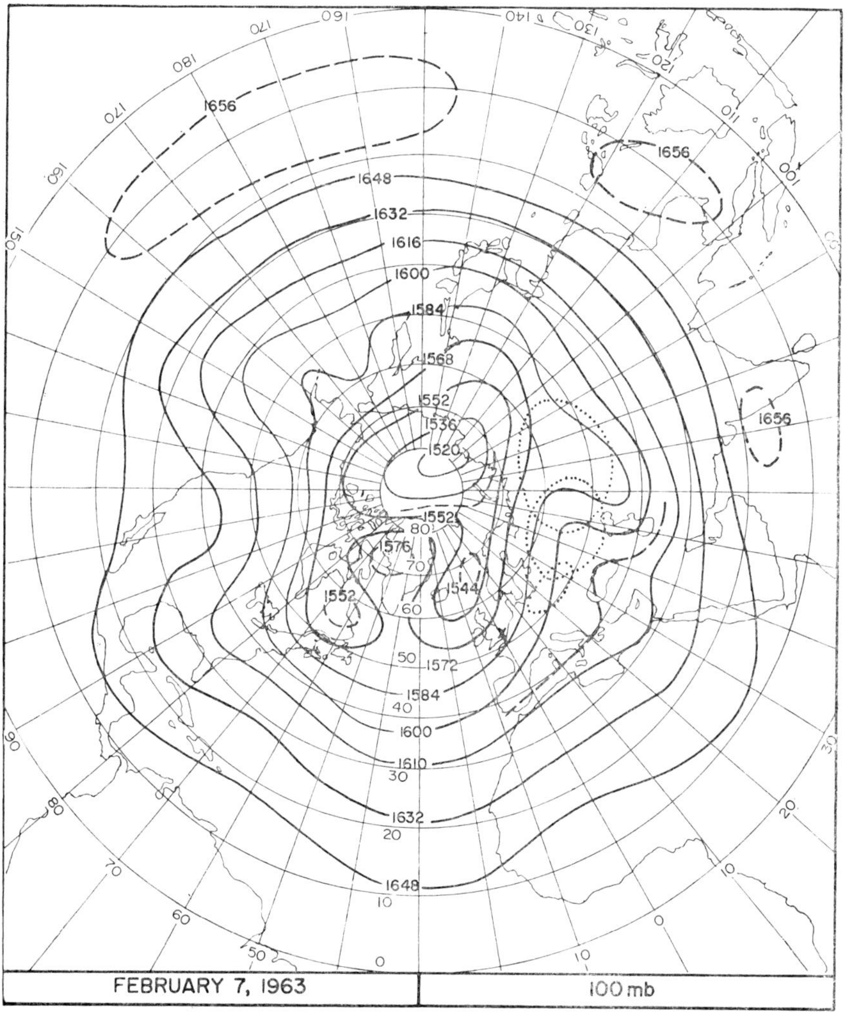

9: 100 mb level, winter

Murgatroyd (1970). On energetic considerations alone Newell (1964) (see Appendix 3) proposed that large scale quasi-horizontal eddy mixing take place at angles related to the mean isentropes as illustrated in Fig. 2.

Qualitative support for this may be drawn from the observation that all radioisotopes whose sources are within the upper atmosphere develop distributions with their surfaces of maximum concentration sloping downwards from tropics to high latitudes at an angle in excess of the isentropes (Machta *et al.* 1970) as shown in Fig. 11.

On the basis of diagnostic studies of ozone from a global network

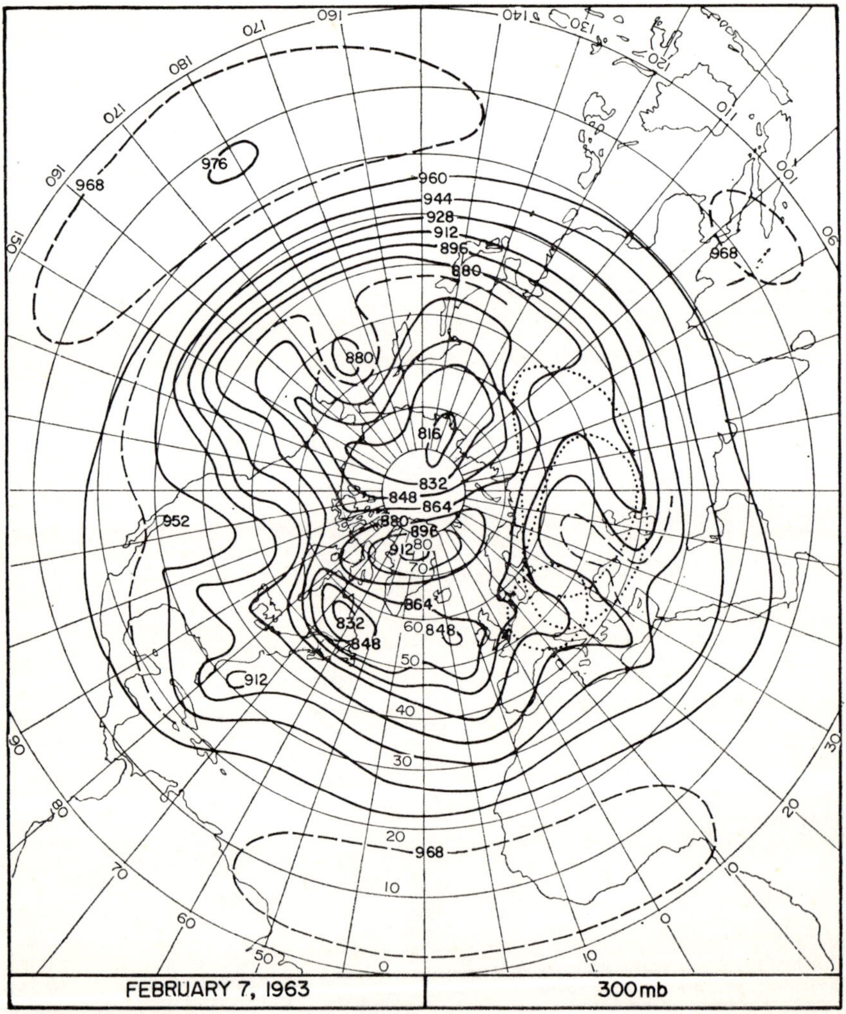

10: 300 mb level, winter

(Newell 1961, 1964) and volcanic dust (Dyer 1970), using continuity equatoins to be discussed in the next section, it has been demonstrated that *while the Brewer-Dobson mean circulation is of major importance in low latitudes the eddy transfers dominate elsewhere.*

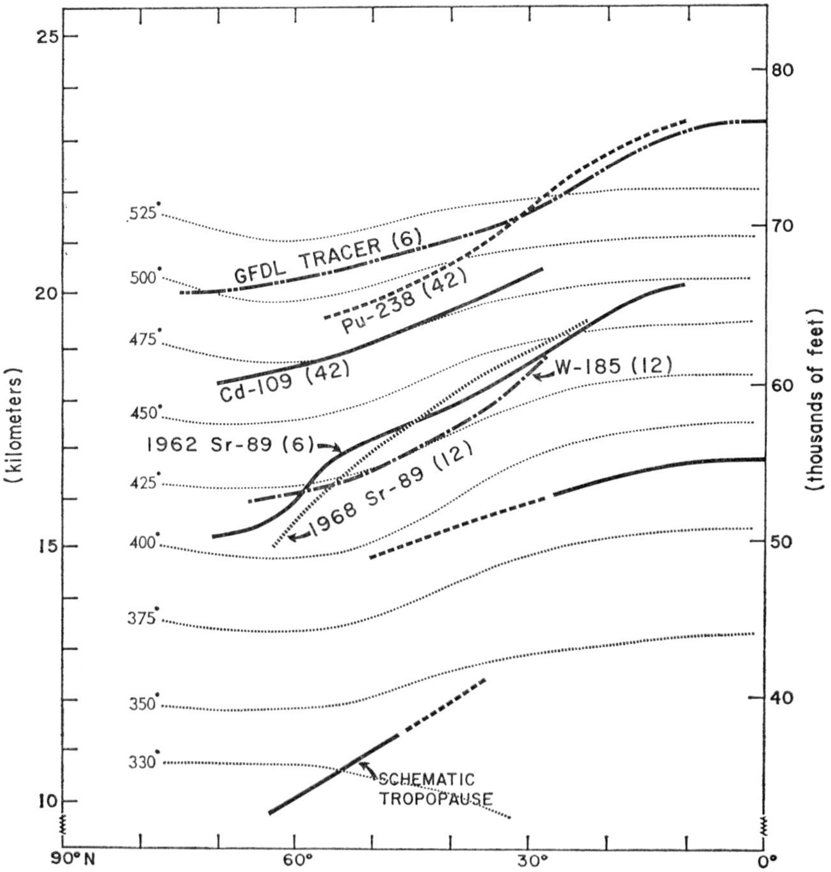

Figure 11. Observed levels of maximum concentration of various radioisotopes injected within the stratosphere and mean annual isentropes. The numbers of parenthesis are the number of months from date of injection. (After Machta *et al.* 1970.)

Studies of ozone and other tracers indicate the complexities of transfer through the tropopause, but suggest strongly that the major region of exchange of air is in the vicinity of the jet axes where the tropopause is ill-defined. Briggs & Roach (1963) analysed cross sections of ozone and water vapour measured by aircraft and demonstrated cases of intrusions of stratospheric air (high ozone, low water vapour) associated with layers of high stability that dipped downwards from the lower stratosphere into the upper troposphere. Danielsen *et al.* (1969) analysed successive distributions of three radioisotopes with long half lives injected into the stratosphere by nuclear explosions and, inferring that air with high concentrations originated in the stratosphere, demonstrated a tendency for such air to enter the troposphere in layers of high stability. Staley (1962), moreover, demonstrated from vertical profiles of gross β activity that where distinctive maxima were noted, they were associated with stable layers containing air of recent stratospheric origin as verified by trajectory analysis. He noted, however, that such activity maxima occurred only on about half the profiles on which stable layers appeared, so that the existence of a layer of above normal stability was not a sufficient condition for downward transport through the tropopause to take place. These early studies have been amply verified by subsequent studies (Reiter 1963; Reiter & Mahlman 1965; Mahlman 1965a) and Mahlman (1965b) has defined a statistical relationship between the transtropopause transport and an index representing the cyclonic activity in the upper troposphere. In specific cases Reiter (1968) has suggested that where relatively small cyclonic and anticyclonic mean shears exist on either side of a jet axis when flow is forced over a 'cold dome' splitting of the jet stream occurs and produces the dynamic conditions appropriate for stratospheric-tropospheric exchange.

Several salient facts established in the preceding comments illustrate the inapplicability of the linear exchange hypothesis.

(a) No specific or non-specific tracer injected into the stratosphere has become well mixed within that region over a sampling period of a few years implying that the transport within the stratosphere is highly anisotropic. Under these circumstances it is unlikely that the lifetime of any molecule in the region will be the same as that in another location as implied by the residence time concept. This same non-uniformity is distribution has also been noted in the case of lead 210 (Feely & Seitz 1970) though the source was in that case natural radioactive decay of radon from a ground source.

(b) Exchange through the tropopause may be by mean or eddy transfers or both but is clearly *not* constant with latitude or season. Studies of meridional and seasonal fallout from nuclear explosions (Staley 1962; Peirson & Cambray 1965) indicate that tracers with stratospheric sources are transferred into the troposphere predominantly in middle latitudes and that such transport is greatest in late spring and least in autumn. It was shown in the previous section that these observations were inconsistent with the linear exchange concept and we proceed to more realistic parameterization in the next section.

(c) Study of individual cross sections quickly leads us to recognize that quasi-horizontal transfer occurs in fairly limited layers bounded by potential temperatures in the range 310° to 350°K. This layer constitutes a layer approximately 100 mb in depth in the polar stratosphere north of 40 latitude and represents only 3% of the net hemispheric mass. While the air in this layer may be exchanged rather easily, exchange with higher levels will be on a quite different time scale as was discovered to be the case when the residence time was derived from bomb injections of debris at different locations.

Having reviewed the nature of the circulation processes we will consider more realistic models by which they may be described.

A fundamental approach to pollution involving the stratosphere

We will discuss the fundamental continuity equation governing a constituent n_p, its relationship to exchange models, and finally identify and discuss problems in which stratospheric contaminants may have ecological impact.

The general form of the continuity equation governing a constituent n_p takes the form:

$$\frac{\partial n_p}{\partial t} = -\nabla \cdot (n_p \mathbb{V}) + \triangle_p \qquad (6)$$

where $\triangle_p = P_p - Q_p n_p - R_p n_p^2$ represents the chemical change, and $\mathbb{V} = iu + jv + k\omega$ represents the vector velocity.

The equations may be simply converted to the more familiar mixing ratios $\chi_p = \rho_p/\rho = \{km_p/R^*\} n_p/\rho$ where:

ρ_p = density of constituent
ρ = net density
k = Boltzman constant
R^* = universal gas constant
m_p = molecular weight of n_p

giving: $$\frac{\partial \rho\chi_p}{\partial t} = -\nabla \cdot (\rho\chi_p \mathbb{V}) + \{km_p/R^*\} \triangle_p$$

If we now express the equation in terms of an average value over time interval τ,

$$(\chi_p)_t = 1/\tau \int_{t-r/2}^{t+r/2} \chi_p dt$$

and an instantaneous deviation from the mean value, $(\chi_p)^t = \chi_p - (\chi_p)_t$ assuming that $(\rho)^t \ll (\rho)_t$, and average all terms in the equation we obtain:

$$(\rho)_t \frac{\partial(\chi_p)_t}{\partial t} + (\chi_p)_t \frac{\partial(\rho)_t}{\partial t} = -\nabla \cdot [(\rho)_t \{(\chi_p)_t (\mathbb{V})_t + ((\chi_p)^t (\mathbb{V})^t)_t\}] + \{km_p/R^*\}(\Delta_p)_t$$

Subtracting the smoothed net continuity equation:

$$(\chi_p)_t \frac{\partial(\rho)_t}{\partial t} = (\chi_p)_t \nabla \cdot \{(\rho)_t(\mathbb{V})_t\}$$

gives:

$$\frac{\partial(\chi_p)_t}{\partial t} = -\mathbb{V} \cdot \nabla(\chi_p)_t - \{1/(\rho)_t\}\nabla \cdot \{(\rho)_t((\chi_p)^t(\mathbb{V})^t)_t\} + \\ + \{km_p/R^*(\rho)_t\}(\Delta_p)_t \qquad (9)$$

The first and second terms on the right side of equation (9) represent advection and eddy mixing respectively.

It is immediately evident that the linear flux approximation implied in the residence time concept does not represent well the dynamic processes as terms of form $(v)_t\ \partial(\chi_p)_t/\partial y$ and $\partial/\partial y\{((\chi_p)^t(v)^t)_t\}$, where the velocity components are limited by constraints implicit in the set of six equations necessary to predict the flow field, suggest strongly non-linear exchange.

Prediction of the wind field requires the solution of:
Three Eulerian equations of momentum.
The *net* continuity equation
The thermodynamic energy equation.
The gas law.

These differential equations are not soluble in general form, but have been integrated in general circulation models (Smagorinsky 1963; Smagorinsky *et al.* 1965; Kasahara & Washington 1967). The derived instantaneous value of the vector wind can be used with the trace constituent continuity equation to predict the changes. In the special cases where the trace gas is one which enters into the atmospheric heating through the radiative transfers, this feedback should be taken into account in the computation of the evolving flow field. Such feedbacks have been introduced in relation to the processes involved in the tropospheric hydrologic cycle (Manabe *et al.* 1970; Washington & Kasahara 1970; Holloway & Manabe 1971) but not in the case of stratospheric ozone (Manabe & Hunt 1968; Hunt & Manabe 1968; Hunt 1969) where only mean distributions have been considered in radiation computations.

Various simplifications of the general continuity equation (9) have been used for specific problems:

In the case where horizontal homogeneity may be assumed $\partial(\chi_p)_t/\partial x = \partial(\chi_p)_t/\partial y = ((u)^t(\chi_p)^t)_t = ((v)^t(\chi_p)^t)_t = 0$, under steady state conditions $\partial(\chi_p)_t/\partial t = 0$ equation (9) reduces to:

$$\partial/\partial z((\omega)^t(\chi_p)^t)_t + (\omega)_t\ \partial(\chi_p)_t/\partial z = \{km_p/R^*(\rho)_t\}(\Delta_p)_t \qquad (10)$$

This equation has been used by Brewer (1949) to interpret water vapour profiles in mid-latitudes assuming $(\Delta_p)_t \equiv 0$, and representing the eddy flux by an eddy diffusion term $((\omega)^t(\chi_p)^t)_t = -K_z\partial(\chi_p)_t/\partial z$, with K_z independent of height. Using the same value of $(\omega)^t$ Stebbins (1960) demonstrated that to explain the profile of tungsten 185 would require a value of K_z an order of magnitude greater than for the case of water vapour. This inconsistency together with the evidence reviewed in the previous section leads us to conclude that the simplest description of the real atmosphere is one consistent with Fig. 2.

When zonal symmetry can be assumed but vertical and meridional transports are finite mean meridional models may be used. When zonal average values are taken that remove the transient waves in the circulation, approximate zonal symmetry results and the valid form of equation (9) to treat such problems is:

$$\partial(\chi_p)_t/\partial t = -(v)_t\partial(\chi_p)_t/\partial y - (\omega)_t\partial(\chi_p)_t/\partial z - \partial/\partial y((v)^t(\chi_p)^t)_t - \\ - \partial/\partial z((\omega)^t(\chi_p)^t)_t + \{km_p/R^*(\rho)_t\}(\Delta_p)_t \tag{11}$$

Provided that the eddy transport terms can be parameterized as functions of the average fields $(v)_t$, $(\omega)_t$, $(\chi_p)_t$, then equation (11) may be used in a predictive model for $(\chi_p)_t$. This type equation, in which the eddy fluxes are assumed proportional to the mean gradient of the parameter to be transported $[((v)^t(\chi_p)^t)_t = -Ky\partial(\chi_p)_t/\partial y]$ has been used to interpret ozone distributions in the stratosphere (Prabhakara 1963; Gebhart 1968).

This parameterization can not account for the counter gradient transports that have been shown to be significant in transports of heat (White 1954; Murukami 1962; Peng 1963; Newell 1964) and trace substances (Newell 1961, 1963; Newell *et al.* 1966) without arbitrarily making K_y a function of latitude and season.

An improvement is possible if we assume that the quasi-horizontal large scale mixing takes place along a preferred mixing slope as was discussed in section 3. Assuming, as proposed by Reed & German (1965), that the mixing may be considered with confidence to occur over a uniform path length (l) (See Appendix 4) then the transfer terms may be represented by the sum of two component terms

$$(\rho)_t((v)^t(\chi_p)^t)_t = -(\rho)_t\left\{Kyy\frac{\partial(\chi_p)_t}{\partial y} + Kyz\frac{\partial(\chi_p)_t}{\partial z}\right\}.$$

This function simplies readily to a form that can be interpreted as implying that the horizontal meridional flux is downgradient or counter gradient depending on whether or not the mixing slope exceeds that of the mean isopleths. Gudiksen *et al.* (1968) have used such a model to simulate transfer of ozone and radioisotopes injected by nuclear bursts into the lower stratosphere.

The use of equation (11) does require a knowledge of the mean velocity components for use in the advection terms. The use of fields computed from observations or simple models incorporating assumptions not included in equation (11) may result in inconsistencies. The appropriate fields to use are those derived from a consistent closed set of equations with similar parameterization. Such a model has been used by Rao & Christie (1973) to interpret ozone changes to be expected in an atmosphere in which excess water vapour and nitrogen oxides are introduced analogous to the problem of S.S.T. exhausts.

Three limitations are implicit in this type model:

The parameterisation of *Kyy*, *Kyz* and *Kzz* is dependent on past history and, as we know that these are related to the thermal field, the assumption that the transfer coefficients will not change over the long term in response to modified radiative heating rates appears questionable.

The estimation of the chemical change term in such a model is possible only in the special cases where the chemical lifetime turns out to be long and insensitive to fluctuations in other constituents. Considerable work remains to be done on parameterizing (Δp).

An effective model for climatic prediction should be capable of providing information on the variability as well as the mean values of parameters. This is particularly so in relation to applications to the ecology where flooding, drought, frost damage, and wind extremes may take more toll than minor changes in the mean values.

Improvement in the parameterization of the eddy fluxes to relate the dispersion and transport to changes in the energy distribution are imperative. Murgatroyd (1969) has demonstrated that the horizontal diffusivities vary with the prevailing synoptic situation and has, moreover, shown, that to a first approximation, diffusivities can be extrapolated from one situation to another mainly in terms of the ratio of the variance of the corresponding wind components. The limits on the mixing angles acceptable in the Reed & German (1965) representation of quasi-horizontal mixing based on energy conversion limitations (Eady 1950) have been outlined and the relationship of the mixing angle to wave number, as well as the large scale characteristics of the thermal, flow and diabatic heating fields require clarification.

Information from both rotating 'dishpan' experiments (Fultz *et al.* 1958; Fultz & Kaylor 1959; Hide 1958; Fowles & Hide 1965) and from climatic studies (News & Views 1973) suggests strongly that in an atmosphere on an earth rotating at constant speed the major factors controlling the circulation regime are the depth of the fluid layer and the thermal characteristics, meridional thermal gradient, $\nabla_y T$, and stability $\partial T/\partial Z$, and the meridional gradient of the coriolis parameter, $\beta = \partial f/\partial y$. There is little doubt that improved parameterization of the transfer coefficients can develop from empirical studies of the above parameters and from their application in the continuity equations of a variety of trace gases whose evolution in time and space is known.

The major restriction in the application of the diffusion parameterization remains in the transition region between one thermal regime and another where the correlation between the horizontal and vertical velocity components is small (the mixing angle has a bimodal distribution). In these regions further sophistication must be introduced into the eddy flux representation if transports between different regimes are to be evaluated in a physically realistic manner.

The gigantic scale of computation necessary to carry out climatic prediction by general circulation models makes it essential to overcome the problems of parameterizing the meridional models as expeditiously as possible for it appears that these offer the best current compromise between the unrealistic 'box' models that fail to represent the appropriate processes, and the physically complete but mathematically cumbersome four dimensional models.

What pollution involving the stratosphere may have ecological impact and can such pollution be investigated by residence time methods?

First toxic chemicals injected at the surface will be almost totally unaffected by the existence of the stratosphere.

$$\frac{dn_T}{dt} = -\{\lambda + a_{TS}\}n_T + a_{ST}n_S + S_T$$

Converting the equation to concentration $n_T = m_T c_T$, and introducing the steady state assumption for net mass, $a_{ST}m_S = a_{TS}m_T$

$$\frac{dc_T}{dt} = -\{\lambda + a_{TS}\}C_T + a_{TS}C_T + S_T/m_T$$

Now $C_S \ll C_T$ and $\lambda > a_{TS}$ [$\lambda \simeq 1/1mo.$, $a_{TS} \simeq (14/100)(1/6\ mo.)$]
$\therefore \lambda \gg a_{TS}\,[1 - C_S/C_T]$

So C_T is almost unaffected by the existence of the stratosphere.

Another form of pollution whose impact on the biosphere might be considered directly detrimental is the introduction of radioisotopes into the atmosphere by nuclear testing. The absorption of radioisotopes by vegetation and animal cells has been widely discussed, but the meteorological effects will receive only cursory comment here as testing is now generally carried on underground with less atmospheric contamination. Some success has been had in predicting long term fallout using the linear transfer equations though artificial means must be resorted to if seasonal and meridional variations are to be derived as outlined in section 2.

On the global, long term scales chronic effects on the biosphere may result from:

1. Increased exposure to solar ultraviolet radiation resulting from chemical depletion of the ozone shield.
2. Alteration of the global climate by changing the distribution of radiatively active constituents and the potential energy distribution through the radiative transfers.

It is clear from Fig. 3, where the relative importance of the contribution of various radiative heating processes above 15 km are shown, that the heating is related to the distribution of the relevant constituents and information on these distributions is necessary for studies of both global climate and the more specific ozone depletion problem.

Ozone depletion, if it takes place, might have some ecological impact by contributing to climate modification, as well as through the more obvious reduction in filtering efficiency to erythemogenic radiation, since the ozone exerts a major influence on the radiative heat budget of the stratosphere (Fig. 3).

The complete chemistry of ozone (Crutzen 1972a, 1972b; Newell 1972) has yet to be resolved but the effects of catalytic removal by odd hydrogen derived from water vapour:

$$\left.\begin{array}{l} OH+O_3 \rightarrow HO_2+O_2 \\ HO_2+O_3 \rightarrow OH+2O_2 \end{array}\right\} 2O_3 \rightarrow 3O_2$$

and nitric oxide and nitrogen dioxide

$$\left.\begin{array}{l} NO+O_3 \rightarrow NO_2+O_2 \\ NO_2+O \rightarrow NO+O_2 \end{array}\right\} O_3+O \rightarrow 2O_2$$

have been under intensive study in relation to the introduction of these pollutants directly into the stratosphere by *S*uper *S*onic *T*ransports. The complete chemistry is, of course, much more complex when sinks and chemical intermediaries are considered, however, the basic questions are:

(a) Can the increases in water vapour and nitrogen oxides be estimated for use in the ozone depletion equation by means of 'box' models?

(b) Can a linear depletion equation be applied to the ozone?

As regards (a), a *rough* estimate of the water vapour increase can be derived from knowledge of the rate of production by the S.S.T. engines together with the residence time *appropriate to the general region of injection* since water vapour is almost conservative at these levels. (Any increase should be fairly easy to distinguish as current thoughts are that stratospheric water vapour originates from ground sources and the major influx region is through the cold 'vapour trap' of the tropical stratosphere).

The situation regarding the nitrogen oxides is much more complex as they are much less chemically inactive and their natural backgrounds are not unequivocally determined. They may originate from naturally occurring nitrous oxide or ammonia from biological sources.

The ozone depletion then clearly cannot be studied by residence time methods since the chemical depletion cannot be represented by a linear factor, moreover, numerical models have shown that the ozone distribution can not be modelled with any degree of success without the introduction of representative dynamical transports. A preliminary steady state mean meridional model (Rao 1973) has already been used to study the ozone depletion that would result from an arbitrary increase in the water vapour and nitrogen oxides (Rao & Christie 1973) but considerable work must yet be done to simulate the chemistry and transports in the transition region between troposphere and stratosphere and the model made time dependent if climatic change is to be studied effectively.

The global climate depends on the distribution of heating and, to the extent that pollution sources on the earth's surface may alter the distribution of radiatively active constituents, they may indirectly alter climate with subsequent effects on many ecosystems. Climatic impact may take a variety of forms such as:

Overall change in the mean temperature.

Change in the regional distribution of winds and temperatures.

Changes in the statistical distribution of winds and temperatures (e.g. standard deviations, extremes etc.) on a regional or global scale.

It is clear that only the first factor could hope to be studied by the residence time approach as there is no dynamical information implicit in the linear exchange model.

In discussing the global distribution of heat sources and sinks three factors may be noted in addition to the ozone noted above. These are cloud cover, carbon dioxide and atmospheric aerosols (SMIC 1971).

Clouds are almost entirely restricted to the troposphere and will not be considered further.

Carbon dioxide is conservative in the atmosphere and consequently relatively well mixed, so radiation computations are insensitive to the small changes in concentration with height and the important parameter is average CO_2 concentration. The major effect of CO_2 in the radiative computations is due to the strong absorption band around 15 μ which absorbs and then emits longwave infrared radiation downwards to the ground and upwards to space. As the CO_2 concentration is increased the net downward longwave flux will be increased diminishing the heat loss of the earth's surface (Moller 1963; Manabe & Wetherald 1969; Rasool & Schneider 1971) until a point of saturation is reached when all available energy in the 15 μ band has been absorbed. An increase of about 0·7°K tropospheric temperature resulting from a doubling of the CO_2 concentration. Preliminary comments on changes in the generation of zonal available potential energy (Lorenz 1955) that might arise as a result of altered heat source and sink distributions due to increased CO_2 have been made by Newell *et al* (1972). The residence time concept has been applied with some success to long term changes in the carbon dioxide resulting from the increased use of fossil fuel over the past several decades (Craig 1957; Bolin & Eriksson 1959; Machta 1971).

The aerosol effects are more difficult to interpret as solution of the radiative transfer equation requires a knowledge of the aerosol concentration profile, in addition to the refractive index and size distribution of particles (Yamamoto & Tanaka 1972). Current estimates of the climatic effects of aerosols are based on an observed profile (Elterman 1964) and the formative processes are not yet fully understood. The profile falls steadily by four orders of magnitude from the earth's surface to the tropopause above which it increases by almost an order of magnitude to a broad maximum centred about 20 km. The stratospheric aerosol, which consists mainly of ammonium sulphate and sulphuric acid is thought to be formed by gas phase reactions involving sulphur dioxide, sulphides and ammonia while those in the troposphere are predominantly the result of reactions between these gases in the presence of liquid water. Increases in the stratospheric concentrations in the northern hemisphere in the sixties have been attributed to gases of volcanic origin (Lazrus *et al.* 1971; Cadle 1972) though increases in the tropospheric turbidity attributable to increased industrial use of fossil fuels has been noted (McCormick & Ludwig 1967; Peterson & Bryson 1968; Joseph & Manes 1971). This increase is not to be confused with the drop in sooty particles that has resulted from increasing use of filtering devices.

To date estimates of the global climatic surface temperature change have been based on a model where the aerosol concentration profile is augmented by a proportional increase at all levels. A proper model to determine the aerosol distribution will require integration of the relevant continuity equations.

These examples may illustrate that while the residence time approach has merit in some problems for an atmosphere degraded by industrial activity, many more potentially ecologically significant changes will require the more subtle fundamental approach.

Conclusion

We can now evaluate the success of the residence time concept under the three criteria stated in the third section.

We have shown that it has useful application in some relatively simple problems such as changes in quasi-conservative constituents where the gross change in reservoir concentration is all that is required. The application to assessment of climatic change due to CO_2 increases has been cited as an illustration. The method has also had some success in interpretation of radioactive fallout distributions, particularly when the location of input of the debris is known and statistical corrections are applied to the trend values of global mean fallout to simulate seasonal and meridional variations.

In general it is safe to say that chemically inert contaminants of *tropospheric* origin will be largely unaffected by the stratospheric reservoir since the latter has small net mass capacity.

Consideration of both the circulation pattern presented graphically in Fig. 1 and the general trace gas continuity equation illustrates plainly the inadequacy of the basic assumption that a linear transfer coefficient describes adequately transfer between stratosphere and troposphere. Moreover only in the case where $R_p = 0$ and Q_p is constant is the linear chemical depletion coefficient a justifiable assumption. Most chemically active trace constituents have highly non-uniform spatial distributions so that the recent trajectory of air in a specific location can contribute to non uniformity of Q_p through the $\sum k_{lmp}\, n_l n_m$ and $\sum k_{lp}\, n_l$ terms, in addition to variability resulting from changes in photodissociation J_p. In studies of chemically interactive trace constituents, then, the use of a valid continuity equation that takes proper account of the representative atmospheric transports is mandatory.

Ideally the integration should be carried out for each of the Z species involved in the S reactions. In practice the set may be substantially reduced using simplifying procedures based on reaction rate considerations, but improvements in the parameterization process are necessary.

In the cases involving ecological impact through ozone depletion or climatic change the feedbacks between the radiative, chemical and transport processes are of paramount importance and it is clear that linear models are almost totally inadequate.

We may conclude that, in the study of problems of major global ecological importance, application of the residence time approach must be viewed with considerable scepticism. Models suitable for attacking these questions are slowly being developed and are of formidable complexity. Interdisciplinary science is finally coming of age with problems of an ecological nature involving the interactions of biology, oceanography, volcanology, industry and meteorology.

Summary

The impact of pollution in upper troposphere and stratosphere on the global environment is discussed in the context of the observed atmospheric structure.

Models that have been used to interpret trace constituent distributions are critically evaluated.

It is concluded that the major influence of the stratosphere on the biosphere will derive from climatic change attributable to redistribution of radiatively active constituents.

References

ANGELL J.K. (1960) An analysis of operational 300 mb transconde flights from Japan in 1957–8. *J. Met.* 17, 20–35.

ANGELL J.K. (1961) Use of constant level balloons in meteorology. *Adv. Geophys.* 8, 137–219.

BJERKNES J. & SOLBERG H. (1922) Life cycle of cyclons and the polar front theory of atmospheric circulation. *Geo. Publ.* 3, 1–18.

BOLIN B. & ERIKSSON E. (1959) Changes in the CO_2 content of the atmosphere and sea due to fossil fuel combustion. In *The Atmosphere and Sea in Motion.* O.U.P., London.

BOVILLE B.W., WILSON C.V. & HARE F.K. (1961) Baroclinic waves of the polar-night vortex. *J. Met.* 18, 567–80.

BREWER A.W. (1949) Evidence for a world circulation provided by measurements of helium and water vapour distribution in the stratosphere. *Q. Jl. R. met. Soc.* 75, 351–63.

BRIGGS J. & ROACH W. (1963) Aircraft observations near jet streams. *Q. Jl. R. met. Soc.* 89, 225–365.

BROWN J.A. (1969a) A numerical investigation of hydrodynamic instability and energy conversions in the quasi-geostrophic atmosphere. I. *J. atmos. Sci.* 26, 352–65.

BROWN J.A. (1969b) A numerical investigation of hydrodynamic instability and energy conversions in the quasi-geostrophic atmosphere. II. *J. atmos. Sci.* 26, 366–75.

CADLE R.D. (1972) Composition of the stratospheric 'sulphate layer'. Proc. of Survey Conf., C.I.A.P., Springfield, Va., U.S.A.

CHARNEY J.G. (1947) The dynamics of long waves in the upper westerlies. *J. Met.* 5, 44–57.

CRAIG H. (1957) The natural distribution of radio carbon and the exchange time of CO_2 between atmosphere and sea. *Tellus* 2, 1–17.

CRUTZEN P.J. (1969) Determination of parameters appearing in the 'dry' and the 'wet' photochemical theories for ozone in the stratosphere. *Tellus*, 21, 368–88.

CRUTZEN P.J. (1971) Ozone production rates in an oxygen-hydrogen, nitrogen oxide atmosphere. *J. geophys. Res.* 76, 7311–27.

CRUTZEN P.J. (1972) SST's—A threat to the earth's ozone shield. *Ambio* 1, 41–51.

DANIELSEN E.F. (1959) The laminar structure of the atmosphere and its relation to the concept of a tropopause. *Arch. Met. Geoph. Bioklim. A* 11, 293–332.

DOBSON G.M.B., HARRISON D.N. & LAWRENCE J. (1929) Measurements of the amount of ozone in the earth's atmosphere and its relation to geophysical conditions. III. *Proc. R. Soc.* A 122, 546–61.

DOBSON G.M.B. (1956) Origin and distribution of the polyatomic molecules in the atmosphere. *Proc. R. Soc.* A 236, 72–81.

DURST C.G., CROSSLEY A.F. & DAVIS N.E. (1959) Horizontal diffusion in the atmosphere as determined by geostrophic trajectories. *J. Fluid Mech.* 6, 401–22.

DUTSCH H.V. (1956) Atmospheric ozone as an indicator of currents in the stratosphere. *Arch. Met. Geophys. Bioklim. A* 11, 240–51.

DYER A.J. (1970) Anisotropic diffusion coefficients and the global spread of volcanic dust. *J. geophys. Res.* 75, 3007–12.

EADY E.T. (1949) Long waves and cyclone waves. *Tellus* 1, 34–52.

EADY E.T. (1950) The cause of the general circulation of the atmosphere. *Centenary Proc. of Roy. met. Soc.* 156–72.

ELTERMAN L. (1964) Rayleigh and extinction coefficients to 50 km for the region ·27μ to ·55μ. *Appl. Opt.* 3, 1139–47.

ENGELMANN R.J. (1968) The calculation of precipitation scavenging. *Meteorology and Atomic Energy*, 208–21, U.S.A.E.C., TID-24190.

FEELY H.W. (1960) Strontium 90 content of the stratosphere. *Science N.Y.* **131**, 645–649.

FEELY H.W. & SPAR J. (1960) W^{185} from nuclear bomb tests as a tracer for stratospheric meteorology. *Nature, Lond.* 188, 1062–4.

FEELY H.W., FRIEND J.P., KREY P.W. & RUSSELL C.A. (1965) Flight data and results of radiochemical analysis of filter samples collected during 1961 and 1962. U.S.A.E.C., HASL 153.

FEELY H.W. & SEITZ H. (1970) Use of lead 210 as a tracer of transport properties in the stratosphere. *J. geophys. Res.* 75, 2885–94.

FOWLES W.W. & HIDE R. (1965) Thermal convection in a rotating annulus. *J. atmos. Sci.* **22**, 541–60.

FULTZ D., LONG R.R., OWENS G.V., BOHAN, W., KAYLOR R. & WEIL J. (1958) Studies of thermal convection in a rotating cylinder and large scale atmospheric motions. *Meteor. Monog. A.M.S.* 4, 21.

FULTZ, D. & KAYLOR R. (1959) The propagation of frequency in experimental baroclinic waves in a rotating annular ring. In *The Atmosphere and Sea in Motion.* O.U.P., London.

GEBHART R. (1968) Photochemical, advective and turbulent effects on the meridional distribution of ozone. *Arch. Met. Geophys. Bioklim A* 17, 301–35.

GREEN J.S.A. (1960) A problem in baroclinic stability. *Q. Jl. R. met. Soc.* 86, 237–51.

GUDIKSEN P.H., FAIRHALL A.W. & REED R.J. (1968) Roles of mean meridional circulation and eddy diffusion in the transport of trace substances in the lower stratosphere. *J. geophys. Res.* 73, 4461–80.

GUSTAFSON P.H., BRAR S.S. & KERRIGAN M.A. (1962) Airborne radioactivity due to nuclear weapons tests. *J. geophys. Res.* 67, 4641–51.

HAGEMANN F., GRAY J. & MACHTA L. (1965) Carbon 14 measurements in the atmosphere 1953 to 1964. U.S.A.E.C., HASL 159.

HAGEMANN F., GRAY J., MACHTAL L. & TUKEVICH S.L. (1959) Stratospheric carbon 14, carbon dioxide and tritium. *Science, N.Y.* **130**, 542–52.

HESSTVEDT E. (1965) Some characteristics of the oxygen-hydrogen atmosphere. *Geofys. Publr* 26–1.

HESSTVEDT E. (1968) On the photochemistry of ozone in the ozone layer. *Geofys. Publr* 27–5.

HIDE R. (1958) An experimental study of thermal convection in a rotating liquid. *Phil. Trans. R. Soc.* A **250**, 441–78.

HOLLOWAY J.L. & MANABE S. (1971) Simulation of climate by global general circulation model. I. Hydrologic cycle and heat balance. *Mon. Weath. Rev. U.S. Dep. Agric.* 99, 335–70.

HOLTON J.R. & LINDZEN R.S. (1968) A note on Kelvin waves in the atmosphere. *Mon. Weath. Rev. US. Dep. Agric.* 96, 385–6.

HUNT B.G. & MANABE S. (1968) Experiments with a stratospheric general circulation model. II. Large-scale diffusion of tracers in the stratosphere. *Mon. Weath. Rev. U.S. Dep. Agric.* 96, 503–39.

HUNT B.G. (1969) Experiments with a stratospheric general circulation model. III. Large scale diffusion of ozone including photochemistry. *Mon. Weath. Rev. U.S. Dep. Agric.* 97, 287–306.

JOSEPH J.H. & MANES A. (1971) Secular and seasonal variations of atmospheric turbidity at Jerusalem. *J. appl. Met.* 10, 453–62.

KAO S.K. & BULLOCK W.S. (1964) Lagrangian and Eulerian correlations and energy spectra of geostrophic velocities. *Q. Jl. R. met. Soc.* 90, 166–74.

KAO S.K. (1965) Some aspects of large-scale diffusion and turbulence in the atmosphere. *Q. Jl. R. met. Soc.* 91, 10–17.

KAO S.K. (1968) Relative dispersion of particles in a stratified, rotating atmosphere. *J. atmos. Sci.* 25, 481–7.

KAO S.K. & GAIN A.A. (1968) Large-scale dispersion of particles in the atmosphere. *J. atmos. Sci.* 25, 214–21.

KAO S.K. & HILL W.R. (1970) Characteristics of the large-scale dispersion of particles in the southern hemisphere. *J. atmos. Sci.* 27, 126–32.

KAO S.K. & HENDERSON D. (1970) Large-scale dispersion of clusters of particles in various flow patterns. *J. geophys. Res.* 75, 3104–13.

KASAHARA A. & WASHINGTON W.M. (1967) N.C.A.R. Global general circulation model of the atmosphere. *Mon. Weath. Rev. U.S. Dep. Agric.* 95, 389–402.

KREY P.W. & KRAJEWSKI B. (1970) Comparison of atmospheric transport model calculations with observations of radioactive debris. *J. geophys. Res.* 75, 2901–8.

KUO H.L. (1952) Three dimensional disturbances in a baroclinic zonal current. *J. Met.* 10, 235–43.

LAZRUS A.L., GANDRUD B. & CADLE R.D. (1971) Chemical composition of air filtration samples of the stratospheric sulphur layer. *J. geophys. Res.* 76, 8083–96.

LIST R.J. & TELEGADAS K. (1969) Using radioactive tracers to develop a model of the circulation of the stratosphere. *J. atmos. Sci.* 26, 1128–36.

LIST, R.J., SALTER L.P. & TELEGADAS K. (1966) Radioactive debris as a tracer for investigating stratospheric motions. *Tellus* 18, 345–54.

LORENZ E. (1955) Available potential energy and the maintenance of the general circulation. *Tellus* 7, 157–67.

MACHTA L. (1971) The role of the oceans and the biosphere in the carbon dioxide cycle. Nobel Symposium 20, Gothenburg.

MACHTA L., TELEGADAS K. & LIST R.J. (1970) The slope of surfaces of maximum tracer concentration in the lower stratosphere. *J. geophys. Res.* 75, 2279–89.

MACHTA L. & LIST R.J. (1959) Analysis of stratospheric strontium 90 measurements. *J. Geophys. Res.* 64, 1267–76.

MAHLMAN J.D. (1965a) Relation of stratospheric—tropospheric mass exchange mechanisms to surface radioactivity peaks. *Arch. Met. Geophys. Bioklim. A* 15, 1–25.

MAHLMAN J.D. (1965b) Relation of upper air hemisphere index patterns to seasonal fallout fluctuations. Proc. 2nd Conf. Radioactive Fallout from Nuclear Weapons Test, Germantown, Md.

McCLAIN E.P. & DANIELSEN E.F. (1955) Zonal distribution of baroclinity for three Pacific storms. *J. Met.* 12, 314–23.

McCormick R.A. & Ludwig I.H. (1967) Climate modification by atmospheric aerosols. *Science, N.Y.* 156, 1358–9.

Manabe S., Smagorinsky J., Holloway J.L. & Stone H.M. (1970) Simulated climatology of a general circulation model with a hydrologic cycle. III. Effects of increased horizontal computational resolution. *Mon. Weath. Rev. U.S. Dep. Agric.* 98, 175–212.

Manabe S. & Hunt B.G. (1968) Experiments with a stratospheric general circulation model. I. Radiative and dynamic aspects. *Mon. Weath. Rev. U.S. Dep. Agric.* 96, 477–502.

Manabe S. & Wetherald R.T. (1967) Thermal equilibrium of the atmosphere with a given distribution of relative humidity. *J. atmos. Sci.* 24, 241–9.

Martell E.A. (1959) Atmospheric aspects of Sr-90 fallout. *Science, N.Y.* **129**, 1197–1206.

Martell E.A. (1968) Tungsten radioisotope distribution and stratospheric transport processes. *J. Atmos. Sci.* 25, 113–25.

Maruyama T. (1967) Large scale disturbances in the equatorial lower stratosphere. *J. met. Soc. Japan.* 45, 391–408.

Maruyama T. (1969) Long-term behaviour of Kelvin waves and mixed Rossby-Gravity waves. *J. met. Soc. Japan* 47, 245–54.

Maruyama T. & Yanai M. (1967) Evidence of large-scale wave disturbances in the equatorial lower stratosphere. *J. met. Soc. Japan* 45, 196–9.

Miles J.W. (1964a) Baroclinic instability of the zonal wind. *Rev. Geophys.* **2**, 155–76.

Miles J.W. (1964b) Baroclinic instability of the zonal wind. II. *J. atmos. Sci.* **21**, 500–6.

Moller F. (1963) On the influence of changes in CO_2 concentration in air on the radiation balance of the earth's surface and on the climate. *J. Geophys. Res.* 68, 3877–86.

Murgatroyd R.J. (1969) Estimations from geostrophic trajectories of horizontal diffusivity in the midlatitude troposphere and lower stratosphere. *Q. Jl. R. met. Soc.* 95, 40–62.

Murgatroyd R.J. (1970) The structure and dynamics of the stratosphere. In *The Global Circulation of the Atmosphere*, Royal Met. Soc., London.

Murukami T. (1962) Stratospheric wind, temperature and isobaric height conditions during the I.G.Y. period. I. Rept. 5, Planetary Circulations Project., M.I.T.

Namias J. & Clapp P.F. (1944) Studies of the motion and development of long waves in the westerlies. *J. Met.* **1**, 57–77.

Neiburger M. & Angell J.K. (1956) Meteorological applications of constant-pressure balloon trajectories. *J. Met.* 13, 166–94.

Newell R.E. (1961) The transport of trace substances in the atmosphere and their implications for the general circulation of the stratosphere. *Geofis. pura appl.* 48, 137–58.

Newell R.E. (1963) Transfer through the tropopause and within the stratosphere. *Q. Jl. R. met. Soc.* 89, 167–204.

Newell R.E. (1964) Stratospheric energetics and mass transports. *Geofis. pura appl.* 58, 145–56.

Newell R.E., Wallace J.M. & Mahoney J.R. (1966) The general circulation of the atmosphere and its effects on the movement of trace substances. II. *Tellus* **18**, 363–380.

Newell R.E. & Dopplick T.G. (1970) The effect of changing CO_2 concentration on radiative heating rates. *J. appl. Met.* 9, 958–9.

Newell R.E., Herman G.F., Dopplick T.G. & Boer G.J. (1972) The effect of changing CO_2 concentration on radiative heating rates: Further comments. *J. appl. Met.* **11**, 864–7.

NEWELL R.E., VINCENT D.G., DOPPLICK T.G., FERRUZZA D. & KIDSON J.W. (1970) The energy balance of the global atmosphere. In *The Global Circulation of the Atmosphere*, Roy. Meteor. Soc., London.

News and Views (1973) Global perspectives on climate. *Nature, Lond.* **242**, 295–6.

PALMER E. (1948) On the distribution of temperature and wind in the upper westerlies. *J. Met.* **5**, 20–7.

PEDLOSKY J. (1965) On the stability of baroclinic flows as a functional of the velocity profile. *J. atmos. Sci.* **22**, 137–45.

PEDLOSKY J. (1970) Finite amplitude baroclinic waves. *J. atmos. Sci.* **27**, 15–30.

PEIRSON, D.H. & CAMBRAY R.S. (1965) Fission product fallout from the nuclear explosions of 1961 and 1962. *Nature, Lond.* **497**, 433–40.

PENG L. (1963) Stratospheric wind, temperature and isobaric height conditions during the I.G.Y. period. II. Rept. 10 Planetary Circulations Project, M.I.T.

PRABHAKARA C. (1963) Effects of non-photochemical processes on the meridional distribution and total amount of ozone in the atmosphere. *Mon. Weath. Rev. U.S. Dep. Agric.* **91**, 411–31.

PETERSON I.T. & BRYSON R.A. (1968) Atmospheric aerosols: Increased concentration during the last decade. *Science, N.Y.* **162**, 120–1.

RAMANATHAN K.R. & KULKARNI R.N. (1960) Mean meridional distributions of ozone in different seasons calculated from umkehr observations, and probable vertical transport mechanisms. *Q. Jl. R. met. Soc.* 86, 144–55.

RAO V.R.K. (1973) Numerical experiments on the steady state meridional structure of the stratosphere. *Mon. Weath. Rev. U.S. Dep. Agric.* **101** (in press).

RAO V.R.K. & CHRISTIE A.D. (1973) The effects of water vapour and oxides of nitrogen on ozone and temperature structure of the stratosphere. *J. atmos. Sci.* **30** (in press).

RASOOL S.I. & SCHNEIDER S.H. (1971) Atmospheric carbon dioxide and aerosols: effects of large increases on global climate. *Science, N.Y.* **173**, 138–41.

REED R. (1955) A study of a characteristic type of upper level frontogenesis. *J. Met.* **12**, 226–37.

REED R. & DANIELSEN E.F. (1959) Fronts in the vicinity of the tropopause. *Arch. met. Geophys. Bioklim. A* **11**, 1–28.

REED R.J. (1964) A tentative model of the 26 month oscillation in tropical latitudes. *Q. Jl. R. met. Soc.* 90, 441–66.

REED R.J. (1966) Zonal wind behaviour in the equatorial stratosphere and lower mesosphere. *J. geophys. Res.* 71, 4223–33.

REED R.J. & GERMAN K.E. (1965) A contribution to the problem of stratospheric diffusion by large-scale mixing. *Mon. Weath. Rev. U.S. Dep. Agric.* 93, 313–21.

REITER E.R. (1963) A case study of radioactive fallout. *J. appl. Met.* **2**, 691–705.

REITER E.R. & MAHLMAN J.D. (1965) Heavy radioactive fallout over the southern U.S., Nov. 1962. *J. geophys. Res.* **70**, 4501–20.

REITER E.R. (1968) The behaviour of Jet streams in potential fallout situations. *Arch. met. Geophys. Bioklim. A* 17, 8–16.

ROSSBY C.G. (1939) Relation between variations in the intensity of the zonal circulation of the atmosphere and the displacement of the semi-permanent centers of action. *J. mar. Res.* **2**, 38–55.

ROSSBY C.G. (1949) On a mechanism for the release of potential energy in the atmosphere. *J. Met.* 6, 163–80.

SHEPPARD P.A. (1963) Atmospheric tracers and the study of the general circulation of the atmosphere. *Repts. Progress in Physics* **26**, 213–67.

SMAGORINSKY J. (1963) General circulation experiments with the primitive equations. I. The basic experiment. *Mon. Weath. Rev. U.S. Dep. Agric.* 91, 99–164.

SMAGORINSKY J., MANABE S. & HOLLOWAY J.L. (1965) Numerical results from a nine-level general circulation model of the atmosphere. *Mon. Weath. Rev. U.S. Dep. Agric.* 93, 727–68.

SMIC (1971) *Report of the Study of Man's Impact on Climate.* M.I.T. Press, Cambridge, Mass.

STALEY D.O. (1962) On the mechanisms of mass and radioactivity transport from stratosphere to troposphere. *J. atmos. Sci.* 19, 450–67.

STALEY D.O. (1960) Evaluation of potential-vorticity changes near the tropopause and transfer of radioactive debris from stratosphere to troposphere. *J. Met.* 17, 591–620.

STEBBINS A.K. (1960) Special report on high altitude sampling program. DASA 532 B, Washington D.C.

TELEGADAS K. & LIST R.J. (1969) Are particulate radioactive tracers indicative of stratospheric reactions? *J. geophys. Res.* 74, 1339–50.

WALLACE J.M. & KOUSKY V.E. (1968) Observational evidence of Kelvin waves in the tropical stratosphere. *J. atmos. Sci.* 25, 900–7.

WILSON C.V. & GODSON W.L. (1963) The structure of the arctic winter stratosphere over a 10-year period. *Q. Jl. R. met. Soc.* 89, 205–24.

WASHINGTON W.M. & KASAHARA A. (1970) A January simulation experiment with the two layer version of the NCAR global circulation model. *Mon. Weath. Rev. U.S. Dep. Agric.* 98, 559–80.

WHITE R.M. (1954) The countergradient flux of sensible heat in the lower stratosphere. *Tellus* 6, 177–82.

YAMAMOTO G. & TANAKA M. (1972) Increase of global albedo due to air pollution. *J. atmos. Sci.* 29, 1405–12.

YANAI M. & MARUYAMA T. (1966) Stratospheric wave disturbances propagating over the equatorial Pacific. *J. met. Soc. Japan* 44, 291–4.

YANAI M. & HAYASHI Y. (1969) Large-scale equatorial waves penetrating from the upper troposphere into the lower stratosphere. *J. met. Soc. Japan* 47, 167–82.

Appendix 1 Static stability and potential temperature

Air is defined as statically stable to *vertical* convection if a parcel is displaced upwards under adiabatic conditions and assumes a temperature cooler than that of the ambient air since it will then be denser than its surroundings and sink again.

If diabatic heating is neglected in a parcel of dry air of unit mass its thermodynamic properties may be described by the identity: $0 = dU + dW$ where dU represents the change in internal energy and dW the work done on the unit mass of air

$$\text{i.e. } 0 = C_u dt + p d\alpha = (C_u + R)dT - \alpha dp \qquad \text{(i)}$$

Since the gas law gives $p\alpha = RT$ or $pd\alpha = RdT - \alpha dp$

T = temperature

p = pressure

α = specific volume

ρ = density

C_u = specific heat at constant volume
R = gas constant for unit mass of dry air
C_p = specific heat at constant pressure
g = gravitational acceleration

Introducing further the hydrostatic assumption $dp = -g\rho dZ$

$$C_p dT = -gdz \text{ or } -\frac{dT}{dZ} = g/C_p \qquad \text{(ii)}$$

$-dT/dZ = g/C_p$ is defined as the dry adiabatic lapse rate and wherever $-dT/dZ < g/C_p$ the air is statically stable to vertical convection.

The potential temperature is the temperature that a parcel of air at pressure 'p' would assume if moved adiabatically to a pressure of 1,000 mb. Returning to equation (i)

$$C_p\, dT = \alpha dp \text{ or } d\, lnT = (R/C_p)\, d\, lnp$$

$$\theta = T(p_o/p)^k \qquad \text{(iii)}$$

where $k = R/C_p$
$p_o = 1000$ mb.

The thermodynamic equation may be written

$$\frac{dT}{dt} = (1/C_p)\left[\alpha\frac{dp}{dt} + q\right] \text{ or by substitution of (iii)}$$

$$\frac{d\theta}{dt} = (1/C_p)(p_o/p)^k q$$

where $q = q_{sr} + q_{ar} + q_L + q_D$ is rate of heating per unit mass of air.

q_{sr} = rate of heating due to absorption of solar radiation.
q_{ar} = rate of heating due to net flux convergence of atmospheric infrared radiation
q_L = rate of heating due to latent heat release.
q_D = rate of dissipative heating.

Appendix 2. WMO definition of thc Tropopause

1. The first tropopause is defined as the lowest level at which the thermal lapse rate decreases to 2°K km^{-1} or less, provided also that the average lapse rate between this level and all higher levels within 2 km does not exceed 2°K km^{-1}.

2. When, above the first tropopause, the average lapse rate between any level and all higher levels within 1 km exceeds 3°K km^{-1} then a second tropopause can occur and is defined by the same criteria as in paragraph 1. This tropopause can either be within or above the 1 km lapse.

Appendix 3. Slantwise convection: energetics and heat fluxes

In the lower stratosphere the meridional and vertical gradients of potential temperature $(\theta)_t$ are directed poleward ($\partial(\theta)_t/\partial y > 0$) and upward ($\partial(\theta)_t/\partial z > 0$) respectively, as shown in Fig. 12.

Neglecting dynamical constraints, we can see how heat may be transferred by air parcels exchanged along slopes greater or less than the isentropic surface in the statically stable atmosphere implied by the positive vertical gradient of potential temperature.

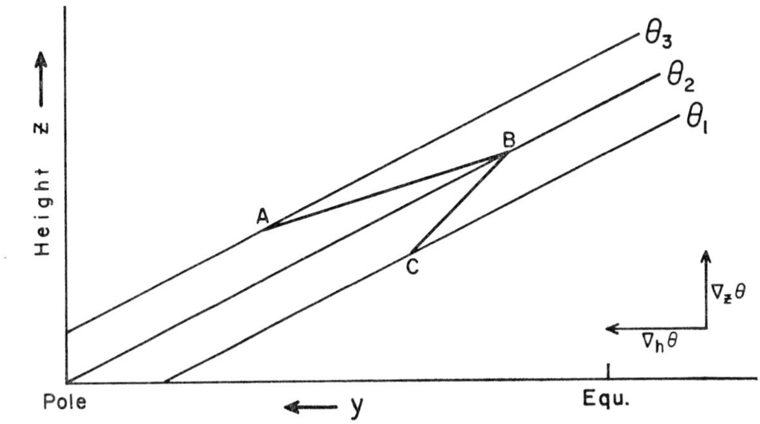

Figure 12. Schematic drawing of mixing in a meridional plane at angles greater and less than the slope of the mean isentropes. Potential temperature increases with height and polewards as in the lower stratosphere.

In the following treatment we use the standard meteorological notation:

y = Sec ϕ: is northward directed coordinate
Z: is height
v: is meridional velocity component
w: is vertical velocity component

and mean values and deviations are denoted by subscripts and superscripts such that $(\theta)_t = 1/\tau \int_{t-\tau/2}^{t+\tau/2} \theta dt$ and $(\theta)^t = \theta - (\theta)_t$ where Γ represents the smoothing interval.

In Fig. 12 we consider an air parcel at A displaced to B. This parcel will be warmer than its new environment and so further displacement in that sense will be encouraged. Similarly the cooling of air displaced to A from B will also serve to release potential energy through the subsidence of potentially cold air.

Expressed symbolically:

A→B implies $(w)^t > 0$, $(v)^t < 0$, $(\theta)^t = \theta_3 - \theta_2 > 0$,
B→A implies $(w)^t < 0$, $(v)^t > 0$, $(\theta)^t = \theta_2 - \theta_3 < 0$.

The net heat fluxes resulting from mixing at the slope given by AB in an atmosphere with thermal distribution $\partial(\theta)_t/\partial y > 0$ and $\partial(\theta)_t/\partial z > 0$ are $((v)^t(\theta)^t)_t > 0$ and $((w)^t(\theta)^t)_t > 0$. *The heat fluxes associated with a release of potential energy then are down and against the thermal gradient in the meridional and vertical directions respectively.*

When motion is along a slope greater than that of the isentropes, as implied by mixing along BC, the process must increase the potential energy so must be forced. The heat fluxes, evaluated as before, are $((v)^t(\theta)^t)_t > 0$ (countergradient) and $((w)^t(\theta)^t)_t < 0$ (down-gradient) in this case.

Similar results obtain in the troposphere.

Appendix 4. Parameterization of eddy fluxes

If the mixing path is allowed to have any orientation in space the deviation of the conservative quantity, χ_p, from its mean value is given to a first approximation by:

$$(\chi_p)^t = -l.\nabla(\chi_p)_t = -\{l_y\ \partial(\chi_p)_t/\partial y + l_z\ \partial(\chi_p)_t/\partial z\}$$

l mixing length vector.

$$\therefore F_y = -(\rho)_t\{(l_y(v)^t)_t\ \partial(\chi_p)_t/\partial y + (l_z(v)^t)_t\ \partial(\chi_p)_t/\partial z\}$$

Assume velocity and displacement vectors are in the same direction:

$$(w)^t/(v)^t = l_z/l_y = \tan\alpha$$

$$(v)^t = (\mathbb{V})^t \operatorname{Cos}\alpha = (\mathbb{V})^t\{1-\alpha^2/2+\ldots\} \simeq (\mathbb{V})^t,$$

$$l_y = |l| \operatorname{Cos}\alpha \simeq |l|,$$

$$(w)^t = (\mathbb{V})^t \operatorname{Sin}\alpha = (\mathbb{V})^t\{\alpha+\alpha^3/\underline{|3}+\ldots\} \simeq (\mathbb{V})^t\alpha,$$

$$l_z = |l| \operatorname{Sin}\alpha \simeq |l|\ \alpha.$$

$$F_y = -(\rho)_t\{(\ |l|\ (\mathbb{V})^t)_t\ \partial(\chi_p)_t/\partial y + (\ |l|\ \alpha(\mathbb{V})^t)_t\ \partial(\chi_p)_t/\partial z\}$$

Assuming that α independent of $(\mathbb{V})^t$:

$$F_y = -(\rho)_t\ \{K_{yy}\ \partial(\chi_p)_t/\partial y + K_{yz}\ \partial(\chi_p)_t/\partial z\}$$

where

$$K_{yz} = (\alpha)_t\ K_{yy}.$$

Similarly:

$$F_z = -(\rho)_t\ \{K_{zy}\ \partial(\chi_p)_t/\partial y + K_{zz}\ \partial(\chi_p)_t/\partial z\}$$

where

$$K_{zy} = (\alpha)_t K_{yy},$$

and

$$K_{zz} = \{(\alpha)_t{}^2 + (\alpha)^{t2}\}K_{yy}$$

Effects of combinations of atmospheric pollutants upon vegetation

R. J. H. WILLIAMS and G. R. RICKS *University of Wales Institute of Science and Technology, Cardiff, Wales, U.K.*

Introduction

Many investigations of pollutant injury to plants, both in the field and the laboratory, have been restricted to a consideration of the effects of individual gases; either as a result of difficulties in simultaneously monitoring a wide spectrum of ambient pollutants or from the need to define experimentally the plant response to particular phytotoxicants.

Although these approaches (e.g. Katz 1939; Middleton *et al.* 1950; Hill *et al.* 1961) have provided valuable knowledge of the levels of toxicity of particular gases some inconsistencies have been observed particularly when defining threshold concentrations causing injury in the field, for under natural conditions plants may be subjected to a variety of phytotoxicants that can vary in concentration both in space and with time. For instance, in Canada and the North Eastern United States, field levels of ozone (O_3) in excess of 5 pphm (parts per hundred million) usually produce symptoms of 'weather fleck' in certain varieties of tobacco (Macdowall *et al.* 1964). However symptoms have been observed (Macdowall *et al.* 1964; Menser & Heggestad 1966; Cole & Katz 1966) when ambient concentrations of O_3 were as low as 2 pphm; results which could not be explained in terms of greater plant sensitivity or associated climatological factors, but suggested a positive interaction between the ozone and other pollutants carried in the weather front.

Thus an interaction between gaseous pollutants poses an insidious threat to vegetation, in order to recognize and comprehend the full effects of air pollutants on plants it is necessary to examine the interactions between pollutants and the vegetative response to them.

Interaction between pollutants

It is convenient to categorize pollutant interaction into two general groups: (1) primary pollutants that react chemically in the atmosphere to produce secondary *de novo* phytotoxicants which may or may not be more potent than the original gases and (2) gas mixtures that coexist without antagonism at normal ambient concentrations but can produce enhanced injury symptoms on exposed vegetation.

In addition the effect of soot and dust particles upon stomatal behaviour must be considered as stomatal interference which can increase gas exchange resulting in greater rates of uptake of atmospheric gases (Ricks & Williams 1974).

(1) INTERACTION TO PRODUCE SECONDARY PHYTOTOXICANTS

The classic example of pollutants reacting at a site remote from the vegetation is the Los Angeles type of oxidant smog, where the primary substances, oxides of nitrogen and unsaturated olefinic hydrocarbons derived from automobile exhausts combine photochemically in the atmosphere to produce secondary toxic pollutants such as O_3, nitrogen dioxide (NO_2) and the peroxyacetyl nitrates (Darley 1969). In addition sulphur trioxide may be produced from the oxidation of sulphur dioxide (SO_2) by peroxy radicals to give sulphuric acid mist—the visible component of the smog. The atmospheric photochemical reaction systems are extremely diverse, partly because of the multiplicity of organic pollutant species, which can include aldehydes, ketones, olefins, aromatics and saturated hydrocarbons, and partly from the difficulties of observation and analysis of the various components in the complex (Haagen-Smit & Wayne 1968).

Despite this there is considerable evidence of the effects of some of the individual components of the secondary pollutant complex upon vegetation (e.g. Middleton 1961; Heggestad *et al.* 1964; Hindawi *et al.* 1965; Heggestad 1968) even down to the effects upon cellular components and biochemical metabolism (Dugger & Ting 1970) although the overall effects of any gas mixtures from within the complex are not, as yet, completely clear.

(2) GAS MIXTURES

During recent years evidence of the importance of mixtures of pollutant gases that do not typically interact to produce secondary phytotoxicants has been accumulating from studies on the injury sustained by plants subjected to fumigation with pairs of phytotoxicants. The pollutant mixture may induce (a) antagonism (reduced effects), (b) synergism (enhanced effects), (c) simple additive effects or (d) potentiation, an enhanced effect produced when one of the components in the mixture is typically non-toxic (Stokinger & Coffin 1968). Mixtures of phytotoxicant pollutants generally produce effects of types (b) and (c). Reports of an antagonism between sulphur dioxide and the oxidant complex in photochemical smog (Thomas *et al.* 1952) have not been substantiated (Tingey *et al.* 1971) although it is possible that the toxic effects of SO_2 could be ameliorated by ammonia, itself a phytotoxicant, for Eggleton & Atkins (1972) have shown that SO_2 in droplets can be oxidized by ammonia to give an ammonium sulphate aerosol. This may be beneficial to vegetation since ammonium sulphate is a plant fertilizer. No examples of potentiation have, as far as we know, been reported.

Sulphur dioxide and ozone

Of all mixtures of pollutant gases studied, the interaction between SO_2 and O_3 has been investigated in most detail, especially in response to the 'weather fleck' syndrome shown by tobacco in North America, which can be caused by the synergistic action of the two gases. This effect on tobacco was first shown by Menser & Heggestad (1966) who obtained injury symptoms, tiny white flecks randomly scattered over the upper surface of mature leaves, that closely resembled typical O_3 injury in plants that had been fumigated with a mixture of SO_2 and O_3 with both gases at sub-threshold concentrations (Table 1). Their results have since been confirmed for a number of

Table 1. Injury to tobacco leaves after exposure to ozone and sulphur dioxide singly and in combination.

Duration of fumigation	Toxicant (pphm)		No. of damaged leaves			Damaged leaf area %		
(hr)	O_3	SO_2	Bel-W_3	Consolation	Bel-B	Bel-W_3	Consolation	Bel-B
2	3·0		0	0	0	0	0	0
2		24·0	0	0	0	0	0	0
2	2·7	24·0	38	37	25	15	12	9
4	3·1		0	0	0	0	0	0
4		26·0	0	0	0	0	0	0
4	2·8	28·0	75	76	48	41	43	23

(Each value is the average of 4 experiments. Total 18 plants; 7 leaves per plant scored.) From Menser & Heggestad (1966).

other varieties of tobacco (Menser & Hodges 1970; Hodges *et al.* 1971; Grosso *et al.* 1971) although Macdowell & Cole (1971) in a kinetic study of the interaction between the two gases suggested that synergism did not occur below the individual threshold dose for O_3.

Synergism between these gases has also been reported to affect other species and has been implicated (Jaeger & Banfield 1970; Dochinger *et al.* 1970; Houston & Stairs 1972) in the Chlorotic Dwarf Disease of Eastern White Pine (*Pinus strobus*), a physiogenic condition characterized by mottling of the needles, premature foliar abscission and stunted roots and shoots. The results of Dochinger *et al.* (1970) in Table 2 show that the gas combination produced more than twice the needle mottling caused by each gas alone. They suggested that SO_2 was more reactive than O_3 in producing disease symptoms, an effect supported by Costonis (1971) who found that in field trials the degree of leaf mottling in sensitive ramets subjected to a mixture of 6 pphm SO_2 with O_3 at a fluctuating concentration of not more than 4 pphm over 4 hour periods was not at all correlated to the changes in O_3 concentration.

Table 2. Mottling of needles of sensitive ramets of white pine after exposure to ozone and sulphur dioxide singly and in combination.

Toxicant (pphm)		% area of needles showing mottling	No. Replications
O_3	SO_2		
9·9		3	9
	9·6	4	6
9·9	9·8	16	9
0·0	0·0 (Control)	0	6

(Resistant ramets exhibited no mottling of the needles.) From Dochinger *et al.* (1970).

The injuries sustained by tobacco and white pine from the SO_2–O_3 mixture are generally similar to those produced by O_3 alone although Grosso *et al.* (1971) found that gas mixtures produced flecking of the upper surface together with lower surface glazing of tobacco leaves whereas the same O_3 concentration only injured the lower surface, and Menser & Hodges (1970) reported that the upper surface lesions produced by this gas mixture were slightly larger than those produced by O_3 alone. Applegate & Durrant (1969) however found that when Spanish peanuts cv. Starr (*Arachis hypogaea*) were fumigated with SO_2 and O_3, singly and in combination, the injuries to the leaves produced by the gas mixture were completely different from those produced by either of the gases alone. At O_3 concentrations of 2–3 pphm the leaves turned chlorotic, starting at the midrib, 24–48 hours after beginning fumigation, and in SO_2 concentrations from 5–12 pphm the youngest leaves turned necrotic around the leaf margins some 4–8 hours after commencing the gas treatment. When chlorosis or necrosis was complete the entire leaf abscissed. In mixtures at much lower gas concentrations (ozone 0·8–1·0 pphm and sulphur dioxide 2·0–3·0 pphm) symptoms appeared earlier, after 4–5 hours, but took the form of an initial marginal chlorosis spreading to the midvein, and eventually after 48 hours the whole leaf became necrotic. The symptoms produced by the synergistic action of the gas mixture in the controlled fumigation were identical to those initially observed in field grown plants after the passage of weather fronts. The subsequent bronzing of leaflets in the field plants observed by Applegate & Durrant (1969) suggests that phytotoxic levels of peroxyacetyl nitrate (PAN) may also have affected the plants (Taylor 1969).

In all the examples described above SO_2 and O_3 have acted synergistically to produce foliar injury, but recently Tingey *et al.* (1971a) showed that reductions in growth and yield in radish cv. Cherry Belle (*Raphanus sativus*) produced by an SO_2–O_3 mixture were at most additive and in several cases significantly less than additive. The reduction in the leaf fresh weight was additive but the overall plant fresh weight and the root fresh and dry weights were not.

Nitrogen dioxide and Sulphur dioxide

The effects of mixtures of NO_2 and SO_2 on a number of species have been studied by Tingey *et al.* (1971b). They found the form of the injury sustained by the leaves of the plants exposed to the NO_2–SO_2 mixtures varied considerably. Symptoms representative of injury from O_3, the oxidant complex, SO_2–O_3 mixtures and SO_2 alone were noted to some degree in all of the species examined; these similarities would make field diagnosis of injury due to specific phytotoxicants extremely difficult!

Their results illustrate an important concept involving the injury threshold concentrations of a gas mixture which hitherto has not been discussed. Table 3 shows the degree of leaf injury induced in tobacco by mixtures of NO_2 and SO_2 at varying concentrations. In preliminary experiments the threshold injury concentrations for NO_2 and SO_2 in 4-hour exposures were found to be 200 and 50 pphm respectively and clearly the gas mixtures induce leaf injury at much lower concentrations than these. However, the results suggest that there exists a combination threshold injury concentration below which no symptoms are observed on the leaf. In this case the concentration lies between 5 and 10 pphm, for only when the concentration of each gas in the mixture exceeds this level does injury occur. The combination or mixture threshold injury concentration is presumably directly analogous to the threshold injury concentration of individual phytotoxic gases although in a gas mixture injury is induced at much lower concentrations.

Table 3. Injury to tobacco exposed to mixtures of nitrogen dioxide and sulphur dioxide for 4 hours.

Toxicant pphm		Leaf injury %	No. observations
NO_2	SO_2		
5	5	0	8
5	25	0	4
10	10	26	16
20	10	60	4
25	5	0	4
25	25	68	4
25	50	100	4

Leaf injury is percentage injury to the three most severely injured leaves/plant. From Tingey *et al.* (1971b).

The results of fumigation on a number of other species are equivocal however (Table 4) for although all species except tomato developed a trace of injury when the concentration of each gas was 5 pphm, injury varying between 0 and 16% occurred when one of the toxicants was held at 5 pphm

Table 4. Injury to the upper surface of plants exposed to mixtures of nitrogen dioxide and sulphur dioxide for 4 hours.

Toxicant (pphm)		Leaf injury %					
NO_2	SO_2	Pinto bean	Oats	Radish	Soy bean	Tomato	No. observations
5	5	2	1	1	2	0	7
5	10	0	0	0	0	0	4
5	20	1	0	0	6	0	3
5	25	1	3	0	7	1	4
10	5	0	0	1	1	0	3
10	10	11	27	27	35	1	3
15	10	24	12	24	20	17	4
15	25	4	0	4	1	0	3
20	20	16	10	6	9	0	2
25	5	0	0	13	2	0	2

Threshold injury concentrations for NO_2 and SO_2 in 4 hr exposures were 200 and 500 pphm respectively. Leaf injury is the percentage injury to the three most severely injured leaves/plant except for Pinto bean where the 2 primary leaves were used. From Tingey *et al.* (1971b).

and the other raised from 5 to 25 pphm. Maximum injury was observed on all species at concentrations between 10 and 15 pphm but as the concentration of both gases increased foliar injury decreased, an effect the authors could not explain.

Sulphur dioxide and hydrogen fluoride

Matsushima and Brewer (1972) studied the effects of this gas mixture on citrus trees in California and found that although leaf area was decreased and leaf chlorosis increased the degree of injury sustained was proportional to simple addition effects.

(3) PARTICULATE MATTER

The deposition of particulate matter on to leaves can affect them by adsorption of toxic gases (Bjorkman 1970) or by altering rates of gas exchange (Berge 1970; Ricks & Williams 1974).

Bjorkman (1970) powdered alder and spruce seedlings with wood ash and after exposure to SO_2 for 18 days found the leaves were damaged more than unpowdered control plants. Powdering with activated charcoal saturated with SO_2 was particularly harmful.

Particles can also occlude stomatal pores (Mansfield & Majernik 1970) preventing their complete closure resulting in gas exchange during the night (Berge 1970). In a recent study of *Quercus petraea* the number of stomata with particles occluding their pores was found to be proportional to the amount of particulate fallout from an industrial source (Williams *et al.* 1971). Stomatal occlusion significantly reduced the stomatal diffusion resistance during the night and the maximal diffusion resistance measured was found to be proportional to the degree of stomatal occlusion (Ricks & Williams 1974). The overall effect of the particulate deposit was to increase rates of gaseous exchange at night which led to significant increases in the levels of leaf sulphur from uptake of SO_2. Therefore increased rates of foliar uptake of phytotoxicants, either single gases or mixtures, would result in an apparent lowering of threshold injury concentration.

Discussion

The realization that interactions between atmospheric pollutants can result in enhanced injury symptoms, new injury symptoms and injury at concentrations far below those at which the individual phytotoxicants cause injury has opened a new field of pollutant injury diagnosis. Work on these aspects has not been extensive but it is apparent that gas mixtures may cause serious effects to vegetation, although very little is yet known of the mechanisms of the plant response to particular mixtures of pollutants.

It appears that the response of individual species varies with different gas mixtures and some plants are more resistant to injury than others e.g. tomato is less sensitive than tobacco to NO_2–SO_2 mixtures (Tables 3 and 4). The importance of the role of the constituent gases in the mixture can also differ with different species, for instance, with tobacco O_3 appears to be the dominant toxicant in an SO_2–O_3 mixture (Macdowell & Cole 1971) whereas with white pine SO_2 has the greater influence (Costonis 1971). Some gas mixtures, especially SO_2–O_3 and NO_2–SO_2, induce a synergistic injury response in most of the plant species so far studied (Tables 1–4), although this may not always be so, for the injury response of radish to an SO_2–O_3 mixture is only additive (Tingey *et al.* 1971a). The other gas mixture (SO_2–HF) which has been investigated in some detail (Matsushima & Brewer 1972) shows no evidence of synergism and the overall effect is at most additive.

The biochemical effects of phytotoxicants involve many processes that occur at different times and at different levels of organization, but the results of fumigation with gas mixtures suggest that when the plant response is additive the constituent phytotoxicants are acting individually, perhaps attacking different initial target loci within the tissues, whereas with the synergistic response the pollutants may influence related metabolic processes resulting in their accelerated disruption. This is not inconsistent with the evidence that whenever synergism has been reported, the threshold values for injury of the gas mixture are always much lower than the threshold values of the constituent gases alone. As yet little or no work has been accomplished on this aspect at the metabolic level.

Unpublished work by Matsushima and Taylor (quoted by Matsushima & Brewer (1972) has emphasized the importance of alternate successional exposure of plants to pairs of gases. They found that injury resulting from alternate exposure to SO_2 and NO_2 was not additive and depended on the order of exposure. SO_2 followed by NO_2 produced severe damage whereas SO_2 following NO_2 was no more damaging than either gas alone. It seems that NO_2 predisposed the tissue to severe damage by the SO_2 whereas an initial exposure to SO_2 had no such complementary effect on subsequent NO_2 fumigation. Further investigations of this nature may help to clarify the relative roles of each phytotoxicant in the synergistic action of gas mixtures.

Under field conditions the importance of alternate exposures must not be minimized for they may easily occur when wind direction fluctuates around several sources of pollutants. In addition, the role of particulate pollution may be of importance in long term exposure of plants to low levels of pollutants for over long periods quite small reductions in stomatal diffusion resistance could result in substantial accumulation of gaseous phytotoxicants.

Plants in the field are subjected to a range of different conditions, where they may experience alternate and/or simultaneous exposure to both gaseous and particulate pollutants often at low concentrations, and this suggests that future investigations must include the study of the effects of low levels of pollutants over the long term particularly upon the growth and yield of economic species. Before realistic emission control standards can be suggested fundamental research must be conducted into these aspects of plant response to pollutants.

Summary

Many studies of the effects of air pollutants upon vegetation have been restricted to the examination of individual pollutants. In recent years it has become evident that important interactions between pollutants may occur. Examples from the literature together with recent experimental work are reviewed.

References

APPLEGATE H.G. & DURRANT L.C. (1969) Synergistic action of ozone—sulphur dioxide on peanuts. *Envir. Sci. Technol.* 3, 759–60.

BERGE H. (1970) Immisionsschaden (Gas-, Rauch-, und Staubschaden). *Handbuch der Pflanzenkrankheiten* 1 (4), 1–69. P. Parey, Berlin and Hamburg.

BJORKMAN E. (1970) The effect of fertilisation on sulphur dioxide damage to conifers in industrial and built-up areas. *Stud. for. Suec.* 70, 1–50.

COLE A.F.W. & KATZ M. (1966) Summer ozone concentrations in Southern Ontario. *J. Air Pollut. Control Ass.* 16, 201–6.

COSTONIS A.C. (1971) Effects of ambient sulphur dioxide and ozone on Eastern White Pine in a rural environment. *Phytopath.* 61, 717–20.

DARLEY E.F. (1969) The role of photochemical air pollution on vegetation. *Air Pollution. Proceedings of the First Congress on the Influence of Air Pollution on Plants & Animals*, pp. 137–42. Centre for Agricultural Publishing and Documentation, Wageningen.

DOCHINGER L.S., BENDER F.W., FOX F.L. & HECK W.W. (1970) Chlorotic Dwarf of Eastern White Pine caused by an ozone and sulphur dioxide interaction. *Nature, Lond.* 225, 476.

DUGGER W.M. & TING I.P. (1970) Air Pollution Oxidants—Their effects on metabolic processes in plants. *Ann. Rev. Plant Physiol.* 21, 215–34.

Eggleton A.F.J. & Atkins D.H. (1972) *Results of the Tees-side Investigation.* U.K.A.E.A. Research Group report, H.M.S.O., London.

Grosso J.J., Menser H.A., Hodges G.H. & McKinney H.H. (1971) Effects of air pollutants on *Nicotiana* cultivars and species used for virus studies. *Phytopath.* 61, 945–53.

Haagen-Smit A.J. & Wayne L.G. (1968) Atmospheric Reactions and Scavenging Processes. *Air Pollution* Vol. 1 *Air Pollution and its effects* (Ed. by A.C. Stern) 2nd Ed. pp. 149–86. Academic Press, N.Y. and London.

Heggestad H.E., Burleson T.R., Middleton J.T. & Darley E.T. (1964) Leaf injury on tobacco varieties resulting from ozone, ozonated hexene-1 and ambient air of metropolitan areas. *Int. J. Air Wat. Pollut.* 8, 1–10.

Heggestad H.E. (1968) Diseases of Crops and Ornamental Plants incited by air pollutants. *Phytopath.* 58, 1089–97.

Hill A.C., Pack M.R., Treshow M., Downs R.F. & Transtrum L.G. (1961) Plant injury induced by ozone. *Phytopath.* 51, 316–63.

Hindawi T.J., Dunning J.A., Brandt C.S. (1965) Morphological and microscopical changes in tobacco, bean and petunia leaves exposed to irradiated auto exhaust. *Phytopath.* 55, 27–30.

Hodges G.H., Menser H.A. & Ogden W.B. (1971) Susceptibility of Wisconsin Havana tobacco cultivars to air pollutants. *Agron. J.* 63, 107–11.

Houston D.B. & Stairs G.R. (1972) Physiological and Genetic response of *Pinus strobus* L. clones to sulphur dioxide and ozone exposures. *Mitt. forstl. Bund Vers Aust. Wien* 97 (2), 387–98.

Jaeger J. & Banfield W. (1970) Response of Eastern White Pine to prolonged exposure to atmospheric levels of ozone, sulphur dioxide or mixture of these pollutants. *Phytopath.* 60, 575.

Katz M. (1939) *Effects of sulphur dioxide on vegetation.* Natural Research Council, Canada. Ottawa.

Macdowell F.D.H., Mukammal E.I. & Cole A.F.W. (1964) Direct correlation of air polluting ozone and tobacco weather-fleck. *Can. J. Plant Sci.* 44, 410–17.

Macdowell F.D.W. & Cole A.F.W. (1971) Threshold and synergistic damage to tobacco by ozone and sulphur dioxide. *Atmos. Envir.* 5, 553–9.

Mansfield T.A. & Majernik O. (1970) Can stomata play a part in protecting plants against air pollutants? *Environ. Pollut.* 1, 149–54.

Matsushima J. & Brewer R.F. (1972) Influence of sulphur dioxide and hydrogen fluoride as a mix or reciprocal exposure on citrus growth and development. *J. Air Poll. Control Ass.* 22 (9), 710–13.

Menser H.A. & Heggestad H.E. (1966) Ozone and sulphur dioxide synergism injury to tobacco plants. *Science* 153, 424–5.

Menser H.A. & Hodges G.H. (1970) Effects of Air Pollutants on Burley tobacco cultivars. *Agron. J.* 62, 265–9.

Middleton J.T., Kendrick J.B. & Schwalm H.W. (1950) Injury to herbaceous plants by smog or air pollution. *Plant Dis. Reptr.* 34, 245–52.

Middleton J.T. (1961) Photochemical air pollution damage to plants. *Ann. Rev. Plant Physiol.* 12, 431–48.

Ricks G.R. & Williams R.J.H. (1973) Effects of Atmospheric Pollution upon Deciduous Woodland 2: Effects of particulate matter upon stomatal diffusion resistance in leaves of *Quercus petraea* (Mattuschka) Leibl. *Environ. Pollut.* 6, 87–109.

Stokinger H.E. & Coffin D.L. (1968) Biological Effects of air pollutants. *Air Pollution*

Vol I. *Air Pollution and its effects* (Ed. by A.C. Stern) 2nd Ed. pp. 446–546, Academic Press, N.Y. and London.

TAYLOR O.C. (1969) Importance of peroxyacetyl nitrate (PAN) as a phytotoxic air pollutant. *J. Air Poll. Control Ass.* 19, 347–51.

THOMAS M.D., HENDRICKS R.H. & HILL G.R. (1952) Some impurities in the air and their effects on plants. *Air Pollution* (Ed. by L.C. McCabe), pp. 41–7. McGraw-Hill, N.Y.

TINGEY D.T., HECK W.W. & REINER R.A. (1971a) Effects of low concentrations of ozone and sulphur dioxide on foliage, growth and yield of radish. *J. Amer. Soc. Hort. Sci.* 96, 369–71.

TINGEY D.T., REINERT R.A., DUNNING J.A. & HECK W.W. (1971b) Vegetation Injury from the Interaction of Nitrogen Dioxide and Sulphur Dioxide, *Phytopath.* 61, 1506–11.

WILLIAMS R.J.H., LLOYD M.M. & RICKS G.R. (1971) Effects of Atmospheric Pollution on Deciduous Woodland 1: Some effects on leaves of *Quercus petraea* (Mattuschka) Leibl. *Environ. Pollut.* 2, 57–68.

Biological survey in the detection and assessment of pollution

R. W. EDWARDS, B. D. HUGHES and M. W. READ
University of Wales Institute of Science and Technology, Cardiff, Wales, U.K.

If one regards pollution as the release, by Man, of substances (or energy) to the environment in quantities which damage his resources, then clearly the extent of pollution is determined not only by what Man regards as resources but also the sensitivity of methods in establishing causal connections between them and substances released. Man is acknowledging a wider resource responsibility and his techniques of detection of damage must maintain a matching sensitivity.

Biological survey has two primary functions in relation to pollution, firstly in detecting temporal and spatial patterns in populations and communities and suggesting causes for such patterns, and secondly, in determining the extent of resource damage when both the resource and its relation to pollution have been clearly defined and established. These may be described as the 'detection' and 'assessment' functions of survey and they form part of the observational pathway in the regulation and control of pollution sources. When biological survey techniques are used primarily in the assessment of pollution the analysis of data is standardized and frequently formalized into a system of classification.

Figure 1 demonstrates the stages necessary to establish and control pollution. Whilst it is possible to assess the likely effects of substances introduced by Man into the environment using experimental procedures, our current models of the physical and chemical behaviour of substances with respect to dispersion, concentration and transformation are generally crude and, furthermore, understanding of the action of such substances on biological systems, of different hierarchical levels, is grossly inadequate. Although we must seek to make predictive models of pollutant behaviour more comprehensive and reliable, inevitably our primary signals of such behaviour will originate for a long while from observational studies, namely surveys.

As Fig. 1 shows, survey techniques are directed at establishing (a) the distribution of substances within ecosystems, and these include distributions within biotic as well as abiotic components, and (b) the distribution of effects. Biological surveys in this paper are regarded as having the second function although the term 'biological' survey sometimes refers to the chemical analysis of organisms and must in this sense be regarded as providing distributional data on pollutants or their degradation products which may either form the basis for models of pollutant dynamics or, where dose-effect relationships have been established, for an assessment of likely damage.

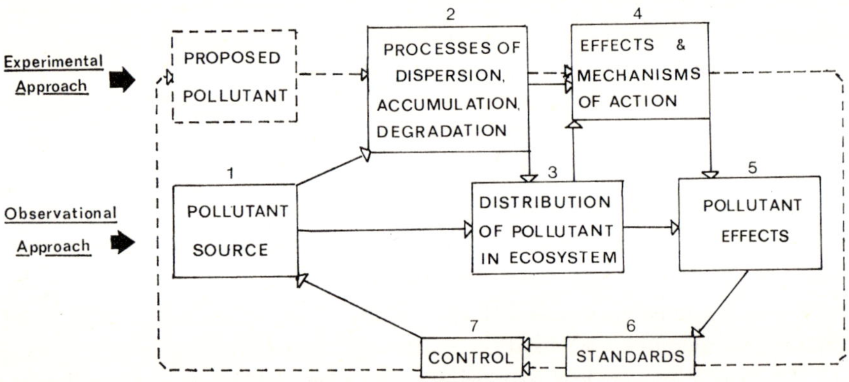

Figure 1. General scheme of pollution assessment and regulation pathways.

An analysis of biological research in the assessment of pollution damage reveals considerable variations in the role of biological survey. In most cases the changing status of populations or communities has led to investigations of the distribution of pollutants and then, where relationships seem significant, to rigorous experimental studies (Fig. 1, Box 5→3→4); this might be regarded as the classical pathway in establishing a causal relationship between a pollutant and its receptor. The work of Ratcliffe (1970) and others on the possible effects of organochlorine pesticides on the status of populations of certain birds of prey developed in this manner, firstly with observations on the changing population density and increased incidence of broken eggs leading to an examination of temporal changes in egg-shell thickness, then to an association of these changes with environmental inputs of certain organochlorine pesticides, and finally to a series of experiments establishing effects of such pesticides on the calcium metabolism of some bird species (Cooke 1973). Although the research development had a classical simplicity there is still some uncertainty about the effect of egg-shell thinning on population density, an uncertainty which is frequently found in the projection of observable biological effects of pollution at the organismic level in the laboratory on the status of natural populations and communities (Sprague *et al.* 1965).

The role of biological survey in some research strategies has been different however. Much of the work in the United Kingdom, concerning the effects of pollution on fresh-water fisheries, over the past twenty-five years, has been orientated towards establishing pollutant concentration—time relationships for fish death, the effects of environmental variables on such relationships and the combined action of poisons. These have been laboratory-based studies and inevitably the ecological relevance of such studies, frequently short-term, has been questioned. Such relevance has been established subsequently (Alabaster *et al.* 1972) in at least two river systems, by examining the distribution of fish in relation to the distribution of toxicity of pollutants as assessed by field analysis of pollutant concentrations and toxicity data derived from laboratory studies (Fig. 1 Box $4\genfrac{}{}{0pt}{}{\nearrow 3}{\searrow 5}$). This role of survey, both biological and chemical, might be considered as an assessment of the adequacy of laboratory data in predicting damage.

In yet other cases the distribution of substances within ecosystems, whose toxic effects have been suspected by their general chemical structure, has led to an intensification of experimental studies on dose-effect relationships and then an examination of possible biological effects using survey techniques (Fig. 1, Box 3→4→5). This pattern of development is shown in relation to the PCBs although there have been considerable problems in evaluating their ecological significance, the principal reasons being that this family of pollutants, ranging in toxic and other properties, generally occurs, at sub-acute levels, with others which have similar effects and that temporal changes in PCB concentrations have not been as dramatic as for some of the chlorinated hydrocarbon pesticides.

The problems of designing surveys to provide signals of environmental damage will be discussed later but these problems must be considered against a general background of changing environmental regulation and pollution control technology. In the past pronounced spatial gradients of pollutants associated with 'hot-spots' were common and local damage was relatively easy to detect and relate to such spatial gradients. Hot-spots are now increasingly regulated and effluents more efficiently dispersed. Whilst this may be effective in preventing acute damage, one consequence is the elimination of pronounced spatial gradients and the attendant difficulty of detecting effects, serious perhaps at the sub-acute level, and relating them to pollutant sources. This situation has developed in relation to SO_2 regulation, particularly in the United Kingdom, with its policy of dispersion from high stacks rather than removal. These problems are exacerbated by the increasing range of pollutants being released.

It may be suggested however that whereas in the past spatial gradients have been useful in establishing damage, rates of technological change are now so rapid that temporal change in ecological patterns is equally useful. Whilst some pollution studies have demonstrated the usefulness of analysing such temporal changes (Ratcliffe 1970), retrospective environmental analysis is generally difficult.

In search of biological pattern

Temporal and spatial patterns in the distribution of individuals, species etc., pattern merely implying non-randomness, may develop in homogeneous environments as well as heterogeneous or changing ones and problems arise not only in establishing causation of such patterns with respect to environmental variables but also with respect to intrinsic biological behaviour leading to non-randomness and the interactions between such extrinsic factors and intrinsic behaviour. Much has been written in recent years concerning the analysis of population and community pattern or structure (Southwood 1966; Williams 1964; Greig-Smith 1964) and such analyses have become increasingly sophisticated with increasing computational facilities. The authors do not intend to intrude into this field except to indicate the practical consequences of collecting information for various forms of analysis and ways in which various analyses may be used to detect relationships with pollutant distributions.

When surveys have a detection function it is generally not possible to establish at the outset what analytical approach will subsequently be most valuable. Nevertheless it is essential to estimate the costs of contemplating different forms of analysis the reliability of which will depend in part on the intensity of data collection.

In pollution surveys sampling sites which have similar basic features are generally selected to reduce inter-site variation and frequently sampling areas with minimum in-site variation are also selected, both being attempts to reduce noise to signal ratios. In lichen surveys related to air pollution studies, specific substrates (asbestos, particular tree species etc.) are generally chosen at sites having similar topographical features. River surveys have generally been carried out wherever possible by sampling only 'uniform riffles'. Nevertheless variations between biological samples taken from such substrates are considerable (Needham & Usinger 1956). Analyses of 50 samples (0·1 m^2) collected in a standardized manner from a uniform riffle on the River Cynon, a polluted river in South Wales, demonstrate some of the difficulties in obtaining adequate quantitative and qualitative data at survey sites. Table 1 shows the number of samples required to obtain an estimate of the average density ($\pm 40\%$) of some of the most abundant invertebrates at the site. For most of these species an accurate estimate of abundance would represent insuperable problems if attempted on a routine basis. Table 1 also shows the wide confidence limits for such average densities with what might be considered a tolerable sampling programme.

Table 1. Analysis of riffle data on the River Cynon.
A. Number of samples for estimation of mean density (± 40%) and
B. Percentage variation from the mean with 4 samples—both with 95% confidence limits.

Taxon	A	B
Ecdyonurus dispar	27	(102)
Ephemerella ignita	15	77
Caenis rivulorum	26	(102)
Baetis scambus	17	80
Baetis rhodani	29	(106)
Leuctra fusca	66	(162)
Hygrobates fluviatilis	67	(163)
Potamopyrgus jenkinsi	84	(183)
Nais elinguis	24	98
Nais alpina	38	(123)
Ancylus fluviatilis	98	(197)
Chironomidae	10	62
Total invertebrates	10	62

Similarly an expression of qualitative variation between samples was obtained for this riffle using various distance coefficients such as that proposed by Jacquard (Southwood 1966) and at this site the average Jacquard coefficient was only 0·44 (S.E. ±0·12) indicating that only 44% of species in two random samples are likely to be common to both. Little of this quantitative and qualitative variation can be accounted for in terms of variations in substrate and stream-velocity at the site and must, for practical purposes, be considered intrinsic biological variation.

Figure 2 shows the increasing number of species collected from this riffle with increasing number of samples and, although this rate of increase may be described mathematically, qualitative comparisons between sites seem meaningful only for the most common species except where intensive sampling programmes can be maintained.

Hierarchical expressions based on the number of organisms and species in samples, such as those proposed by Margalef (1956) and Menhinnick (1964) are also highly dependent on sample size or sample number when the customary few samples are taken, although there is much more stability in the Shannon-Wiener function (see Southwood 1966) in the macro-invertebrate community of this riffle.

These analyses indicate that the analytical treatment of site data is fairly restricted unless a substantial sampling programme is undertaken.

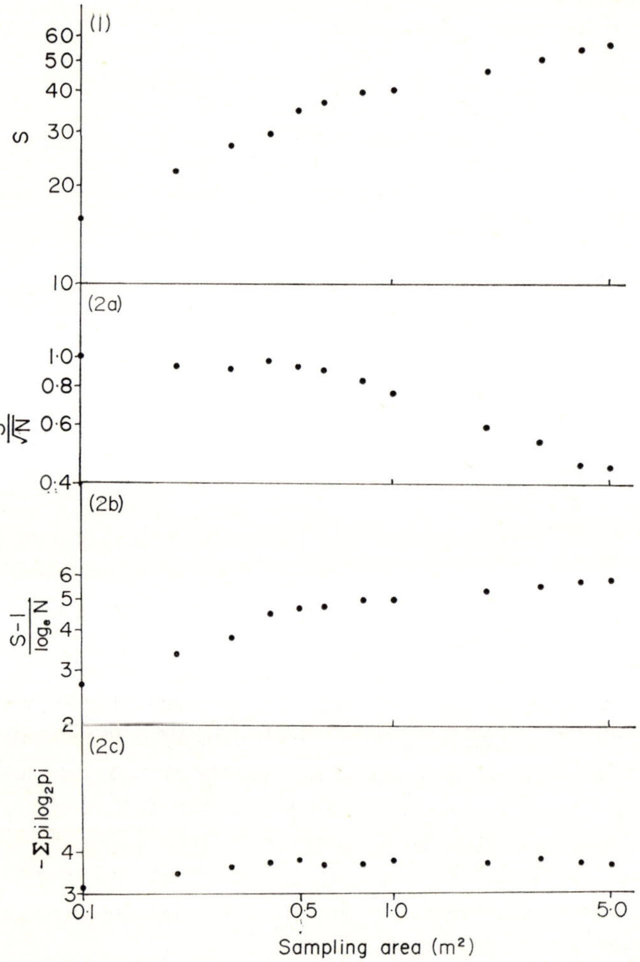

Figure 2. Relation between number of samples (reflecting sample size) and (1) the number of species collected (2) diversity indices a. Menhinick b. Margalef and c. Shannon-Wiener.

It might be assumed that spatial distributions of communities suspended in water-bodies are more homogeneous than terrestrial or aquatic-benthic communities and whilst this may be so for some neutrally buoyant components with no powers of locomotion, most planktonic species show, at times, extremely contagious distributions where even life-stages and sexes are spatially separate (George 1972) and where spatial patterns may be small-scale (George & Edwards 1973).

Various artificial substrates have been developed, sometimes to reduce sampling difficulties, but generally to reduce variation between samples and so facilitate inter-site comparisons (Dickson & Cairns 1972; Hester & Dendy 1962). Their effectiveness in reducing variability has frequently not been tested and the ecological relevance of their results has been questioned.

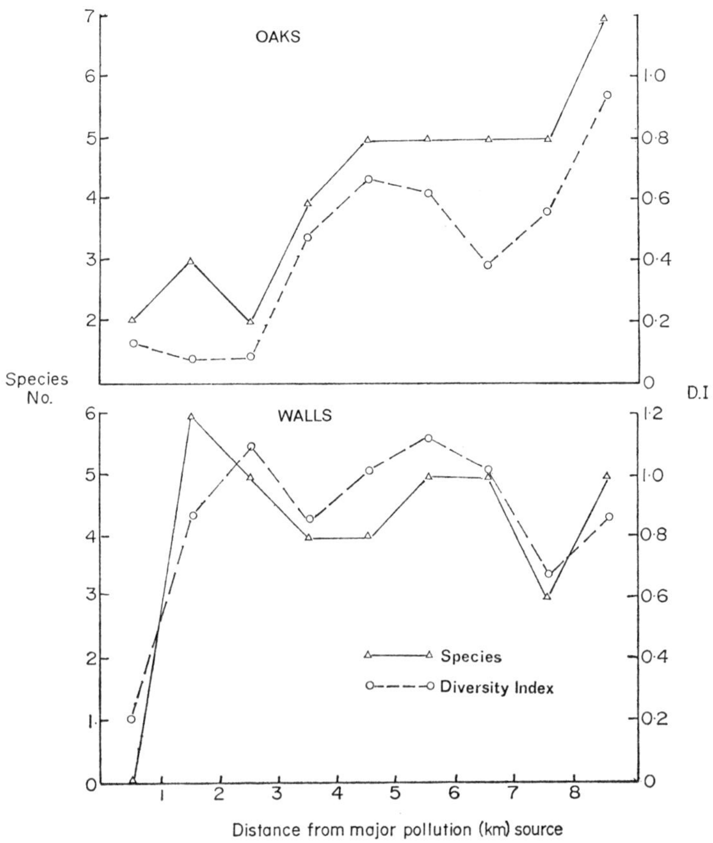

Figure 3. Number of lichen species and their diversity indices (Shannon-Wiener) on two substrates around an industrial and urban source of pollution in S. Wales.

Even when artificial substrates are not used, conclusions may be very dependent on the substrate or habitat selected. Figure 3 shows the average number of lichen species and their indices of diversity (Shannon-Wiener) on defined sampling areas of stone walls and sessile oaks at various distances from an urban and industrial source of SO_2 and other pollutants in South Wales. The substrate chosen greatly influences not only the qualitative composition of the lichen flora but also the spatial changes in species abundance and clearly affects interpretations of spatial patterns of pollutant effects (see also Gilbert 1965). Pyatt (1970) also surveyed the lichens in this area and showed yet another spatial pattern by listing all the species found per km² irrespective of substrate. Such an approach may lead to confusion between species-richness on standard substrates and spatial patterns of substrate diversity, this diversity being generally low, but not necessarily related to pollution, around urban and industrial areas.

Whilst the time, and sometimes disruption, involved in sampling provide constraints in the analysis of ecological survey data, generally the laboratory analysis of samples, particularly taxonomic, frequently constitutes the major constraint on the number of samples analysed. The level of taxonomic identification may considerably influence the detection of patterns, but where the role of survey is primarily in the assessment of pollution it is sometimes possible to reduce the time devoted to identification either by selecting key components of communities or by reducing the level of identification to the genus or family rather than the species. Confusion arises however where the key or indicator species approach is used to detect effects of unrecognized pollutants from unknown sources: this approach, the validity of which is based on established dose-effect relationships, has limited diagnostic value.

Figure 4 demonstrates the close correlations obtained for several commonly-used expressions of community structure, using information either at the species or family level, at several macro-invertebrate sampling stations

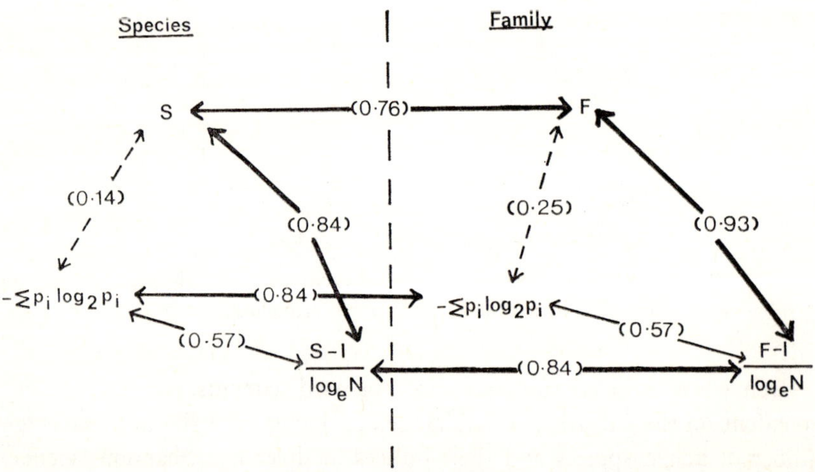

Figure 4. Correlations between expressions of species and family diversity for macro-invertebrates in River Cynon. (Significant correlations shown with continuous lines; N, S and F are numbers of organisms, species and families respectively.)

on the River Cynon (data taken from Learner *et al.* 1971). Clearly if these were the only expressions of community structure required there would be little point in devoting considerable time to species identification and certain of these expressions, closely correlated, could be omitted.

Where the role of survey is predominantly detection, a more open-ended approach to data analysis is required and reduced taxonomic analysis may result in a crucial loss of information in detecting pattern and relating this to possible causal factors.

In search of causative factors

There seem to be broadly two ways in which biological survey data may be used in attempts to detect pattern and establish causal factors. These, using a short-hand notation, may be described as the A, B, C . . . and N_1, N_2, N_3 . . . approaches.

Firstly where a body of information exists on the ecological requirements and tolerance limits of species (A, B, C, . . .) then survey data on the distribution of such species may indicate the factors primarily responsible for the distribution. This may be described as the 'qualitative' input of survey and although potentially a most powerful approach it is limited in its usefulness at present by the relatively scanty information on the basic ecological requirements of many species and their reactions to pollutants.

Secondly, suggestions of causation come from direct correlations between biological patterns involving the distribution of separable components (N_1, N_2, N_3 . . .) of communities and environmental variables. The qualitative differences between components are ignored and they are merely used to derive numerical expressions of community structure. Changes in such expressions, in themselves, give no specific clues on causation although it has been suggested that certain of them give a general indication of the degree of stress. Whilst studies have shown temporal and spatial changes in the value of such expressions as the Shannon-Wiener function, which can be related to the distribution of toxic substances (Brooker & Edwards 1974; Wilhm 1970), the general validity of this method of detecting patterns of stress has not been established and furthermore the approach gives no indication of the kind of stress involved.

Clearly attempts should be made, accepting the practical constraints described earlier, to obtain biological data which can be analysed qualitatively (A, B, C, . . .) and, in conjunction with other environmental information, quantitatively (N_1, N_2, N_3, . . .) to provide reinforcement of analytical approaches. These complementary approaches have been used on macro-invertebrate data of the River Cynon, given in Learner *et al.* (1971), to demonstrate some of the problems in detecting pattern and its possible causation, particularly in relation to pollution. The study of this river is being continued more intensively to test initial conclusions.

Figure 5 shows some of the common numerical expressions, namely the number of species and organisms, diversity index and redundancy, at sampling stations down the river (Station 1→11). The most marked spatial change is clearly between stations 4 and 5, although there is no change in the number of species between these stations. The low diversity of the upstream station is probably caused by physical instability and is common in headstreams, but the sharp biological discontinuity between station 4 and 5 cannot be similarly explained.

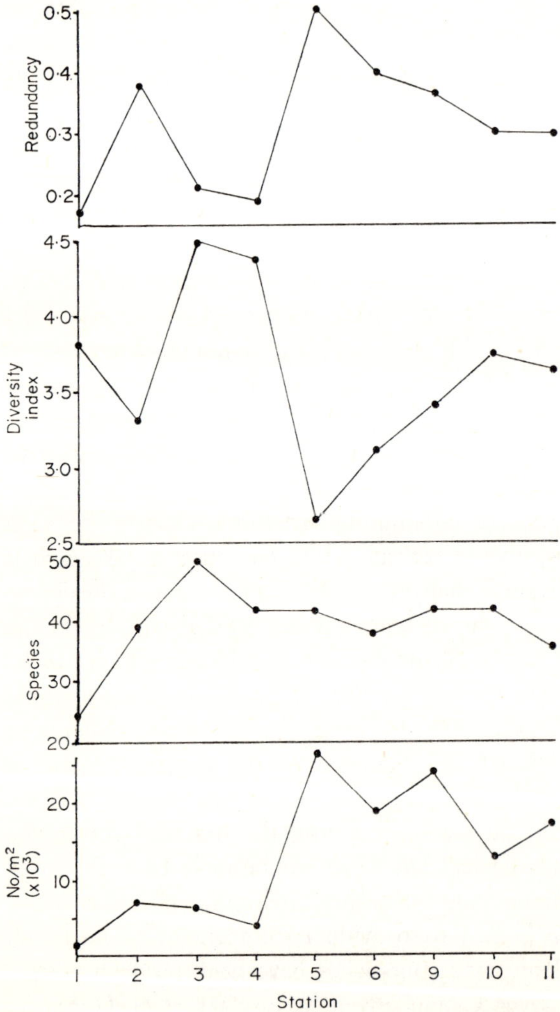

Figure 5. Expressions of community structure at stations in the River Cynon.

When one compares the similarity of stations using the presence or absence of biological components (species) at these stations using a distance measurement such as the Jacquard coefficient (Table 2), the similarity of the fauna downstream of station 5 clearly emerges. These Jacquard coefficients are derived from bulked samples at each station and are not comparable with those given earlier in the paper for faunal variation within a riffle.

Biological discontinuities can also be detected by establishing which

Table 2. Similarity of macro-invertebrate fauna at stations on River Cynon based on Jacquard coefficient.

2	19·6							
3	20·0	42·6						
4	8·2	28·6	29·6					
5	18·2	19·4	19·7	41·7				
6	7·1	17·2	19·4	35·6	49·0			
9	14·6	16·4	20·5	23·9	43·8	48·0		
10	18·9	12·8	21·6	27·3	40·6	47·2	58·8	
11	13·7	12·3	20·0	27·9	40·0	44·0	56·2	66·1
Station	1	2	3	4	5	6	9	10

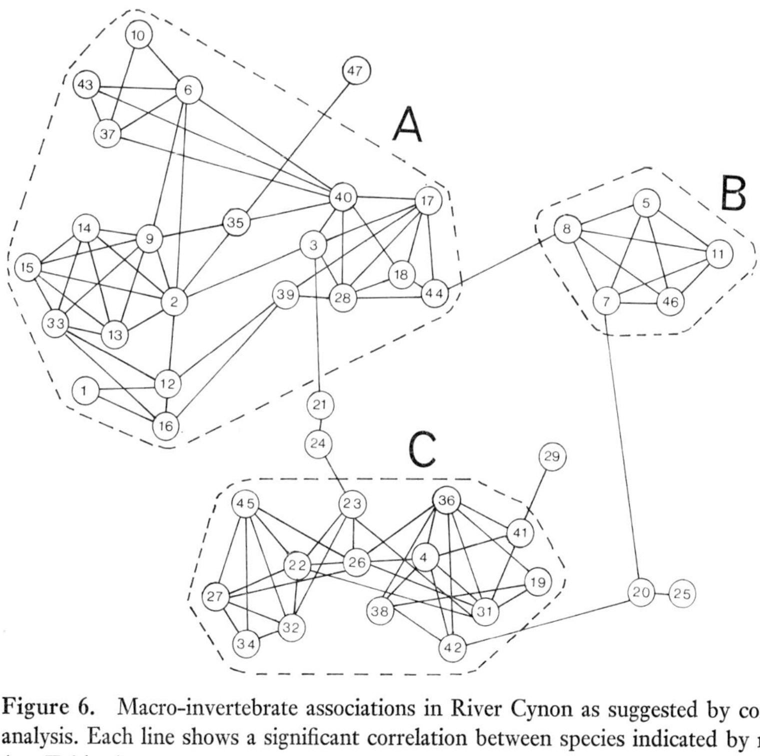

Figure 6. Macro-invertebrate associations in River Cynon as suggested by correlation analysis. Each line shows a significant correlation between species indicated by numbers (see Table 3).

groups of components are spatially associated. Figure 6 shows the results of a correlation analysis in which each component, in this case species, existing at more than four stations, including tributaries, and exceeding a density of at least $30/m^2$ at one station, is shown if its abundance is significantly correlated with that of another component. Such correlations are shown by lines joining species denoted by numbers. The species tend to sort themselves into three primary groups: sub-groups are apparent if higher values of correlation coefficients are taken. Other more rigorous clustering techniques have been used on these data and different distance measurements adopted e.g. euclidian, but the same basic pattern emerges.

In Fig. 7 the total density of each group is shown for all the stations,

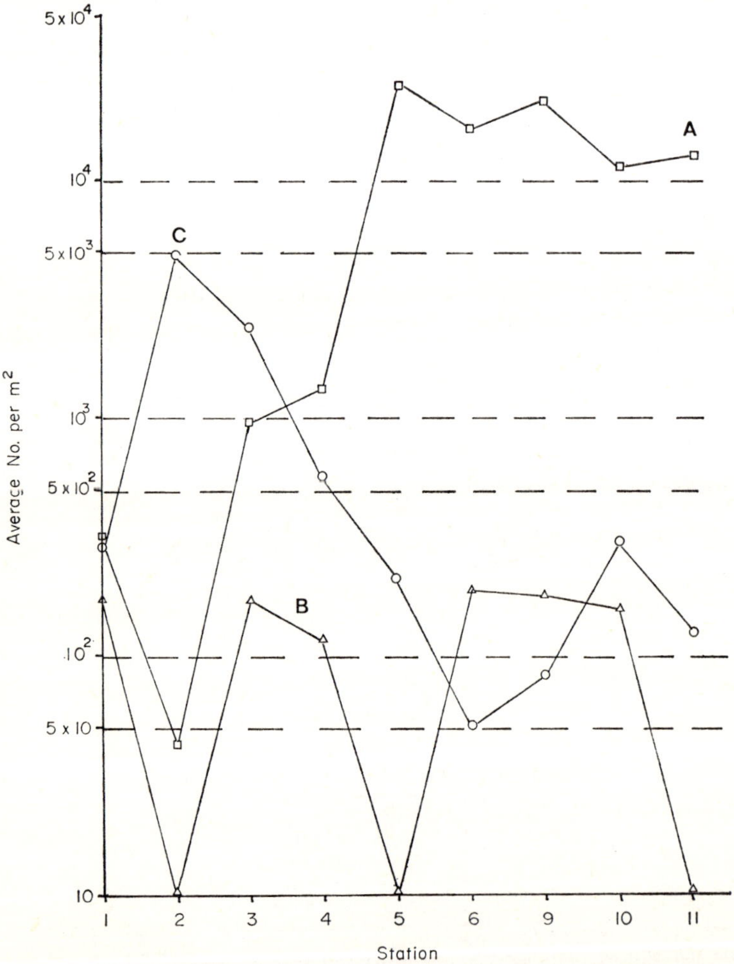

Figure 7. Changes in abundance of major macro-invertebrate associations in River Cynon (see Fig. 6).

excluding tributaries: there is a clear decline in Group C principally downstream of station 3 and a sharp increase in Group A downstream of station 4. Group B, shown for completeness, is only abundant in certain tributaries but never exceeds 200/m^2 in the main river.

Some of the faunal differences between stations will result from changes in physical factors such as temperature, basic substrate and water velocity although sampling stations were selected to minimize differences in the last two factors. When certain faunal elements change there will be consequential biological interactions and secondary changes. Nevertheless the sharp spatial discontinuities in community structure within the reaches 3–4 and 4–5, shown in Fig. 7, suggest important environmental discontinuities.

The search for causal factors of these biological discontinuities, as has been suggested earlier, can take broadly two courses:

1. examination of ecological requirements and effects of pollutants with respect to those species whose distributions show spatial changes, and
2. establishment of relationships between species distributions and environmental variables.

Table 3 shows the lists of species spatially grouped in Figs. 6 and 7 and although some associations probably have a biological basis, for example that of *Eukiefferiella calvescens* which frequently lives in the mantle cavity of *Ancylus fluviatilis* (Hynes 1972), other distributions are probably related to

Table 3. Macro-invertebrate associations on the River Cynon.

A	B	C
Hydra sp.	*Nais alpina*	*Chaetogaster langi*
Chaetogaster crystallinus	*Nais communis*	*Leuctra fusca*
Chaetogaster diaphanus	*Nais elinguis*	*Ephemerella ignita*
Nais barbata	*Pristina idrensis*	*Ecdynurus dispar*
Nais variabilis	*Potamopyrgus jenkinsi*	*Oreodytes septentrionalis*
Pristini foreli		*Limnius volckmari*
Stylaria lacustris		*Trissopelopia longimana*
Limnodrilus hoffmeisteri		*Polypedilum acutum*
Tubifex tubifex		*Brillia modesta*
Lumbricillus rivalis		*Eukiefferiella calvescens*
Helobdella stagnalis		*Orthocladius* sp. A.
Acanthocyclops vernalis		*Bezzia-Palpomyia gp.*
Eucyclops speratus		*Wiedemannia* sp.
Prodiamesa olivacea		*Ancylus fluviatilis*
Brillia longifurca		
Cricotopus bicinctus		
Orthocladius rubicundus		
Rheocricotopus foveatus		
Syncricotopus rufiventris		
Hygrobates fluviatilis		
Lebertia porosa		

environmental factors. Certain components of Group A, abundant below station 4, are commonly associated with organic enrichment e.g. *Tubifex tubifex*, *Limnodrilus hoffmeisteri*, *Helobdella stagnalis*. Several components of Group C, which declines in abundance downstream of station 3 are sensitive to substrate siltation.

Although biological analysis suggests perhaps two sources of pollution, organic pollution upstream of station 5 and settleable solids upstream of station 4, river quality data provided by the Glamorgan River Authority did not confirm pronounced changes attributable to undoubted potential pollution sources within these reaches, namely storm—sewer outlets and a coal washery. Subsequent chemical studies have shown that the latter gives rise to marked changes in the flow of certain cations (Mg^{++}, Na^{+}, K^{+}) and anions (SO_4^{--}, Cl^{-}) in the river (Fig. 8 for Mg^{++}), which probably have no pronounced effects on species distributions. This more frequent chemical sampling programme has now detected intermittent discharges of coal particles also.

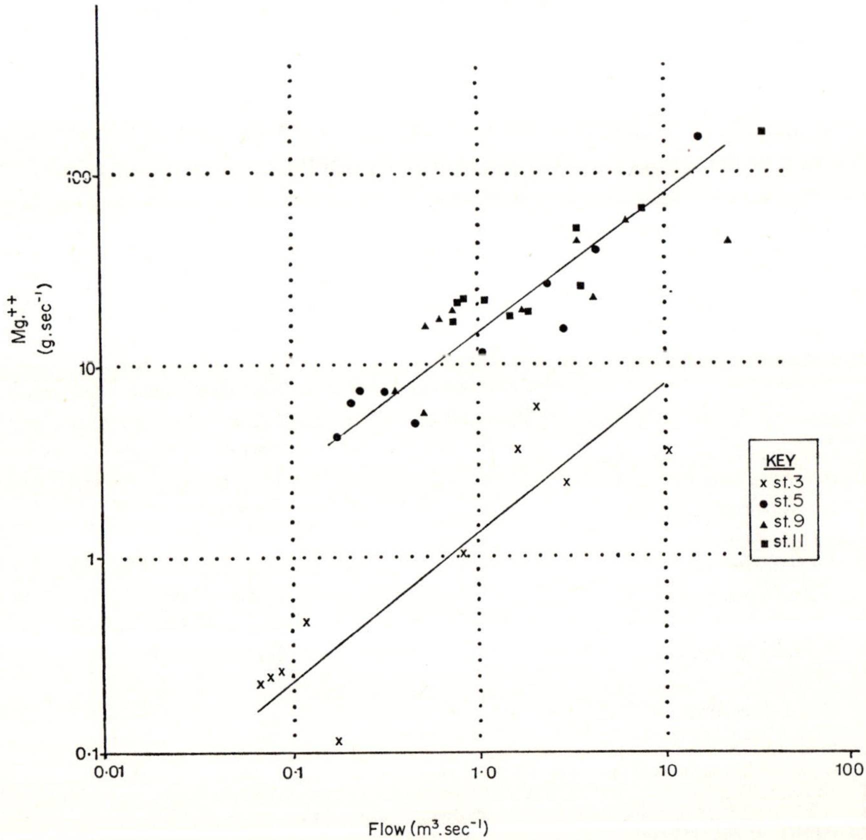

Figure 8. Relationship between water and magnesium flows at four stations on River Cynon.

The analysis of the Cynon data demonstrates the power of biological survey in detecting pollution sources, particularly intermittent discharges which are so easily missed by periodic chemical sampling. A variety of multivariate analytical techniques is now available which enables one to detect biological pattern and establish relationships much more readily. The major need seems to be background information on species requirements and their response to pollutants and other environmental stresses.

The Cynon example also demonstrates that quite striking changes in chemical quality spatially coincident with community discontinuities may not be primary causative factors of the latter. The continuous substantial discharges of magnesium, sodium, potassium, sulphate and chloride, are probably in this category (Sutcliffe *pers. comm.*). Nevertheless their detection suggested that other materials of more profound ecological consequence might be discharged intermittently from the same source.

Rivers represent the simplest systems for the detection of pollution sources, pollutants are dispersed unidirectionally and act on a linear receptor system. Air pollutants are dispersed three dimensionally and the terrestrial receptor system is predominantly two-dimensionally distributed. In lentic water-bodies, such as lakes and most marine systems effluent dispersal is three-dimensional and acts on both three-dimensionally distributed planktonic and nektonic and two-dimensionally distributed benthic receptors. What is more, dispersal patterns in both air and most water-bodies are variable. These differences give rise to considerable sampling, analytical and interpretive problems.

The role of survey in pollution assessment

When there are clearly defined effects of specific kinds of pollution on receptor species or communities, which in themselves may not be regarded as resources, such effects can be used as a measure of the level of this pollution. These may be regarded as 'indicator' species or communities. Their value depends however on established dose-effect relationships or on the relationship of their response with that of other receptors which may be regarded as resources.

In the field of river pollution, biological assessment, using formal descriptions of communities or expressions of their structure, have been used to describe levels of organic pollution. The first such system, a classification proposed by Kolkwitz and Marsson (1967) and based primarily on lists of species existing under different quality conditions has been widely used and subsequently modified (Fjerdingstad 1964). Another system used in the Trent catchment (U.K.) is based on the spectrum of benthic animal species or groups from which is derived a numerical value or 'biotic index'. Woodiwiss has amply demonstrated the relationship between this biotic index and aspects of organic enrichment e.g. B.O.D., NH_3, in the catchment. Some formal biological systems of classification of organic pollution take into account not only qualitative data but also the numerical abundance of components (Chandler 1970). With all these schemes however the biological data are collected and treated in such a manner that the detection function, with respect to other kinds of pollution, is sacrificed: the search for signals is replaced by the measurements of the strength of a defined signal at minimum cost.

In recent years the level of air pollution, and frequently, by implication, SO_2 pollution, has been assessed using lichen associations (Hawkesworth & Rose 1970). Such implication is of dubious validity for adequate data on dose-effect relationships are not available for most species and SO_2 is one of many pollutants which affect lichens. Moreover, in this particular instance, cheap integrative chemical techniques are available to determine the average levels of SO_2 (Ricks & Williams in press).

So frequently biological survey techniques are prematurely focussed on assessment, through the use of indicator species and communities, when the scientific basis for such assessment has not been established. The simplification and formalization of data collection and handling involved often precludes their use in detecting new sources of pollution, a function which is of primary importance with the increasingly complex array of pollutants being discharged and our inadequacy in predicting, experimentally, their effects.

Acknowledgments

The authors wish to thank Dr R. J. H. Williams and Mrs A. H. James for lichen data and the Glamorgan River Authority for flow data. Mr B. D. Hughes is a N.E.R.C. Research Student and Mr M. W. Read receives a grant from the British Steel Corporation.

Summary

The role and problems of surveys, both with respect to the distribution of pollutants and the detection of biological effects is discussed. Examples are given which demonstrate how the analysis of spatial and temporal distributions of communities enable the detection of pollution sources and the assessment of damage to be made.

References

ALABASTER J.S., GARLAND J.H.N., HART I.C. & SOLBÉ J.F. DE L.G. (1972) An approach to the problem of pollution and fisheries. *Symp. zool. Soc. Lond.* 29, 87–114.

BROOKER M.A. & EDWARDS R.W. (1974) Effects of the herbicide paraquat on the ecology of a reservoir. III. Fauna and general discussion. *Freshwat. Biol.* 4, 311–35.

CHANDLER J.R. (1970) A biological approach to water quality management. *Wat. Pollut. Contr. London.* 69, 415–22.

COOKE A.S. (1973) Shell thinning in Avian eggs by environmental pollutants. *Environ. Pollut.* 4, 85–152.

DICKSON K.L. & CAIRNS J. (1972) The relationship of freshwater macroinvertebrate communities collected by floating artificial substrates to the MacArthur–Wilson equilibrium model. *Amer. Midl. Nat.* 88, 68–75.

FJERDINGSTAD E. (1964) Pollution of streams estimated by benthal phytomicroorganisms. 1. A saprobic system based on communities of organisms and ecological factors. *Int. Revue. ges. Hydrobiol.* 49, 63–131.

GEORGE D.G. (1972) *Zooplankton Studies on a Eutrophic Reservoir*. Ph.D. thesis, Univ. of Wales.

GEORGE D.G. & EDWARDS R.W. (1973) *Daphnia* distribution within Langmuir circulations. *Limnol. & Oceanogr.* 18, 798–800.

GILBERT O.L. (1965) Lichens as indicators of air pollution in the Tyne valley, pp. 35–47. In *Ecology and the Industrial Society* (Ed. by G.T. Goodman, R.W. Edwards & J.M. Lambert), Blackwell, Oxford.

GREIG-SMITH P. (1964) *Quantitative Plant Ecology*. Butterworths, London.

HAWKSWORTH D.L. & ROSE F. (1970) Qualitative scale for estimating SO_2 air pollution in England and Wales using epiphytic bryophytes. *Nature, Lond.* 227, 145–8.

HESTER F.E. & DENDY J.J. (1962) A multiple-plate sampler for aquatic macroinvertebrates. *Trans. Amer. Fish. Soc.* 91, 420–1.

HYNES H.B. (1972) *The Ecology of Running Water*. Liverpool Univ. Press, Liverpool.

KOLKWITZ R. & MARSSON M. (1967) Ecology of animal saprobia. In *Biology of Water Pollution*, U.S. Department of the Interior. Federal Water Pollution Control Administration.

LEARNER M.A., WILLIAMS R., HARCUP M. & HUGHES B.D. (1971) A survey of the macrofauna of the River Cynon, a polluted tributary of the River Taff (South Wales). *Freshwat. Biol.* 1, 339–67.

MARGALEF R. (1956) Diversidad de especies en las commundades naturales. *Pub. Inst. Biol. Apl.* (Barcelona) 9, 5–27.

MENHINICK E.F. (1964) A comparison of some species-individuals diversity applied to samples of field insects. *Ecology* 45, 859–61.

NEEDHAM P.R. & USINGER R.L. (1956) Variability in the macrofauna of a single riffle in Prosser Creek, California, as indicated by the Surber Sampler. *Hilgardia* 24, 383–409.

PYATT F.B. (1970) Lichens as indicators of air pollution in a South Wales steel town. *Environ. Pollut.* 1, 45–56.

RATCLIFFE D.A. (1970) Changes attributable to pesticides in egg breakage frequency and eggshell thickness in some British birds. *J. appl. Ecol.* 7, 67–115.

RICKS G.R. & WILLIAMS R.J.H. (1974) Effects of atmospheric pollution on deciduous woodland. 2. Effects of particulate matter upon stomatal diffusion resistance in leaves of *Quercus petraea* (Mattuschka) Liebl. *Environ. Pollut.* 6, 87–109.

SOUTHWOOD T.R.E. (1966) *Ecological Methods with Particular Reference to the Study of Insect Populations*, Methuen, London.

SPRAGUE J.B., ELSON P.F. & SAUNDERS R.L. (1965) Sublethal copper-zinc pollution in a salmon river—a field and laboratory study. *Int. J. Air Wat. Pollut.* 9, 531–43.

WILHM J. (1970) Range of diversity index in benthic macroinvertebrate populations. *J. Wat. Pollut. Contr. Fed.* 42, R221–3.

WILLIAMS C.B. (1964) *Patterns in the Balance of Nature*. Academic Press, London.

Geochemical parameters in the assessment of estuarine pollution

I. THORNTON *Applied Geochemistry Research Group, Imperial College, London, U.K.*

Introduction

The Applied Geochemistry Research Group at Imperial College, London was formed in the early 1950s with the brief of investigating and developing methods for purposes of mineral exploration. For the past ten years we have also been concerned with the application of geochemistry to the life sciences (Webb 1964), initially in relation to agriculture and more recently to estuarine fisheries and pollution studies. In agriculture for instance, it has been possible to relate the regional distribution of trace elements, based on stream sediment sampling, to the distribution of recognized deficiencies and toxicities in crops and livestock. In addition further more extensive areas have been defined where less severe but nonetheless potentially economic subclinical or latent problems may exist and in which more costly studies on the soil, plant and animal may then be concentrated (Thornton & Webb 1970).

Recent studies have shown an encouraging degree of correlation between land-based geochemical patterns and chemical parameters measured in selected estuaries and at the same time provide valuable base-line information on the distribution of heavy metal pollutants. This paper summarizes this aspect of the Group's activities together with some parallel studies on the effect of sewage sludge dumping on trace metal distribution in the Firth of Clyde and illustrates both the use and potential of geochemical studies in the overall field of estuarine pollution.

Geochemical maps of England and Wales

Following trial surveys over some 17,000 square miles of England, Wales and Ireland, a major project to prepare a geochemical atlas of England and

Wales was commenced in 1969 under a grant from the Wolfson Foundation. Using the active stream sediment as the sampling medium, nearly 50,000 samples were taken from tributary/road intersections, giving a density of around one sample per square mile. The minus 204 μm fraction was analysed for 20 elements by direct-reading emission spectrography, atomic absorption spectrophotometry and colorimetric procedures. Provisional trace element maps were then plotted by computer (Fig. 1). These and similar data have been used as a basis for the selection of estuarine study areas.

The concept of producing geochemical maps by the analysis of stream sediment material is based on the premise that the active drainage sediment represents Nature's closest approximation to a composite sample of the weathered products of rock and soil upstream from the point of sampling. The geochemical reconnaissance map thus reflects on a regional scale the chemical composition of both rock and soil and as such provides a useful catalogue of base-line information on the distribution of those trace metals, such as cadmium and lead, which have become rather emotive household terms in the context of pollution today. Of course, not all metal excess is man-made and knowledge of the natural variation is essential as a foundation on which to base and assess the relative significance of contamination by man.

In addition geochemical reconnaissance maps have been shown to highlight areas of regional metal contamination by past and present-day industry, including mining and smelting going back to Roman times (Thornton & Webb in press). In South-West England and Wales, for instance, geochemical maps reflect extensive contamination of agricultural land with one or more of the elements arsenic, cadmium, copper, lead and zinc.

The provisional geochemical map for cadmium in England and Wales (Fig. 1) illustrates these points. Anomalous or high patterns of cadmium relate to three main sources of the metal:

(a) Areas associated with sulphide mineralization, contaminated by base-metal mining activity, including parts of South-West England, Wales and Central England.
(b) Areas contaminated by present-day and recent industrial processes such as the Avonmouth industrial complex and the Swansea smelting activities.
(c) Areas underlain by cadmium-rich black shales of Carboniferous age in which stream sediments and soils derived from the rock are enriched in the metal. Further studies have in fact shown that cadmium may be present in concentrations up to 100 ppm in marine black shales (Lowenstein & Holmes *pers. comm.*).

All three sources of excess metal would seem to be of importance in assessing their potential contribution to estuarine pollution.

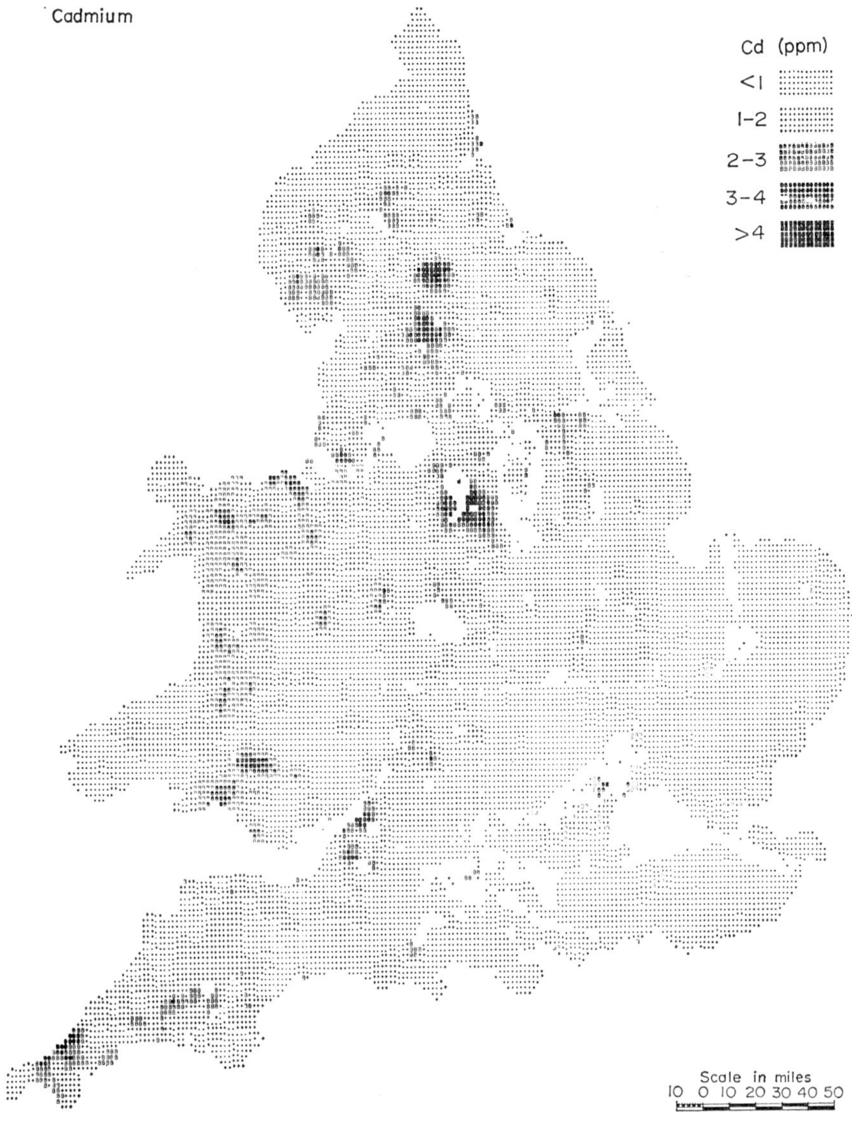

Figure 1. Provisional map showing the distribution of cadmium in stream sediment in England and Wales, based on a modification of the grey-scale mapping programme described by Howarth (1971) and compiled by Dr. P.L. Lowenstein as part of a project financed by the Wolfson Foundation.

Geochemical studies in estuaries

NORTH WALES

Geochemical studies were commenced in 1968 at the invitation of the White Fish Authority following problems encountered during the first three years of a pilot programme of oyster seed production in their hatchery at Conway, North Wales. Intermittent failure of oyster larvae to develop and settle satisfactorily led to irregular productivity particularly in the early months of the year seemingly at times of greatest turbulence and turbidity in the estuary. Lead and zinc have been mined in the western part of the Conway catchment since Roman times with particular activity in the second half of the nineteenth century and heavy metal contamination was considered a possible reason for the poor performance. Widespread contamination of tributary drainage in the vicinity of the disused mines had in fact been shown be previous geochemical reconnaissance (Nichol *et al.* 1970).

A more detailed survey of both tributary sediments and waters was extended to include stations in the Conway River and its estuary. Analytical data confirmed the higher levels of lead and zinc in sediments derived from Ordovician mineralized rocks to the west of the catchment compared to those derived from unmineralized Silurian rocks to the east (Elderfield *et al.* 1971). Peak values of one per cent or more lead and zinc were found in streams draining old mines contaminated with mill tailings and mineralized waste.

Waters in general showed a similar picture with up to 3,000 μg/1 Zn in streams draining old mines compared with 25 μg/1 in uncontaminated streams. Estuarine waters sampled at the location of the intake pipe to the hatchery contained 100 μg/1 Zn at the time of sampling compared with median values of 10 μg/1 quoted in the literature for normal fresh and sea waters, a value similar to that determined in local sea water off Conway (Table 1). Lead showed less contrast than zinc in the water probably because of its relatively low solubility.

Waters from the hatchery were sampled each working day over a six-month period from January to June 1969 (Elderfield *et al.* 1971). Concentrations of soluble zinc (passing through a 0·45 μm millipore filter) showed marked periodic fluctuations over the range 0·5–470 μg/1 in February with a progressive decline to around 50–100 μg/1 in June (Fig. 2). Zinc values showed a tendency towards an antipathetic relationship with those for salinity with peak concentrations corresponding to neap tides. Other elements measured showed no consistent patterns and were present at more normal levels. Undoubtedly the periodic high zinc values are in some way related to contamination from mines and mine waste in the catchment though the cause is almost certainly more complicated than the simple

Table 1. Range and mean lead and zinc content of river, estuarine and sea water at Conway.

	Tributary drainage		River Conway		
	Mineralized area	Unmineralized area	Above tidal influence	Below tidal influence	Local sea water
Zn μg/l	853 94–3260	23·8 1·7–75	33·4 0·8–94·8	114 4·0–172	8·8
Pb μg/l	7·4 2·7–16·3	4·5 0·5–16·5	8·7 2·2–31·8	6·0 0·9–29·8	3·6
No. of samples	12	22	8	16	

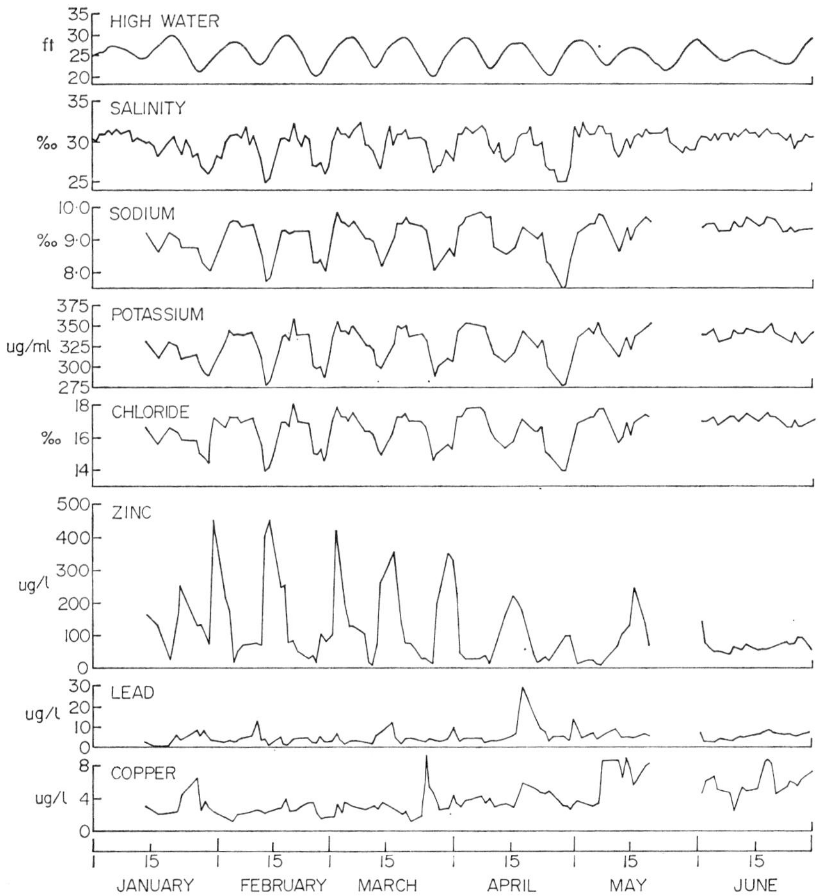

Figure 2. Variation in the composition of the hatchery water in relation to the salinity and high water at Conway, North Wales (reproduced by courtesy of Marine Pollution Bulletin).

mixing of zinc-rich fresh water with sea water in the estuary and is possibly associated with both water layering in the estuary and the removal of zinc into solution from sediments and particulates under certain localized conditions.

Subsequently beaker trials were carried out with the larvae of the Pacific oyster (*Crassostrea gigas*) comparing zinc treatments applied both as zinc sulphate and as zinc-rich mine-adit water (Brereton *et al.* 1973). With increasing zinc from both sources over the range 100–500 μg/1 growth decreased, mortality increased (90% of larvae died within two days at 500 μg/1 Zn) and the incidence of abnormal larvae increased. The range of zinc values tested corresponded to those found periodically in the natural waters used in the hatchery where, as the normal larval cycle is approximately 18 days, each batch was subjected to one or two periods of abnormally high zinc in the first four months of the year.

Further studies have since confirmed these results and have demonstrated that similar amounts of zinc retard growth and development of older larvae, reduce metamorphosis or settling and reduce growth of the young spat (Boyden *et al.* in preparation).

The above trials together with earlier toxicity tests on larvae of *Ostrea edulis* both by the Fisheries Experimental Station of the Ministry of Agriculture, Fisheries and Food and the White Fish Authorities late hatchery at Conway (Walne 1970; Knowles *pers. comm.*) demonstrate the deleterious effect of excess zinc though the experimental conditions do not represent precisely those which occur in the estuarine water. We do consider, however, that they provide sufficient evidence to suggest a causal relationship between excess natural zinc and larval failure; if this is so then amounts of the metal must be taken into consideration in assessing the suitability of other estuaries for the siting of oyster hatcheries.

BASE-LINE INVESTIGATIONS

Following the encouraging results at Conway geochemical studies were extended to four other estuaries of direct concern to the oyster industry, comprising those of the River Helford and Restronguet Creek, known to be contaminated by heavy metals from past mining activity, and that of Poole Harbour and the River Colne in Essex, both thought to be relatively uncontaminated. The main objectives of the programme were twofold: firstly to assess the relationship between the geochemistry of the estuaries and that of the associated catchments and secondly to provide base-line information on trace metals in waters and sediments in different estuarine environments.

It is not within the scope of this paper to discuss the results of this work in detail. In summary, however, the sediments varied appreciably in chemical composition between the selected estuaries and this variation was clearly reflected by the geochemical survey of major tributary drainage in the respective catchments (Table 2). These preliminary data would seem to indicate that trace element maps based on stream sediment analysis would indeed be useful in focussing attention on potentially contaminated estuaries, where such contamination is of a regional nature. Of course maps of this type will not indicate metal pollution by industrial effluent passing directly into the main river or estuary since the sampling is confined to the tributary drainage.

Table 2. Mean composition of tributary and estuarine sediments (ppm in the minus 204 μm fraction).

	Copper		Lead		Zinc	
	Tributary Drainage	Estuary	Tributary Drainage	Estuary	Tributary Drainage	Estuary
Restronguet	615	1,690	1,005	684	1,890	1,540
Helford	51	179	126	177	204	353
Colne	21	35	122	143	102	152
Poole	24	12	102	104	94	60

Geochemical studies of this type are still at an early stage of development and the data need careful and detailed assessment. It is necessary for instance to clarify the effect of rock and soil geochemistry on the trace metal status of the estuaries and similarly to establish criteria for the more detailed interpretation of geochemical maps in areas of urban and industrial contamination.

Naturally it is important to determine the effect, if any, of metal status on estuarine fauna, particularly for economic species where both production and quality are of significance to man. Monitoring of commercial species by the Ministry of Agriculture, Fisheries and Food has shown that, with the exception of shellfish, variations in the metal concentration of inshore seawater has little effect on the composition of edible species, probably due to their mobility (Portman 1972). Shellfish however showed marked accumulation of metals in specific estuarine locations. Indeed workers in the U.S.A. have actually used metal concentrations in the Eastern oyster (*Crassostrea virginica*) to detect sources of heavy metal pollution (Huggett *et al.* 1973).

We have studied the accumulation of metals by the Pacific oyster in the four above-mentioned estuaries and found, for instance, 18 times the rate of copper accumulation over an eight-month period in the partly mineralized

Helford system compared with that in the estuary of the River Colne (initial concentration of 180 ppm Cu in dry matter rising after eight months to 640 ppm in River Helford and 205 ppm in River Colne). Similar oysters in the heavily contaminated Restronguet Creek, containing initially 250 ppm Cu in dry matter, after one month contained 1,500 ppm, two months 2,500 ppm and four months 6,000 ppm Cu. These results would seem to reflect the variations in sediment composition shown in Table 2.

Before a realistic assessment can be made of the significance of metal contamination in these estuaries we need to know far more about both the chemical forms of the metals in the natural water, particulates and sediments and the factors affecting availability, uptake and the food chain. In studying the effect of metal excess on shellfish performance and reproductive capacity it is also necessary to take into account the possible adaptability of particular species to metal-rich environments as found by Bryan & Humerstone (1971) in specimens of ragworm (*Nereis diversicolor*) in the copper-rich Restronguet Creek.

SEWAGE SLUDGE DISPOSAL

Research of a somewhat different nature, though complementary to the above, was commenced in the Firth of Clyde in 1971 in collaboration with the Clyde River Purification Board, to determine the effect of sewage sludge disposal on the trace metal status of the estuary. Sewage from Glasgow and neighbouring areas is treated in a number of plants and the sludge dumped off Garoch Head currently at the rate of 1×10^6 tons per year (Fig. 3). The material includes both domestic and industrial waste and contains quite large amounts of cadmium, chromium, copper, lead and zinc (Mackay *et al.* 1972; Halcrow *et al.* 1973).

Samples of the bottom sediment taken on a grid basis showed appreciable enrichment of all these metals together with silver, mercury and organic matter in the vicinity of the deposit area (Table 3, Figs 3 & 4). The degree

Table 3. Range and mean metal content (ppm) of sandy-mud sediments from the sludge disposal area and nearby control areas in the Firth of Clyde (based on 10 samples from each population: data refer to the oven-dried minus 204 μm fraction).

	Ag	Cd	Cr	Cu	Pb	Zn
Disposal area	4	6	122	269	361	631
	2–6	4–8	87–175	250–300	269–403	437–681
Control area	0·2	2	33	16	42	85
		1–3	10–65	9–20	24–67	60–130

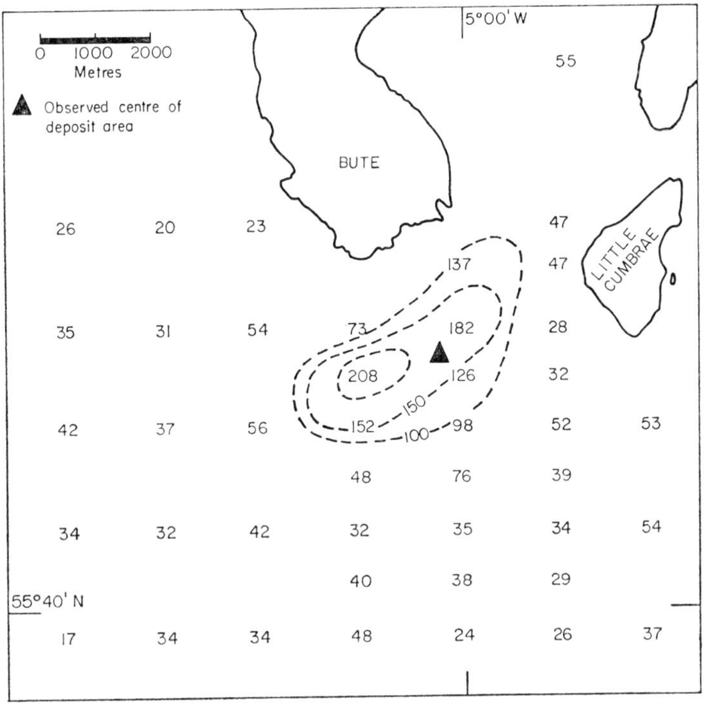

Figure 3. Distribution of copper in superficial sediments in the vicinity of the sewage sludge disposal area in the Firth of Clyde (μg/g Cu in the minus 204 μm fraction).

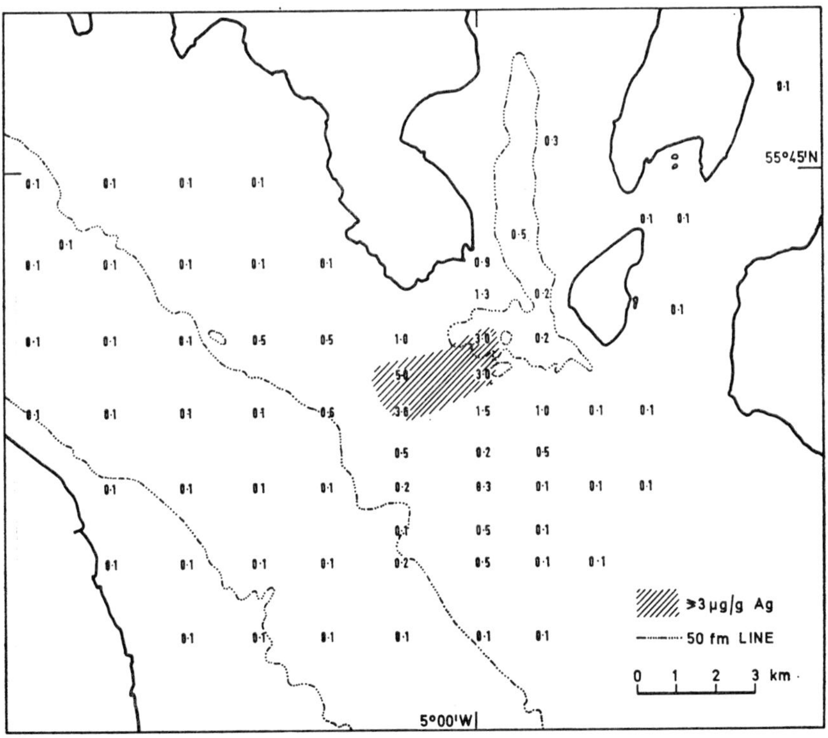

Figure 4. Distribution of silver in superficial sediments in the vicinity of the disposal area in the Firth of Clyde.

of enrichment was greater in the disposal area than in neighbouring areas contaminated by material from the River Clyde and by sediment derived from local mineralized areas in Loch Fyne.

A survey based on Agassiz trawling showed that the distribution of epifaunal species in particular is effected by sludge disposal. *Crangon*, *Pandalus* and *Buccinum* spp. were concentrated in and around the disposal area while *Nucula*, a primitive bi-valve, was notably absent though present in the surrounding areas (Fig. 5). It is not known whether these effects are due to variations in the metal content or organic matter status of the respective environments. Differences in metal content of those species present in both deposit and control areas were mostly small with the exception of the common whelk (*Buccinum undatum*) which accumulated 1,500 ppm Zn and 3 ppm Cd in the deposit area compared to 250 ppm Zn and 1 ppm Cd in the surrounding area (data based on dry matter content). Of the economic species present, prawns and Norwegian lobster are migratory and demersal fish such as herring are naturally mobile, making it difficult to assess any possible effect of sludge dumping either on their numbers or chemical composition.

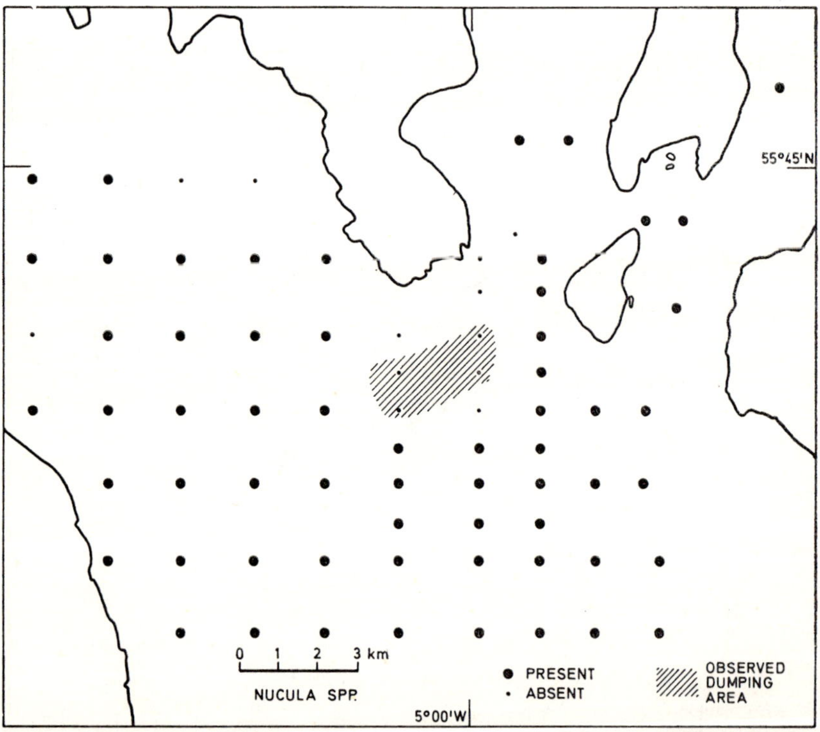

Figure 5. Distribution of *Nucula* spp. in the vicinity of the disposal area in the Firth of Cylde.

It has been concluded (Halcrow *et al.* 1973) that the effects of sludge dumping in this area are local probably affecting some 20 km^2 of sea bed. When compared to the heavily contaminated Restronguet Creek in South-West England, the concentrations of metals in water and sediments are relatively low, though little is known of the forms of the metals and their availability.

Conclusions

Geochemical reconnaissance maps for England and Wales have been shown to reflect natural variations in the chemical composition of rock and soil and at the same time to highlight specific sources of industrial heavy metal pollution when this is on a regional scale. Detailed studies in selected estuaries and their catchment drainage show an encouraging degree of correlation between geochemical patterns based on the analysis of sediments from tributary drainage and the chemical status of estuarine sediments and waters.

In discussing data for British coastal waters obtained by M.A.F.F., Preston (1973) concludes that elevated metal concentrations are 'limited to estuaries and the narrow coastal margin, being associated with drainage from industrial areas or the dumping of sewage sludge and industrial wastes'. Similarly Abdullah *et al.* (1972) point to metal-rich runoff and industrial and domestic waste disposal as prime sources of trace metal enrichment in some of our inshore and estuarine waters. In the light of these remarks and of the above conclusions it is suggested that regional geochemical maps can be used as an aid in assessing the potential metal status of our estuaries supplemented, where possible, by information on the location, composition and degree of waste disposal.

Such information may well be of valuable assistance in determining the suitability of estuaries and near-shore areas for the siting of hatcheries, shellfish stocks and developmental programmes for other forms of aquaculture.

Acknowledgments

The studies described in this paper, financed by a research contract from the Natural Environment Research Council, form part of a continuing programme in environmental geochemistry under the general direction of Professor J. S. Webb. Thanks are due to the many past and present staff members of the Applied Geochemistry Research Group who contributed to the work and in particular to Miss Anne Brereton, Dr C. H. Boyden,

Dr H. Elderfield, Mr W. Halcrow and Miss Helen Lord. Collaboration and assistance from the Clyde River Purification Board, Ministry of Agriculture, Fisheries and Food and White Fish Authority is gratefully acknowledged.

Summary

Trace metals have been determined in sediments and waters of contaminated and relatively uncontaminated estuaries selected on the basis of regional geochemical maps of England and Wales. This baseline information is compared with the distribution of trace metals in a sewage sludge disposal area in the Firth of Clyde.

References

ABDULLAH M.I., ROYLE L.G. & MORRIS A.W. (1972) Heavy metal concentration in coastal waters. *Nature, Lond.* **235**, 158–60.

BOYDEN C.R., LORD H. & THORNTON I. (in preparation) The influence of zinc on settlement of the oyster *Crassostrea gigas.*

BRERETON A., LORD H., THORNTON I. & WEBB J.S. (1973) Effect of zinc on growth and development of larvae of the Pacific oyster *Crassostrea gigas. Mar. Biol.* **19**, 96–101.

BRYAN G.W. & HUMMERSTONE L.G. (1971) Adaptation of the polychaete *Nereis diversicolor* to estuarine sediments containing high concentrations of heavy metals I. General observations and adaptation to copper. *J. Mar. biol. Ass.* **51**, 845–63.

ELDERFIELD E., THORNTON I. & WEBB J.S. (1971) Heavy metals and oyster culture in Wales. *Mar. Pollut. Bull.* **2**, 44–7.

HALCROW W., MACKAY D.W. & Thornton I. (1973) The distribution of trace metals and fauna in the Firth of Clyde in relation to the disposal of sewage sludge. *J. Mar. biol. Ass.* **53**, 721–39.

HUGGETT R.J., BENDER M.E. & SLONE H.D. (1973) Utilizing metal concentration relationships in the Eastern Oyster (*Crassostrea virginica*) to detect heavy metal pollution. *Wat. Res.* **7**, 451–60.

HOWARTH, R.J. (1971) Fortran Iv program for grey-level mapping of spatial data. *Mathemat. Geol.*, **3**, 95–121.

HALCROW W., MACKAY D.W. & THORNTON I. (1972) Sludge dumping in the Firth of Clyde. *Mar. Pollut. Bull.* **3**, 7–11.

NICHOL I., THORNTON I., WEBB J.S., FLETCHER W.K., HORSNAIL R.F., KHALEELEE J. & TAYLOR D. (1970) *Regional Geochemical Reconnaissance of the Denbighshire Area.* Report No. 70/8, Institute of Geological Sciences, London.

PORTMAN J.E. (1972) The levels of certain metals in fish from coastal waters around England and Wales. *Aquaculture* **1**, 91–6.

PRESTON A. (1973) Heavy metals in British waters. *Nature, Lond.* **242**, 95–7.

THORNTON I. & WEBB J.S. (1970) Geochemical reconnaissance and the detection of trace element disorders in animals. In *Trace Element Metabolism in Animals* (Ed. by C.F. Mills), Livingstone, London.

THORNTON I. & WEBB J.S. (in press) Environmental geochemistry: some recent studies in the United Kingdom. *Proc. 7th Ann. Conf. on Trace Substances in Environmental Health*, University of Missouri.

WALNE P.R. (1970) Present problems in the culture of the larvae of *Ostrea edulis*. *Helgoländer. wiss. Meeresunters* **20**, 514–25.

WEBB J.S. (1964) Geochemistry and life. *New Scientist* **23**, 504–7.

SECTION THREE

Terrestrial degradation

Acidic and ferruginous mine drainages

H. G. GLOVER *National Coal Board, Rotherham, U.K.*

Introduction

Springs, streams and rivers draining from outcrops of coal-bearing strata are normally contaminated by small quantities of iron, manganese, aluminium and other compounds derived from the natural weathering of the exposed minerals. In the early history of mining, the presence of iron oxide (ochre) deposits in natural seepages was used as an indicator of the presence of coal seams (Plot 1686; Thomas 1698).

Compared with these natural conditions, mining operations expose relatively large areas of rock to the action of the atmosphere, with the result that abnormal quantities of water-soluble compounds of iron, manganese etc. may contaminate the drainage from the mine, and in turn, the local surface drainage system. The visible effect of such compounds on rivers and streams is the deposition of the highly coloured pigment, ochre. Such deposits are commonly associated with coal mining, and the offending discharges have become known as 'acid mine drainage'.

In the United Kingdom, most of the ochreous discharges from coal mines are not acidic in the chemical sense, the term referring to the supposed state of the chemical conditions at the point of weathering of the parent mineral, iron pyrite (FeS_2). A more correct term for the non-acidic drainages would be 'ferruginous'. It is also necessary to point out that sulphide minerals such as iron pyrite occur in rocks other than coal-bearing strata, and that the mining of minerals other than coal also leads to the discharge of acidic and ferruginous mine drainages (Bilharz 1949; Kinney 1964; Treharne 1962).

In recent years, the pollution caused by acidic and ferruginous mine drainages has achieved prominence in the press and in the technical literature (Anon. 1965), and has led to changes in legislation in some countries. The causes of this increased interest are, chiefly:

(i) A general increase in awareness of pollution;
(ii) the time-lag, often as long as several decades, in the appearance of acid mine drainage after the abandonment of certain types of mines;
(iii) the increase in the total amount of pyritic material exposed in each mining area;
(iv) the availability, at a practical cost, of methods of controlling some of the problems of acidic and ferruginous mine drainage discharges.

The majority of the discourse relating to these drainages during the last decade has come from the United States of America, where it was reported in 1962 that the total acid load from mine drainages was estimated to be 3·5 million tons/year (Anon. 1962). This quantity of acid was small compared with the alkalinity of the river systems as a whole, but had an overwhelming effect on particular rivers and streams within mining areas. In the United States, nearly 6,000 miles of streams and 1,500 acres of impoundments were reported to be polluted by acid mine drainage (Kinney 1964).

In the United Kingdom, individual acidic and ferruginous drainage problems from coal mines are qualitatively similar to those in the U.S.A. but the scale of the overall problem is smaller by two to three orders of magnitude. No major rivers in the United Kingdom are measurably affected by these drainages, but several minor rivers and a large number of streams are affected by the typical ochre deposits and a few are so contaminated as to be acidic.

The formation of acidic and ferruginous mine drainages

The mine drainages to be considered in this text arise from the sources listed below:

Underground coal mine workings

Water from the strata which penetrates into such workings must be removed. Workings below the local surface water table are drained by pumping; workings above the water table may be drained by gravity flow. After abandonment, the former type of workings usually fill to the local water table but seldom overflow, whereas the latter type of workings normally continue to drain indefinitely. The rate of flow of drainages from individual underground coal mines in the United Kingdom averages about 1 million gal/day.

Surface coal mine workings

These are also known as opencast mines, contour or area strip mines or pits. Water may enter such workings from the strata or as rainfall and may be pumped or drained by gravity. The rates of flow of such drainages vary over wide ranges.

Coal mine spoil heaps

These are also known as 'tips', 'rucks', and 'banks' in England and Wales, as 'bings' in Scotland, and as 'refuse banks' in North America. Drainages arise from the action of rainfall and may appear as surface run-off or as seepages at the toe or from perched water tables. Rates of surface drainage may be high in storms, and otherwise negligible. Rates of toe seepages caused by percolation are usually less than 10,000 gal/day but may be higher in districts of very high rainfall.

Coal stock piles

Drainages from the action of rainfall may appear at the toe of such piles. Rates of flow are usually less than 10,000 gal/day.

The factors which determine the extent of contamination of mine drainage waters by acidic and ferruginous salts have been the subject of extensive studies in the last few years. It has been found that a large number of physical, chemical and biological factors may be involved. The more important of these factors are summarized below. A comprehensive review has recently been published (Anon. 1971a).

Chemically, the contamination of coal mine drainages by acids and iron salts starts with the oxidation of the mineral, iron pyrite (FeS_2). This mineral is found throughout the coal seams and the associated shales and occasionally in the sandstones. Pyrite takes the form as brassy crystals and greenish bronze amorphous masses. The former range in size from a few μm to a few cm, and the latter appear in sizes up to over 1 m diameter.

Pyrite begins to oxidize as soon as it is exposed to air, the rate of oxidation increasing progressively with the extent of the oxidation. The sulphates resulting from the oxidation can be detected in drainage water within a few days of first exposure of the pyrite. The immediate products of oxidation of iron pyrite are ferrous sulphate, $FeSO_4$; ferric sulphate, $Fe_2\,(SO_4)_3$; sulphuric acid, H_2SO_4; and hydrated ferric oxide, $Fe_2O_3.xH_2O$.

Many investigations have been made into the processes involved in the oxidation of iron pyrite. The principal rate-controlling factors have been identified, but the interpretation of some of the test results has led to differences of opinion concerning details of the oxidation mechanism.

Increasing temperature leads to an increased rate of oxidation, it having been suggested that the rate doubles for a 10°C rise of temperature. Water appears to be essential for the reaction, the rate of oxidation increasing with the water vapour pressure until, at 100% relative humidity, the rate becomes equal to that for immersed pyrite. It has been suggested that water may be necessary not as a reactant, but as a medium for the transfer of the oxidation products from the reaction sites. The rate of oxidation is generally agreed to be determined by the pressure of the reacting oxygen, approximating to a first order reaction at atmospheric pressures.

It is claimed that the actual electron transfer associated with the oxidation of the pyrite occurs on the surface of the mineral from either adsorbed oxygen or from ferric ion. A reaction mechanism involving ferric ion would require the presence of an inorganic or biological catalyst to oxidize ferrous to ferric, a reaction which occurs at a negligible rate in acidic solutions in the absence of such a catalyst.

The rate of oxidation is also apparently related to the reactive surface area of the pyrite, thus explaining the apparent anomaly that 'brassy, museum grade' pyrite does not appear to oxidize. The various mechanisms which have been postulated for the oxidation of pyrite have recently been reviewed in a report describing a mathematical model for a pyrite oxidation system (Morth *et al.* 1972).

The role of bacteria in the formation of acidic and ferruginous drainages has been studied extensively in the last twenty-five years, partly as a means of elucidating the mechanism of the oxidation reaction and partly in the hope that some means of controlling the formation of acidic and ferruginous mine drainages by bacteriostats would be discovered (Temple & Colmer 1951; Ashmead 1956; Lundgren 1971). Acidophilic ferrous oxidizing bacteria have been found to be ubiquitous in coal mine workings and have been shown to increase the rate of oxidation of pyrite in laboratory experiments (Ashmead 1956; Baker & Wilshire 1970; Smith & Shumate 1970). The effectiveness of bacterial catalysis in the field has been questioned by the observation that the number of organisms which can develop in field conditions is limited to about 10^8 to 10^9/ml, a number which can only achieve the observed rate of production of pyrite oxidation products when the ratio of water to pyrite is very high (Lau *et al.* 1970).

Compared to the attention which has been given to the mechanism of pyrite oxidation, comparatively little interest has been shown in the hydrological and other mass transport aspects of the problem. Careful inspection of mine workings, spoil heaps and coal stock piles in the United Kingdom has revealed that only a small fraction of the total quantity of pyrite oxidation products which are produced are actually discharged in the drainage. For example, it is found that although most underground mine workings produce an iron-free neutral drainage, all of the pyrite exposed on the walls, floors and roofs of the mine roadways becomes more or less oxidized and covered with efflorescences of acidic oxidation products. In such situations, the primary factor which determines the degree of contamination of the drainage is the effectiveness with which the drainage water comes into contact with the oxidation products and not the rate of oxidation of the pyrite.

A second factor which may determine the quality of the drainage from mine workings and spoil heaps is the neutralizing and purifying effect of the clays, carbonates and other minerals in the strata. These minerals may react with pyrite oxidation products to form aluminium, manganese, calcium and magnesium sulphates. Most of the coal mine drainages encountered in the United Kingdom contain high (up to 3,000 mg/l concentrations of these neutral sulphates, which are not present in the parent ground waters, thus indicating the extent of internal neutralization of pyrite oxidation products which has occurred. The extent of this internal purification can be judged by the observation that whereas fresh spoils contain no sulphates, a well weathered spoil may contain as much as 5% of calcium sulphate.

Finally, a drainage which becomes contaminated within a mine working or a spoil heap may be purified naturally before it is discharged. The contaminated drainage may mix with naturally alkaline waters, and subsequent aeration and sedimentation of the mixture may remove appreciable quantities of what would otherwise be polluting components such as iron salts.

The quality of a mine drainage is thus seen to be determined by the following factors:

The comminution and exposure of rocks containing pyrite.

A supply of atmospheric oxygen to the vicinity of the exposed pyrite.

The efficiency of transfer of oxygen to the surface of the pyrite.

An effective catalyst for the oxidation of ferrous to ferric in an acidic medium (assuming oxidation by ferric ion to be significant in the particular situation).

A supply of water as liquid or vapour for the promotion of the oxidation reaction or the removal of reaction products from the immediate oxidation sites.

The availability of carbonate or clay minerals to react with the primary oxidation products.

A supply of drainage water moving to and from the zone of pyrite oxidation.

The efficiency of secondary purification of the drainage within the mine or heap, e.g. by alkaline strata water.

Table 1. The composition of drainage water from a number of coal mining activities.

Source of Discharge	Quality	pH value	Alkalinity to pH 4·5 mg/l $CaCO_3$	Acidity to phenol-phthalein at the boiling point mg/l $CaCO_3$	Calcium mg/l Ca	Magnesium mg/l Mg	Dissolved iron mg/l Fe	Suspended iron mg/l Fe	Manganese mg/l Mn	Chloride mg/l Cl	Sulphate mg/l SO_4
Underground mine working, shallow depth	Naturally purified, low salinity	8·0	290	0	90	93	1·0	0	0·4	90	700
Underground mine working, medium depth	Naturally purified, medium salinity	7·9	580	0	176	137	0	0·5	0·1	6,900	700
Underground mine working, maximum depth	Naturally purified, high salinity	7·5	190	0	2,560	720	0·6	0·2	0·9	30,800	350
Flooded underground mine workings	Alkaline and ferruginous	6·9	340	0	190	130	25	21	6	42	1,720
Shallow underground mine workings, gravity flow	Acidic and ferruginous	2·9	0	480	125	88	122	0	7	50	1,250
Spoil tip seepage	Acidic and ferruginous	4·6	5	580	250	230	23	17	10	95	2,300
Coal stock pile seepage	Acidic and ferruginous	3·1	0	1,100	n.d.	n.d.	160	0	9	80	1,220

n.d. signifies not determined

Faced with this list of conditions, it would seem to be improbable that a mine drainage could become contaminated with pyrite oxidation products. However, by an unfortunate coincidence, sufficient of these conditions occur simultaneously in the Appalachian coalfields of the United States to produce a major problem. The coals and associated rocks which are exposed apparently have a fairly high pyrite content, the mines have largely been driven above the water table and have thus remained filled with air after abandonment, the workings have been shallow and the overlying rocks have fractured easily so that water has penetrated over large areas of the workings, little or no alkalinity has been present in the natural ground waters, and the drainage has continued to flow from the mines after abandonment.

In contrast, conditions in the coal mines of the United Kingdom have been much less likely to cause problems. A few of the earliest mine workings (e.g. those in, South Wales, the Rossendale anticline in Lancashire, the fringes of the Pennine hills, and parts of Scotland), were similar to the mines of Appalachia, and these still yield small quantities of acidic and ferruginous drainages. The majority of coal mines in the United Kingdom have been driven in coal seams of relatively low pyrite content, the workings were below the water table and were covered by relatively soft shales which allowed little water to enter the mines during the active phase, and little or no overflow to occur after abandonment. The natural ground waters encountered in these mines have usually been alkaline, some having alkalinities of 2,000 mg/l (as $CaCO_3$).

Similarly, in the United Kingdom, the coincidence of conditions necessary for the discharge of acidic and ferruginous drainages from spoil heaps and coal stock piles are seldom present.

The composition of drainages from various coal mining activities are shown in Table 1.

The potentially polluting and other adverse effects of acidic and ferruginous mine drainages

From the point of view of modern society, acidic and ferruginous mine drainages cannot be claimed to have any beneficial effects. It should be remembered that these drainages are a living demonstration of a process by means of which iron has been mobilized and concentrated into iron-rich deposits in the course of geological history, but unfortunately, mankind does not have any particular need or sympathy for the continued deposition of iron ores at the present time.

A possible slight advantage of acidic and ferruginous drainages from coal stock piles is that pyritic sulphur removed by leaching is so much less sulphur remaining to pollute the atmosphere on subsequent combustion of the coal (Rogoff *et al.* 1960). However, this advantage is gained only at the risk of causing water pollution. It is estimated that the costs of preventing this pollution would be far greater than the costs of other methods of desulphurizing the coal or the flue gases resulting from its combustion.

Compared with this somewhat dubious advantage, the adverse effects of acidic and ferruginous mine drainages are many and are potentially serious. Some of these effects appear before the drainage is discharged, others appear in the environment external to the mine if the drainage is discharged in a contaminated condition. These effects will be considereed in turn.

ADVERSE EFFECTS WITHIN THE MINE, SPOIL HEAP OR COAL STOCK PILE

In mine workings, the most obvious effect of acidic and ferruginous mine drainages is the corrosion of equipment such as conveyors, roof supports, rails, chains, and steel ropes. These drainages also attack concrete and mortar, and paradoxically, although corrosive, often produce serious fouling of pumps, pipes, etc. with ochreous scale. Fortunately, acidic and ferruginous mine drainages do not contain substances which are particularly toxic to man and animals by ingestion, and do not seriously affect the skin, although clothing may be damaged by repeated immersion and drying. Drainage water contaminated by acidic and ferruginous salts is seldom suitable for use within the mine for purposes such as dust suppression, coal preparation etc.

In spoil heaps, the conditions which lead to the formation of acidic and ferruginous drainages are often directly associated with the fertility of the surface layers of the heap. A solution to a soil fertility problem will normally solve an acidic and ferruginous drainage problem. Ferruginous drainages may deposit ochreous scale in gravel and pipe drains within and beneath a spoil heap and thus affect the mechanical stability (Fischer 1970).

Apart from the obvious difficulties of disposal of a contaminated drainage discharge, acidic conditions within a coal stock pile do not pose immediate problems although concrete underlying a stock pile may be attacked, particularly if the concrete is of low quality. Acidic conditions in a coal stock pile often first reveal themselves by the corrosion of bunkers and fuel-handling equipment when the coal is subsequently utilized.

After the removal of an acidic coal stock pile which has been lying on natural ground, the quality of the surface drainage recovers rapidly provided that the coal and any inferior coal 'carpet' has been completely stripped, the sub-soil well limed, and the top soil replaced.

THE EFFECTS OF DISCHARGES OF ACID AND FERRUGINOUS MINE DRAINAGES ON MUNICIPAL SEWERS

In the United Kingdom, the discharge of acidic and ferruginous drainages to municipal sewers is not a common practice, but, providing that the drainage can be accommodated, certain compensatory beneficial effects may be obtained. In making a discharge to a sewer it is necessary to ensure that sufficient dilution is available to avoid damage to the fabric of the sewer by acids and sulphates, and to avoid interference with the biological sewage treatment processes. Provided that these limitations can be overcome, acidic and ferruginous mine drainages can have the beneficial effects of forming flocs which assist the sedimentation and filtration processes and may precipitate phosphates from the sewage (Benoit *et al.* 1971).

THE POTENTIAL EFFECTS OF ACIDIC AND FERRUGINOUS MINE DRAINAGE DISCHARGES ON RIVERS, STREAMS, CANALS, LAKES AND THE SEA

The effects of acidic and ferruginous mine drainage discharges to surface waters depend on the quality of the drainage, the quality of the receiving water and the relative dilution. The principal pollutants in acidic and ferruginous drainages from coal mines are suspensions of ferric oxide (ochre) and dissolved iron, aluminium, manganese, calcium and magnesium sulphates, and possibly free sulphuric acid. Low concentrations of the heavy metal salts of copper, nickel, zinc etc. may also be present. All of these potential pollutants are present naturally in surface water courses in the geologically exposed coalfields so that the biological systems in the receiving waters have acquired a considerable tolerance. The possible effects of these drainages will be considered under three headings.

The effect on the aesthetic value of the surface water

The principle visual effect of an acidic and ferruginous mine drainage on a surface water is the deposition of the hydrated ferric oxide, ochre, (also known as 'yellowboy' in North America). Chemically-deposited ochre is a pigment of considerable covering power, and as little as 1 mg/l (as Fe) in a stream will cause visible staining of the bed and the accumulation of opalescence in pools.

Discharges of mine drainage containing a few mg/l of iron to still waters such as canals and lakes may acumulate to produce visible coloured patches and opacity of the water. Discharges of ferruginous mine drainages to estuaries are seldom visible because of the naturally suspended silt in such waters, and discharges to coastal waters are similarly seldom perceptible provided that the rate of discharge does not exceed about 100 kg of iron a day at any one point.

In general, the discoloration of surface water by ochre does not produce a favourable reaction from a riparian owner or from the public, although some of the contaminated waters are quite spectacular and have been known to attract visitors.

It is not possible to assess objectively the loss of aesthetic value of a surface water contaminated by ochre.

The effects on the utilization of a surface water

Surface waters may be required for a wide variety of purposes including agriculture, industry and public supply, and recreation, including bathing. Acidic conditions in a surface water, as indicated by a low pH value, must be considered to reduce the value of the water for all these purposes. In the absence of acidic conditions, the neutral salts such as sulphates etc. may also be detrimental. Particular instances of this are the effects of sulphates on concrete structures such as cooling towers, the effect of magnesium salts in the drinking water of cattle, etc. Iron and maganese compounds are detrimental to industrial activities such as paper making and laundering and may increase the costs of treatment of water for public supply.

The effects on the biology of a surface water

The biological effects of an acidic and ferruginous mine drainage discharge are complex and not amenable to completely rational analysis. High concentrations of the acidic drainages are toxic to all normal forms of surface water-life, but in the absence of the mineral acid component, the other contaminants may have a less severe effect than the discoloration by ochre would suggest. It has been shown that the acute toxicity of aged ferric hydroxide is highly problematical and would be in the range of thousands of mg Fe/l (Sykoraj *et al.* 1972) and a fishery has been maintained in a strip mine pond, despite the presence of iron in concentrations from 0·16 to 11 mg/l by the application of lime to maintain a pH value greater than 6 (Davis 1971).

Many accounts of the effects of acidic mine drainages on surface waters have appeared in the technical literature (Weaver & Nash 1968; Herricks & Cairns 1972; Riley 1960; Dugan *et al.* 1968) but little attention has been given to the effects of the alkaline or ferrous bicarbonate type of discharge which is more common in the United Kingdom.

Experiments to determine the effects of specific components of acidic and ferruginous mine drainages on different types of water in the laboratory are seldom conclusive but may help to elucidate some of the observations made in the field. Free mineral acids, as indicated by low pH values, are probably the most toxic components involved (Anon. 1968; Lloyd & Jordan 1964). It has been shown that four species of bacteria used as test organisms should be able to grow at pH values greater than 5·3, in sulphate concentrations up to 500 mg/l and in iron salt concentrations up to 100 mg/l (McCoy & Dugan 1968). Tests over a full generation of juvenile brook trout revealed a trend towards slower growth in increasing concentrations of suspended ferric hydroxide, but egg viability was high in all concentrations except 50 mg/l. Similar tests on fathead minnows showed that the highest concentration of suspended iron which did not seem to affect survival and growth was below 12 mg Fe/l. The safe concentration for reproduction and growth of Gammarus minus larvae was less than 3 mg Fe/l (Sykora *et al.* 1972). Aluminium both in solution and in the form of freshly precipitated suspensions seems to be particularly toxic. A recent report indicated that the maximum safe concentration was 0·1 mg/l to trout, and that concentrations of 1·5 mg/l could be fatal to trout (Freeman & Everhart 1971).

The ferrous salts in mine drainages do not usually have a significant effect on the dissolved oxygen content of a surface watercourse, mainly because 1 mg/l of dissolved oxygen will oxidize 7 mg/l of dissolved ferrous iron, and the rate of oxidation of ferrous iron falls rapidly with a slight fall of pH value, being negligible at pH values (about 5·5) at which the biological life is not seriously disturbed.

Prevention of the formation of acidic and ferruginous mine drainages

Prevention of the formation of an acid and ferruginous mine drainage is basically simple. Either the source of the drainage water can be cut off, contact between the water and the contaminating pyrite oxidation products can be cut, or production of the pyrite oxidation products can be prevented. Ideally, these principles should be adopted in the planning of a mining operation but, in practice, circumstances seldom permit complete control by any of these methods which are discussed below in general terms. A more detailed review of these methods of control has recently been published (Anon. 1971a).

STOPPING OF DRAINAGE WATER AT THE SOURCE

It is almost impossible to prevent surface water from entering underground strata, particularly strata which have been fractured by mine workings (Zaval & Robins 1973; Anon. 1972; Hill & Martin 1972). Occasionally, it may be possible to identify and seal surface fissures, abandoned adits, auger holes etc. and it may be possible to pump from, or to drain, aquifiers, but to attempt to prevent strata water from running into a mine working would usually be expensive and dangerous.

It is not possible to prevent natural precipitation from falling on to surface mine workings, spoil heaps and coal stock piles.

PREVENTION OF CONTACT BETWEEN DRAINAGE WATER AND PYRITE OXIDATION PRODUCTS

Since the majority of the water which flows from the strata into underground mine workings is relatively uncontaminated, it is advantageous to trap this water, if possible, at the point of entry, and to convey it directly from the mine. Unfortunately, the geological conditions in coal-bearing strata are seldom favourable, and it is more usual for water to penetrate the workings in the form of 'drippers' which cannot be trapped.

Contact between drainage water and pyrite oxidation products in spoil heaps can be controlled. In the extreme, water-tight covers of plastic or tar can be placed over the surface of a heap. Such seals seldom last for more than a few months or years due to the action of the weather or to vandalism, and cannot be considered to provide more than a temporary solution. Hydraulic seals which are efficient and permanent can be obtained by placing a layer of well compacted, low permeability spoil on the surface of the heap, but secondary problems of erosion may then be encountered. To control erosion, a comprehensive surface drainage system, with particular attention to the surface contours may be required, and this may involve considerable regrading of an existing heap. It is normally necessary to place a top layer of fertile soil stabilized with vegetation on the outer surface of a heap which has been covered with an impermeable layer of spoil in order to control erosion and the disruptive effects of frost, etc.

Contact between drainage water and pyrite oxidation products in coal stock piles can be controlled by plastic film or tar covers, subject to the limitations previously described. Such covers have other beneficial effects, including partial or complete control of spontaneous combustion and may be highly profitable if their use permits the construction of deeper stock piles than would otherwise be possible (Hall 1967; Kenyeres & Takacs 1972).

PREVENTION OF THE OXIDATION OF IRON PYRITE

It is tempting to think that it should be possible to control the rate of oxidation of pyrite in air by the application of some form of inhibitory reagent, possibly a biological control agent. Various attempts to find such a reagent have been reported but none has been shown to be effective in full scale use. Reagents such as chromates and ammonia have had no effect on acidic drainages running from abandoned mine workings (Braley 1956). Research into the development of a phage for ferrous oxidizing bacteria (Shearer *et al.* 1968) was apparently not successful. Later it was claimed that oxidation could be inhibited by strains of Caulobacters (Shearer *et al.* 1970). This report also discussed the use of 15 different antibiotics, three of which were effective, and the antibiotic-producing species, *Streptomyces aureofaciens*.

Assuming that the oxidation of pyrite cannot be prevented by an inhibitory reagent, it is necessary to consider the possibility of controlling the oxidation by restricting the access of oxygen to the pyrite. During the course of active mining operations it is not possible to prevent air from coming into contact with pyrite in underground mines which are ventilated with air in order to support life.

In an attempt to prevent the oxidation of pyrite, an underground mine is to be operated in the U.S.A. in a sealed condition without ventilation, the miners being supplied with air by individual life-support systems (Anon. 1970). It is unlikely that such systems will be economically viable in Great Britain because of geological difficulties.

Although it is not possible to prevent pyrite oxidation during the active phase of underground mining, it may be possible to control the air supply to the worked-out parts of a mine which are no longer required and there is an even better chance of restricting the air supply when the whole mine, or possibly a group of adjacent mines have been abandoned. Methods of cutting off the air supply, in order of effectiveness, are:

to flood the workings completely,

to fill depressions in connecting tunnels, adits, etc. with water,

to construct air-tight masonry seals in tunnels, shafts, adits etc.

The flooding of workings will often occur naturally once pumping ceases provided that the workings are below the local water table. In order to flood workings above the water table, it would be necessary to construct hydraulic seals, a practice which is seldom feasible due to the low strength of coal-bearing strata, except when very low hydraulic pressures are involved (Foreman & McLean 1973).

Flooding effectively prevents further oxidation of pyrite, but dissolves any pyrite oxidation products which may have formed previously so that it may be necessary to deal with a problem of contaminated water disposal should the flooded workings overflow.

Seals formed by the flooding of depressions and masonry seals may be effective provided that the surrounding strata is undisturbed. However, if the workings are at a shallow depth below the surface, it is seldom possible to prevent the ingress of oxygen through fractures caused by the breathing effect of changes in the atmospheric pressure (Hill & Martin 1972; Braley 1962; Moebs & Krickovic 1970).

It is not possible to control the oxidation of pyrite in surface mine workings, and this is one of many reasons for the method of surface mining used in the United Kingdom which involves rapid extraction of the coal followed by complete restoration of the original site.

The oxidation of pyrite in spoil heaps can be controlled by the use of sealing layers as previously described for the control of water penetration. In order to prevent completely the ingress of oxygen, it may be necessary to ensure that an earth seal is not allowed to dry out. A highly compacted spoil will normally contain about 20% of voids which will be water-filled at the time of placing. The permeability of such a material to oxygen will be negligible, but if the water is displaced by drying, the permeability to oxygen will increase appreciably. However, even though some pyrite may become oxidized in this way, it is unlikely that the rate of leaching of the oxidation products in subsequent wet weather would be significant. Should it be considered necessary to avoid drying out of the surface layers of a spoil heap, large, deep-rooted plants should not be allowed to propagate, particularly on parts of a heap, e.g. steep slopes which do not receive or are unlikely to absorb, a normal proportion of the rainfall.

The oxidation of pyrite in stock piles of graded coals cannot normally be prevented, although the access of air can be reduced by the application of plastic or tar films to the pile. It may be possible to reduce the rate of oxygen penetration into stock piles of the finer grades of coal by heavy compaction.

The purification of acidic and ferruginous mine drainages by treatment

Should it be impracticable to prevent the formation of an acidic and ferruginous drainage in a mine, spoil heap or coal stock pile, it may be necessary to resort to treatment of the drainage before discharge. Technically, these drainages can be treated to any required quality standard, but it should be recognized that the costs of treatment could be so high as to render the continued operation of the mine uneconomic.

The type of treatment required will depend on the quality of the drainage and the standard to which it is to be purified. For the purposes of discussion, methods of treatment suitable for different types of acidic and ferruginous drainages will be considered individually, and finally, certain less orthodox methods of treatment suitable for various types of mine drainage will be described.

DRAINAGES CONTAINING FREE ACID AND INSIGNIFICANT CONCENTRATIONS OF IRON SALTS

Shallow underground mine workings which are well ventilated may provide conditions in which most of the iron originally dissolved from pyrite is deposited leaving essentially a dilute solution of sulphuric acid. Such drainages are almost unknown in the United Kingdom. It is possible to treat such drainages simply by the addition of an alkali such as lime or limestone. Automatic plants have been installed for the treatment of this type of drainage in remote districts in Pennsylvania (Charmbury *et al.* 1968).

DRAINAGES WHICH ARE ACIDIC AND ALSO CONTAIN SIGNIFICANT CONCENTRATIONS OF FERRIC IRON

Drainages from well-ventilated underground mine workings, spoil tips and some coal stock piles may contain ferric iron in solution in concentrations up to a few thousand mg/l. The pH value of such drainages must be less than about 3·5 in order to maintain the ferric iron in solution. Drainages of this type are also relatively uncommon in the United Kingdom.

Treatment by lime neutralizes the free acidity and causes the ferric salts to precipitate as a floc of ferric oxide hydrate. The precipitation is rapid and is effective in giving a neutral supernate of low iron content, although the coagulation of the flocs becomes progressively more difficult the lower the initial iron concentration of the drainage. Lime treatment involves relatively high reagent costs, problems of dosage control, and produces a waste sludge of high water content.

Limestone may be used for the treatment of this type of drainage with the advantage of lower reagent cost, simpler dosage control and more dense sludge. The limestone must either be very finely divided, or mechanically scrubbed during the reaction in order to remove a scale of reaction products. Limestone processes are not generally applicable if the sulphate content of the drainage exceeds about 5,000 mg/l, as SO_4 (Wilmoth & Hill 1970; Lovell 1971; Glover 1967).

DRAINAGES WHICH ARE ACIDIC AND CONTAIN FERROUS SALTS IN SOLUTION

Drainages of this type may be discharged from mine workings, spoil heaps and coal stock piles and although they represent only a small fraction of all discharges from coal-mining operations in Great Britain, a few examples can be found in each coalfield, particularly in association with coals of high (e.g. more than 3% as S) sulphur content. The iron content of such drainages may be as high as 5,000 mg/l (as Fe), and the pH values may be as low as 1·5, although the flow rates of the more contaminated drainages are usually proportionately reduced. These drainages are also liable to contain appreciable concentrations of aluminium and manganese salts and trace concentrations of copper, nickel, zinc etc.

In assessing the type of treatment needed for this type of drainage, it is necessary to measure not only the free acidity revealed by the pH value, but also the concealed acidity due to the aluminium and ferrous components of the drainage which are not indicated by the pH value. For this purpose, the effective acidity can be measured by titration with sodium hydroxide at the boiling point to phenolphthalein. The acidity value so determined is a measure not only of the effective acidity of the drainage from the point of view of treatment, but also of the real alkali demand which the discharge would exhibit if it were to be discharged to a surface water.

These acidic-ferrous drainages are all amenable to treatment by lime as previously described for acidic-ferric drainages, but the sludges produced are liable to be even more voluminous. For example, the treatment by lime of a discharge of 1 m gal/day containing 1,000 mg/l of dissolved ferrous iron would produce about 100,000 to 200,000 gal/day of sludge. In lagoons such sludge would consolidate over a period of several months to a volume of about 50,000 to 100,000 gallons. The de-watering of such sludges is often difficult and expensive, for example, mechanical filtration could cost up to £5/ton of dry solids. Various methods for increasing the density of lime-precipitated sludges have been described, most of which are based on the principle of sludge recirculation to the point of neutralization (Streeter *et al.* 1971; Kostenbader & Haines 1970; Moss 1971). Whenever possible, lime-precipitated sludges are discharged to abandoned mine workings, but the number of such workings to which disposals can be made safely is very limited in the United Kingdom. The costs of treatment of acidic-ferrous drainages by lime are in the range £0.2 to £3/1,000 gallons treated.

Provided that the sulphate content does not exceed about 5,000 mg/l (as SO_4) the acidic-ferrous mine drainages can be treated by limestone as previously described. Ferrous salts will not react directly with limestone, and must be oxidized to ferric salts before a reaction will occur. The cheapest available oxidizing agent is atmospheric oxygen, but unfortunately the rate of oxidation of ferrous salts by air is negligible at pH values of less than about 5·5. In order to use limestone for treatment of drainages containing acidic ferrous salts it is thus necessary either to introduce air with the limestone and to agitate the mixture for periods of up to several hours (Hill 1969; Anon. 1971b), or to oxidize the ferrous salts whilst still acidic by the use of a catalyst. Reagents which have been proposed for this purpose include activated carbon (Ford & Boyer 1973) and gamma rays (Steinberg & Pruznsky 1970), ozone (Beller *et al.* 1970) and electrolysis (Jasinski & Gaines 1972). It is also possible to oxidize ferrous salts in acidic solutions by biochemical means using ferrous oxidizing bacteria (Glover 1967; Whitesell *et al.* 1971; Ueta *et al.* 1972). These bacteria are normally present in acid mine drainages and can be propagated, preferably in an activated sludge type of process, to cell counts sufficiently high for the oxidation of hundreds of mg/l of ferrous salts in a few hours at normal atmospheric temperatures. The overall costs of limestone processes with appropriate ferrous oxidation stages are estimated to be about one half of the equivalent costs of treatment by lime.

DRAINAGES WHICH ARE CHEMICALLY NEUTRAL BUT CONTAIN DISSOLVED FERROUS IRON

Drainages of this type are discharged almost exclusively from the deeper underground mine workings and are the most common form of ferruginous mine drainage in the United Kingdom, although rare in North America because the mine workings are less deep. The iron salts in these drainages originate from the acidic oxidation products of pyrites, which have been neutralized by strata water containing sodium bicarbonate, and by carbonate and clay minerals. These drainages may contain ferrous iron in concentrations up to about 400 mg/l (as Fe) which is present effectively as ferrous bicarbonate ($Fe(HCO_3)_2$). The principle chemical characteristic of these drainages is that on aeration, all the ferrous salts are converted into the insoluble ferric form which precipitates as hydrated ferric oxide (ochre). An efficient and relatively low cost treatment process can thus be based on the use of a series of sedimentation basins with a cascade between each basin. Provided that a suitable site can be found, earth-walled basins can be used and treatment costs may be as low as 5 to 20p/1,000 gallons of water treated. The sludge formed by this method of treatment is of relatively high density and is fairly permeable so that it can be dredged and discharged to a spoil heap without much difficulty. Should it not be possible to construct basins, a more compact plant involving mechanical aeration and filtration may be necessary, the capital and operating costs of which could rise to £1,000 gallons of drainage treated.

DRAINAGES CONTAINING FERRIC IRON IN SUSPENSION

Mine drainages may contain ferric iron in the form of the hydrated oxide (ochre) either alone, or in combination with the dissolved ferric and ferrous salts as previously described. Suspensions of ferric oxide are not normally present in drainages from immobile sources such as abandoned mine workings, spoil heaps, coal stock piles etc. but are fairly common in drainages from active mines where physical disturbances may be present.

Treatment necessary for the removal of dissolved iron will incidentally remove suspended ferric oxides without a significant increase in costs. Drainages containing only ferric oxides can be treated by sedimentation, possibly with added flocculants, at a cost of up to 25p/1,000 gallons in earth-walled lagoons, and up to £1/1,000 gallons in mechanical plant.

UNCONVENTIONAL TREATMENT PROCESSES

A wide range of what may be termed unconventional treatment processes have been tested in recent years and a few are used in the U.S.A. on the full scale:

Reverse osmosis

This process which involves the application of hydraulic pressure to the drainage contained in semi-permeable membranes is applicable to the treatment of the less contaminated acidic and ferruginous mine drainages (Sleigh & Kremen 1971; Mason & Gupta 1972). The disadvantages, apart from the high capital and operating costs are that a concentrated acidic waste may be produced which will still require treatment. An advantage is that a potable water product can be produced, so that, in areas of water shortage, receipts from the sale of water may offset the costs of treatment. No full-scale plant has so far been constructed.

Flash distillation

This process involves the distillation of the drainage under reduced pressure.

Comments similar to the reverse osmosis process are applicable. Although taken to an advanced design stage in Pennsylvania (Maneval & Lemezis 1972) no full-scale plant based on this process has been constructed.

Freezing

It is possible to purify acidic and ferruginous mine drainages by selective freezing (Anon. 1971c). Process evaluation tests have not reached the stage at which cost estimates can be made. It is also possible to de-water, by freezing, the hydroxide sludges produced from alkali treatment of acidic and ferruginous mine drainages, but the costs are high (Streeter *et al.* 1971).

Ion exchange

Various processes involving the use of liquid or solid ion exchange resins in more or less complicated chemical systems have been developed (Holmes & Kreusch 1972). Advantages over some of the other less conventional processes are that the waste products may be produced in solid form and may have a market value. Two full-scale plants have been commissioned (Zaban *et al.* 1972).

Solvent extraction

It is possible to remove iron and other contaminants from mine drainage by the use of organic complexing reagents and solvent extraction, but the costs of such processes would generally be too high to be practicable.

Biochemical reduction

It is possible to cultivate in an acidic and ferruginous mine drainage, after appropriate correction of the pH value, sulphate reducing bacteria, which will produce sulphides, which will in turn precipitate the dissolved iron salts (Dugan *et al.* 1968). It is even possible to form magnetic sulphides by such a process (Freke & Tate 1961). It is considered that corrosion problems due to sulphides, the need to maintain anaerobic conditions and relatively high operating temperatures would lead to very high costs of treatment. No full-scale treatment plants have been constructed.

Acknowledgments

The author wishes to thank the National Coal Board for permission to publish this paper and his several colleagues for assistance with the manuscript. Any opinions expressed are those of the author and not necessarily those of the National Coal Board.

Summary

The origin and potentially polluting effects of acidic and ferruginous drainages from coal mines, spoil heaps and coal stock-piles are described. Procedures available for the control of contamination at the source and for the treatment of these drainages are reviewed.

References

ANON. (1962) Report Prepared for Committee on Public Works. House of Representatives. *Print No.* 18, 87th Congress, 2nd Session. U.S. Dept. Hlth. Ed. & Welfare.

ANON. (1965) *Mine Drainage Abstracts. A Bibliography.* Dept. Environ. Resources, Pennsylvania.

ANON. (1968) *Water Quality Criteria.* Report of Natn. Tech. Adv. Com., U.S. Dept. Int.

ANON. (1970) Feasibility study of mining coal in an oxygen free atmosphere. U.S. Dept. Inter., *Wat. Pollut. Contr. Res. Ser.* 14010 DZM 08–70.

ANON. (1971a) Acid mine drainage formation and abatement. U.S. Environ. Prot. Agency. *Wat. Pollut. Contr. Res. Ser.* 14010 FPR 04–71.

ANON. (1971b) Studies on limestone treatment of acid mine drainage. U.S. Dept. Inter., *Wat. Pollut. Contr. Res. Ser.* 14010 EIZ 01–70 & EIZ 12–71.

ANON. (1971c) Purification of mine water by freezing. U.S. Environ. Prot. Agency, *Wat. Pollut. Contr. Res. Ser.* 14010 DRZ 02–71.

ANON. (1972) Use of latex as a soil sealant to control acid mine drainage. U.S. Environ. Prot. Agency, *Wat. Pollut. Contr. Res. Ser.* 14010 EFK 06–72.

ASHMEAD D. (1956) The influence of bacteria in the formation of acid mine waters. *Colliery Guardian* **192**, 483–7.

BAKER R.A. & WILSHIRE A.G. (1970) Microbiological factor in acid mine drainage formation. U.S.D.I. Fed. Wat. Qual. Adm., *Wat. Pollut. Contr. Res. Ser.* 14010 DKN 11–70.

BELLER M., WAIDE C. & STEINBERG M. (1970) Treatment of acid mine drainage by ozone oxidation. U.S. Environ. Prot. Agency, *Wat. Pollut. Contr. Res. Ser.* 14010 FMH 12–70.

BENOIT R.J., BALAKRISHNAN S. & ATTWATER A.J. (1972) Concentrated mine drainage disposal into sewage treatment systems. U.S. Environ. Prot. Agency, *Wat. Pollut. Contr. Res. Ser.* 14010 09–71.

BILHARZ O.W. (1949) Experiences with acid mine-water drainage in Tri-State Field. *Trans. Amer. Inst. Min. Metall. Engrs.* 181, 282–91.

BRALEY S.A. (1956) Acid coal mine drainage. *Min. Engin.* 8, 314–18.

BRALEY S.A. (1962) An evaluation of mine sealing. *Mellon Inst. Res. Proj.* 370, Pittsburgh, Pennsylvania.

CHARMBURY H.B., BUSCAVAGE J.J. & MANEVAL D.R. (1968) Pennsylvania's abandoned mine drainage pollution abatement program. *Proc. 2nd Symp. on Coal Mine Drainage*, Pittsburgh, Pennsylvania.

DAVIS R.M. (1971) Limnology of a strip mine pond in Western Maryland. *Chesapeake Sci.* 12, 111–14.

DUGAN P.R., RANDLES C.I., TUTTLE J.H., McCOY B. & MACMILLAN C. (1968) *The Microbial Flora of Acid Mine Water and its Relationship to Formation and Removal of Acid.* Water Resources Centre, Ohio State Univ.

FISCHER M. (1971) Ageing symptoms in drainage systems. *Symp. on Opencast Workings.* Leipzig, D.D.R.

FORD T. & BOYER J.F. (1973) Treatment of ferrous acid mine drainage with activated carbon. U.S. Environ. Prot. Agency, *Environ. Protect. Technol. Ser.* EPA-R2-73-150.

FOREMAN J.W. & McLEAN D.C. (1973) Evaluation of pollution abatement procedures, Moraine State Park. U.S. Environ. Prot. Agency, *Environ. Protect. Technol. Ser.* EPA-R2-73-140.

FREEMAN R.A. & EVERHART W.H. (1971) Toxicity of aluminium hydroxide complexes in neutral and basic media to rainbow trout. *Trans. Amer. Fish. Soc.* 4, 644–58.

FREKE A.M. & TATE D. (1961) The formation of magnetic iron sulphide by bacterial reduction of iron solutions. *J. biochem. Microbiol. Tech. Engng.* 3, 29–39.

GLOVER H.G. (1967) The control of acid mine drainage pollution by biochemical oxidation and limestone neutralisation treatment. *22nd Indust. Wastes Conf.*, Purdue Univ., Indiana.

HALL D.A. (1967) The storage of coal: the sealing of coal dumps with road tar. *J. Inst. Fuel* 40, 474–6.

HERRICKS E.E. & CAIRNS J. Jnr. (1972) The recovery of stream macrobenthic communities from the effects of acid mine drainage. *Proc. 4th Symp. on Coal Mine Drainage Res.*, Pittsburgh, Pennsylvania.

HILL D.W. (1969) Neutralisation of acid mine drainage. *J. Wat. Pollut. Contr. Fed.* 41, 1702–15.

HILL R.D. & MARTIN J.F. (1972) Elkins Mine drainage pollution control demonstration project—an update. *Proc. 4th Symp. on Coal Mine Drainage Res.*, Pittsburgh, Pennsylvania.

HOLMES J. & KREUSCH E. (1972) Acid mine drainage treatment by ion exchange. U.S. Environ. Prot. Agency, *Environ. Protect. Technol. Ser.* EPA-R2-72-056.

JASINSKI R. & GAINES L. (1972) Electrochemical treatment of acid mine waters. U.S. Environ. Prot. Agency, *Wat. Pollut. Contr. Res. Ser.* 14010 FNQ 02-72.

KENYERES J. & TAKACS P. (1972) Use of bituminous emulsions for coating stores coals. *Hungan. Min. Res. Inst.*, Publ. 15.

KINNEY E.C. (1964) Extent of acid mine pollution in the United States affecting fish and wildlife. U.S. Dept. Inter. *Fish & Wldlf. Circ.* 191.

KOSTENBADER P.D. & HAINES G.F. (1970) High density sludge treats acid. *Coal Age* 75, 90–7.

LAU C.M., SHUMATE K.S. & SMITH E.E. (1970) The role of bacteria in pyrite oxidation kinetics. *Proc. 3rd Symp. on Coal Mine Drainage Res.*, Pittsburgh, Pennsylvania.

LLOYD R. & JORDAN D.H.M. (1964) Some factors affecting the resistance of rainbow trout (*Salmo Gairdnerii Richardson*) to acid waters. *Int. J. Air Wat. Pollut.* 8, 393–403.

LOVELL H.L. (1971) Limestone treatment of coal mine drainage. *Min. Congr. J.* 57, 28–34.

LUNDGREN D.G. (1971) Inorganic sulphur oxidation by iron-oxidising bacteria. U.S. Environ. Prot. Agency, *Wat. Pollut. Contr. Res. Ser.* 14010 DAY 06–71.

McCOY B. & DUGAN P.R. (1968) The activity of micro-organisms in acid mine Water, II. The relative influence of iron, sulfate and hydrogen ions on the microflora of a non-acid stream. *Proc. 2nd Symp. on Coal Mine Drainage Res.* Pittsburgh, Pennsylvania.

MANEVAL D.R. & LEMEZIS S. (1972) Multistage flash evaporation system for the purification of acid mine drainage. *Trans. Amer. Inst. Min. Engrs.* 252, 42–5.

MASON D.G. & GUPTA M.K. (1972) Reverse osmosis demineralisation of acid mine drainage. U.S. Environ. Prot. Agency, *Wat. Pollut. Contr. Res. Ser.* 14010 FQR 03–72.

MOEBS N.N. & KRICKOVIC S. (1970) *Air-sealing Coal Mines to Reduce Water Pollution.* Report of Investigations No. 7354, U.S. Bureau of Mines.

MORTH A.H., SMITH E.E. & SHUMATE K.S. (1972) Pyritic systems: a mathematical model. U.S. Environ. Prot. Agency, *Environ. Prot. Tech. Ser.* EPA-R2-72-002.

MOSS E.A. (1971) Dewatering of mine drainage sludge. U.S. Environ. Prot. Agency, *Wat. Pollut. Contr. Res. Ser.* 14010 FJX 12–71.

PLOT R. (1686) *The Natural History of Stafford-shire.* The Theater, Oxford.

RILEY C.V. (1960) The ecology of water areas associated with coal strip-mined lands in Ohio. *Ohio J. Sci.* 60, 106–21.

ROGOFF M.H., SILVERMAN M.P. & WENDER I. (1960) The elimination of sulfur from coal by microbial action. Div. of Gas & Fuel Chem., *Amer. Chem. Soc. Preprint* 2, 25–36.

SHEARER R.E., EVERSON W.A. & MAUSTELLER J.W. (1968) Reduction of acid production in coal mines with use of viable antibacterial agents. *Proc. 2nd Symp. on Coal Mine Drainage Res.*, Pittsburgh, Pennsylvania.

SHEARER R.E., EVERSON W.A. & MAUSTELLER J.W. (1970) Characteristics of viable antibacterial agents used to inhibit acid-producing bacteria in mine waters. *Proc. 3rd Symp. on Coal Mine Drainage Res.*, Pittsburgh, Pennsylvania.

SLEIGH J.H. & KREMEN S.S. (1971) Acid mine waste treatment using reverse osmosis. U.S. Environ. Prot. Agency, *Wat. Pollut. Contr. Res. Ser.* 14010 DYG 08–71

SMITH E.E. & SHUMATE K.S. (1970 Sulfide to sulfate reaction mechanism. U.S. Dept. Int., Fed. Wat. Qual. Admin., *Wat. Pollut. Contr. Res. Ser.* 14010 FPS 02–70.

STEINBERG M. & PRUZNSKY J. (1970) Radiation treatment of mine waste waters. *U.S. Patent* 3-537-966.

STREETER R.C., YOUNG R.K. & GLENN R.A. (1971) Studies on densification of coal mine drainage sludge. U.S. Environ. Prot. Agency, *Wat. Pollut. Contr. Res. Ser.* 14010 EJT 09–71.

SYKORA L., SMITH E.J., SHAPIRO M.A. & SYNAK M. (1972) Chronic effect of ferric hydroxide on certain species of aquatic animals. *Proc. 4th Symp. on Coal Mine Drainage Res.*, Pittsburgh, Pennsylvania.

TEMPLE K.L. & COLMER A.R. (1951) The autotrophic oxidation of iron by a new bacterium: *Thiobacillus Ferrooxidans. J. Bact.* 62, 605–11.

THOMAS G. (1698) An account of Pensilvania. In *The First Century and a Quarter of American Coal Industry* (by H.N. Eavenson), Waverley, Baltimore.

TREHARNE W.D. (1962) Pollution problems on the Afon Rheidol. *Wat. & Waste Treat.* 8, 610–13.

UETA E., SASAMOTO K. & FUKUDA K. (1972) Bacterial oxidation treatment of waste water at Hosokura. *Print No. T IIIb* 3, Min. & Metall. Inst. Japan, Tokyo.

WEAVER R.H. & NASH H.D. (1968) The effects of strip mining on the microbiology of a stream free from domestic pollution. *Proc. 2nd Symp. on Coal Mine Drainage Res.* Pittsburgh, Pennsylvania.

WHITESELL L.B. Jnr., HUDDLESTON R.L. & ALLRED R.C. (1971) Microbiological treatment of acid mine drainage waters. U.S. Environ. Prot. Agency, *Wat. Pollut. Contr. Res. Ser.* 14010 ENW 09–71.

WILMOTH R.C. & HILL R.D. (1970) Neutralization of high ferric iron acid mine drainage. U.S. Dept. Int., *Wat. Pollut. Contr. Res. Ser.* 14010 ETV 08–70.

ZABBAN W., FITHIAN T. & MANEVAL D.R. (1972) Converting AMD to potable water by ion exchange treatment. *Coal Age* 77, 107–11.

ZAVAL F.J. & ROBINS J.D. (1973) Water infiltration control to achieve water pollution control. A feasibility study. U.S. Environ. Prot. Agency, *Environ. Prot. Ser.* EPA-R2-73-142.

Estimating the acid potential of coal mine refuse

F. T. CARUCCIO *University of South Carolina, Columbia, U.S.A.*

Introduction

BACKGROUND

When coal is mined the iron disulphides (FeS_2) found in the coal and associated strata, are exposed to the atmosphere and readily oxidized to a series of hydrous iron sulphates. Water coming in contact with these iron sulphates generates acidic mine drainages, commonly containing large concentrations of iron and sulphate.

The chemical reactions involved in this process are as follows:

$$2\ FeS_2 + 2H_2O + 7O_2\ (g) \rightarrow 2\ Fe^{++} + 4H^{+} + 4\ SO_4^{--} \qquad (1)$$

Ferrous ion oxidation produces another increment of acid by the following reaction:

$$4\ Fe^{++} + 10H_2O + O_2\ (g) \rightarrow 4\ Fe(OH)_3 + 8H^{+} \qquad (2)$$

(Barnes & Romberger 1968)

During the past ten years the author has investigated the nature and distribution of pyrite within the strata of the Allegheny Group of the bituminous coal field of Pennsylvania. He has shown that the factors affecting the quality of mine drainages from various strata are:

(i) Alkalinity—a function of the amount of calcareous material present in the stratum;

(ii) pH—which controls the occurrence of iron bacteria that can catalyse the acid-producing chemical reactions (Equation 2);

(iii) The nature and distribution of acid-producing pyrite in the coals and associated strata.

As seen in Fig. 1, in the Hydro-geochemical Environment—I, the pyrite occurring in the strata is coarse-grained and is relatively inert, the pH of the ground water (before the mining phase) is above 6·4 which is not conducive to the support of the iron bacteria and there exists, within the natural system, a ground water with a high neutralizing capacity. In this environment, minor amounts of acidity produced by the pyrite are neutralized by the available alkalinity, the acid reaction is not catalysed (due to the absence of the bacteria) and the mine drainages are chemically neutral and of low sulphate types.

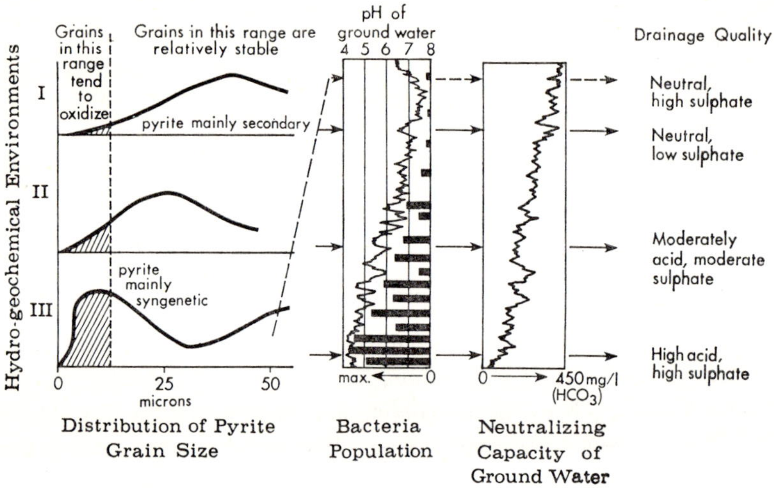

Figure 1. Summary diagram relating mine drainage qualities to various hydro-geochemical factors.

At the other extreme, in Environment III, most of the pyrite in the strata is fine grained (less than 0·25 microns) and readily oxidizes to produce abundant acidity. The pH of the ground water is low and supports iron bacteria which catalyse the acid-producing reactions and generate additional acidity. The buffering capacity of the ground water is extremely low and consequently, drainages from this environment are highly acidic and high in sulphate.

Environment II represents the medium of the two with correspondingly moderate amounts of acidity and sulphate.

Neutral, high sulphate drainages are generated when alkaline waters come in contact with acid waters and neutralization takes place. This situation occurred in north-western Pennsylvania where coals mined from highly acid, high sulphate strata were capped by calcareous glacial drift. Here the highly alkaline waters from the glacial drift effectively neutralized the acidity created in the mining operation, producing a neutral, high-sulphate type of mine drainage.

Of the three factors outlined above, the alkalinity and pH could be more or less controlled by the reclamation technique. For example, chemical compounds or calcareous strata capable of generating alkalinity could be spread over the surface or incorporated in the backfill. If applied correctly an alkaline front can be generated that will migrate with infiltrating water through the reclaimed section, displace the iron bacteria, and in turn reduce and effectively neutralize the acidity.

The mode of pyrite, however, cannot be controlled inasmuch as it is an intrinsic part of the coal and associated strata. Accordingly, it is of interest to the reclamation effort to anticipate the degree of acidity that may be expected from the spoil material and whether or not the acidity can be effectively neutralized. This paper intends to provide a technique that can be used to estimate the amount of acidity and enhance reclamation efforts.

GENERATION OF ALKALINITY

If the degree of acidity that a pyrite-bearing stratum will produce could be predicted then the feasibility of incorporating a calcareous layer in the backfill could be established. This could then guide the mine operator or the reclamation efforts toward appropriate steps to prevent acid mine drainage or enhance revegetation of the reclaimed areas.

It is apparent that in order for the above technique to be successful the amount of acidity generated in the section must be less than the amount of available alkalinity for complete neutralization to take place. The amount of alkalinity that calcareous strata will generate is determined by the solubility of the calcium carbonate. This, in turn, is dependent upon the partial pressure of carbon dioxide through which the infiltrating waters will contact before encountering the carbonate. The alkalinity can be calculated from the following equation:

$$\text{Log Ca}^{++}\ (\text{mg/l}) = 2{\cdot}56 + 0{\cdot}362\ \text{Log pCO}_2 \qquad (3)$$

and for each mole of calcium produced there are two moles of bicarbonate alkalinity (Langmuir, *pers. comm.*).

Under atmospheric conditions where the partial pressures of carbon dioxide is equal to $10^{-3{\cdot}5}$ atmospheres, the alkalinity generated by water flowing over calcareous material is about 60 mg/l (as HCO_3^-). If a good soil-mulch cover is developed the partial pressure of carbon dioxide in the soil could increase to about 10^{-1} (data taken from Black 1957) and increase the available alkalinity of infiltrating waters by a factor of eight. This value assumes that the infiltrating waters move slow enough to allow the chemical reaction to go to completion. The development of a good mulch cover requires many years and the resulting high alkalinities may not be realized for some time. However, even assuming the system to be under conditions far from ideal we could still expect an alkalinity concentration of about 100 mg/l (as HCO_3^-) which can effectively neutralize about 1·6 mg/l of acidity (as H^+).

The dependency of the generation of alkalinity upon the partial pressure of carbon dioxide, which in turn, is a function of a soil mulch cover underscores the necessity of establishing a rapidly growing vegetation cover in order to successfully reclaim an area. To this end, it is urged that there should be an initial planting of carpet grasses and subsequently, after the grasses have been established, the planting of trees. The grasses will provide a good soil mulch as well as effectively reducing the temperatures of the surface rock fragments, thereby providing for the greater retention of soil moisture. Proper management is required to insure that an effective growth medium is provided at the surface and that tree and grass growth be compatible.

Generation of acidity

PREVIOUS WORK

It is generally recognized that the oxidation of iron disulphide (hereafter collectively called pyrite) to hydrous ferrous sulphates and the subsequent hydrolysis of these compounds in water is the cause of the acidity found in most coal mine areas. Because of this relationship it is often assumed that the amount of acidity produced is related to the amount of pyrite present, i.e., the higher the pyrite content the greater the acid production potential. However, many investigators noted that the amount of pyrite present in a sample was not always proportional to the amount of acidity produced. This writer has shown that similar rock types produced acidities that varied more directly with the particle size of the pyrite than with the amount of pyrite present (Caruccio 1968). It was shown that small particles of pyrite ($<0{\cdot}25$ microns) rapidly decomposed when exposed to the atmosphere while coarse grained pyrite particles (greater than 50 microns) remained stable. It was also shown that the variability in acid production was related to the rock's permeability (binders produced less acid than shales and coals) but not related to the mineralogy of the iron disulphide. In some instances pyrite was shown to produce more acidity than marcasite; contrary to what has been reported.

(1)

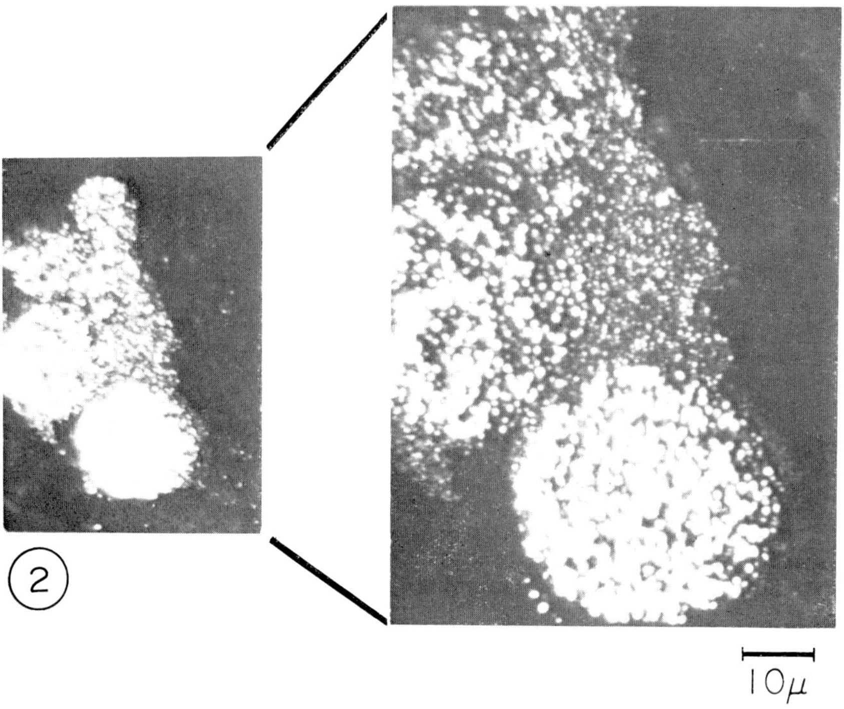

(2)

Plates 1 and 2. Photomicrographs of reactive pyrite.

3

50 μ

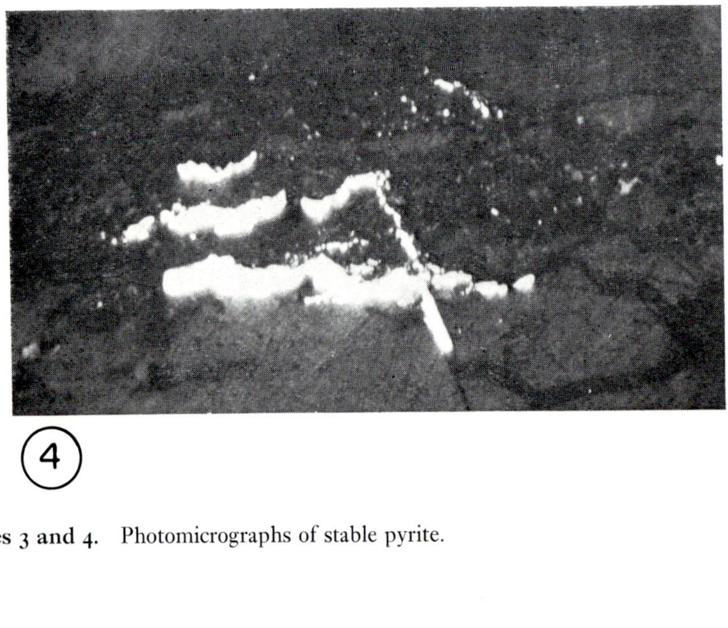

4

Plates 3 and 4. Photomicrographs of stable pyrite.

PYRITE MORPHOLOGY AND PYRITIC SULPHUR ANALYSES

If a polished section of a reactive nodule is compared to one of an inert sample a difference in pyrite type is immediately apparent. The sample that is noted to readily decompose is composed primarily of the pyrite grains seen in Plates 1 & 2. This type of pyrite has been termed 'framboidal' by Gray *et al.* (1963) and will be so used by this writer. The pyrite grains of the inert sample, however, have forms shown in Plates 3 & 4. Not shown but also included in stable pyrite categories are crystals or euhedra of pyrite that commonly have cubical or triangular shapes. These crystals are generally between 5–10 microns and although small sized are inert. Obviously, the pyrite grains of the reactive sample are not only finer grained than the inert sample but also have a different morphology. Large nodules (about 1–2 cm long) were noted to readily decompose and seemingly contradicted this observation. However, by carefully polishing the nodule and examining it with a microscope, it was discovered that the nodule was actually an aggregate of fine grained framboidal pyrite particles.

From the preceding discussion it follows that the morphology of the pyrite (framboidal versus tabular-euhedral) is related to the reactivity of the sample. It may be noteworthy to mention that it is not the size alone that determines the reactivity of the pyrite sample. A stable crystal of pyrite was pulverized in a ball mill and subsequently exposed to conditions that caused other samples to decompose. After two months the pulverized pyrite had not oxidized. This suggests that the framboidal-reactive pyrite has a composition which is different from the stable form and accounts for the differences in reactivity. Recent studies have shown that stable pyrites have trace amounts of titanium whereas reactive pyrites do not. In addition, reactive pyrites contain trace amounts of silver (Caruccio 1972).

Regardless of the actual mineralogical composition the fact that the framboidal pyrite was noted to readily decompose can still be used to evaluate the percentage of reactive pyrite. The technique is to measure the total amount of pyrite in a sample using the Leco (Laboratory Equipment Company) combustion instrument in a method outlined by Neavel (1966). The total sulphur content of a sample is the sum of organic sulphur, pyritic sulphur and sulphate sulphur. The sulphate sulphur constitutes a small portion of the total sulphur content of a sample and can be neglected. A portion of the sample is then digested in a nitric acid solution which effectively removes the pyritic sulphur and the remaining sulphur is in the organic form. The digested sample (containing only organic sulphur) is then analysed for total sulphur content. The difference between the total sulphur content of the raw sample and the total sulphur content of the digested sample is the pyritic sulphur content. The value obtained represents the total pyrite content and is composed of primary-framboidal-reactive pyrite and primary-euhedral and secondary inert pyrite. Of these types only the framboidal pyrite is reactive and will produce acidity.

From these considerations it should be possible to develop a technique which will rate the acid mine drainage 'potential' of a sample by measuring the percentage of primary-fine grained-framboidal pyrite in a sample.

Technique for estimating reactive pyrite

METHODOLOGY

To estimate what percentage of the total pyrite content is reactive, a spoil sample is crushed to pass a 2 mm sieve, riffled to obtain a 10–15 g portion, cast into plastic pellets, cured, cut, and polished by lap wheels. The pellet is then examined with an oil immersion lens and the various types of pyrite are identified. Efforts must be made to insure that the sample cast into the pellet be representative of the field sample.

The pellet is then scanned with respect to a systematic grid network for the purpose of measuring the area of the pellet occupied by various types of pyrite. Pyrite grains that are of the framboidal type and are not euhedral are identified and the area occupied by these grains is measured. Commonly the framboidal pyrite occurs within clusters and an areal measurement of the cluster outline is as valid as the total of individual grain area. The area occupied by secondary pyrite is also measured. After the entire pellet is scanned and the area of the pellet covered by pyrite ascertained, the percentage of the area occupied by the reactive pyrite grains can be calculated by dividing the area of fine-grained pyrite by the total pyrite area. This percentage multiplied by the total pyrite content obtained from the LECO yields the percentage of reactive pyrite contained in the sample. There are various ways to measure the areas of the pellet covered by the various sizes and types of pyrite grains. Some possibilities include:

1. *Planimetric measurement on a photograph* The sample could be photographed through the microscope and a composite photomicrograph obtained. A stage micrometer could also be photographed to determine the scale of the field of view. The photograph is then covered with a piece of tracing paper and the various pyrite areas measured with a planimeter.

2. *Automated microscope electronic data accumulator* (*AMEDA*) This technique has the capacity of measuring reflectivity, the level of which can be controlled to discriminate pyrite from background reflections. The chopped light beam automatically scans the pellet and measures the particle size and records these data electronically. This technique has the poten-

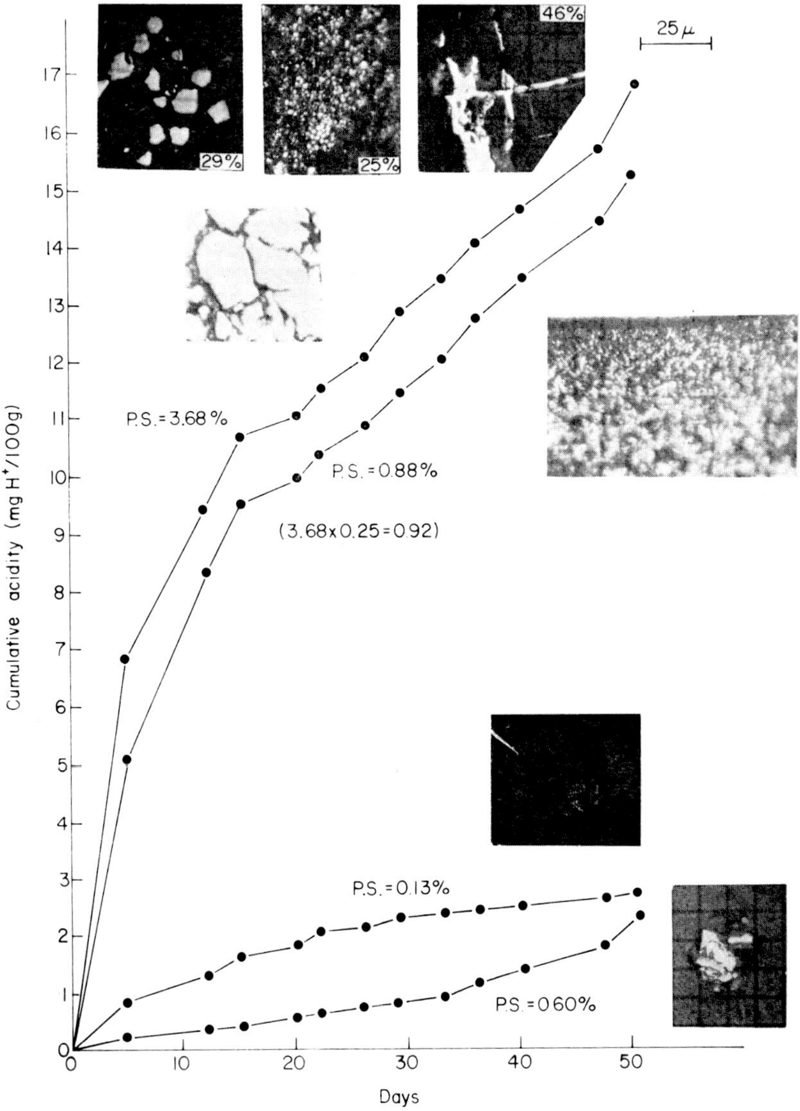

Figure 2. Acid potential of samples related to percentages of various pyrites.

tial of being developed to scan a pellet with a band of light in order to measure areas of various pyrite sizes.

3. *Optical measurement with grid* A grid is inserted into the ocular of a microscope and then calibrated by viewing a stage micrometer. The pellet is then scanned and pyrite types are identified, measured, and the area occupied by the grains estimated from the superposed grid.

EXPERIMENTAL DATA

An experiment was initiated whereby weighed samples of coal of similar particle size were placed in plastic containers which permitted a constant flow of humidified air to pass over the samples. Periodically the samples were flushed with deionized water and the leachates analysed for total hot titratable acidity. The amount of acidity, expressed as milligrams of hydrogen produced at a given time per unit weight of sample was calculated. In the preparation of the experiment a 15–20 g sample was split off of the main sample to be used in making pellets for the microscopic examination. This same split was analysed for pyritic sulphur contents.

The experiment was run for fifty days and the results are shown in Fig. 2. As can be seen the lower curves show the acid production of two samples of coal, one with a pyritic sulphur content of 0·6% and another with 0·13%. Although one sample had almost five times the amount of pyrite than the other, they both produced about equal amounts of acidity. When the polished pellets were examined with a microscope it was found that all of the pyrite in the sample containing 0·6% pyritic sulphur was of the coarse-grained secondary type. In contrast, all the pyrite in the sample containing 0·13% pyritic sulphur was of the framboidal primary type.

Referring to the upper curves, two samples of coal were leached, one containing 0·88% pyritic sulphur and the other containing 3·68%. The rate of acid production was the same for both samples with the 3·68% sample yielding slightly more acid. When the polished pellets of the samples were viewed with a microscope and the areas occupied by various types of pyrite ascertained, it was found that all of the pyrite of the sample containing 0·88% pyritic sulphur was of the framboidal type whereas in the sample containing 3·68% pyritic sulphur, 46% of the pyrite was of the coarse-grained secondary type, 29% was of the primary euhedral type and 25% was of the primary framboidal type. Inasmuch as 25% of the total pyrite found in the latter sample is reactive, the total pyritic sulphur content (3·68%) is multiplied by the percentage of reactive pyrite (0·25%) to yield 0·92%. This estimate of reactive pyrite converts the pyritic sulphur value to a better estimator of acid production potential.

Conclusions

As the experiment in Fig. 2 shows, framboidal pyrite produces about five times the amount of acidity than coarse-grained secondary pyrite. Further, the contention that framboidal pyrite is the primary cause of acidity and is the parameter that should be used as an estimator of acid production potential, was borne out by the experiment.

The technique presented is a valid one, however, additional experiments are needed to relate other variables such as permeability, rock type and particle size to quantitative acidity predictions. As a first order approximation the above technique can be used to obtain coarse estimates of acid potentials and whether or not calcareous materials can be used to ameliorate and neutralize acid problems.

Acknowledgments

This study was underwritten by the Coal Research Board of the Commonwealth of Pennsylvania and by the Research Foundation of the State University of New York (No. 26–7112–A).

Summary

Of the four modes of pyrite occurring in coal strata, only one, *framboidal* pyrite (particles less than 0·5 microns) has been shown to produce abundant acidity. A point counting technique is presented whereby quantitative estimates of acidity can be obtained for use in pre-vegetation treatments of refuse banks.

References

Barnes H.L. & Romberger S.B. (1968) Chemical aspects of acid mine drainage. *J. Wat. Pollut. Contr. Fed.* 40, 371–84.

Black C.A. (1957) *Soil-Plant Relationships.* Wiley, New York.

Caruccio F.T. (1968) An evaluation of factors affecting acid mine drainage production and the ground water interactions in selected areas of Western Pennsylvania. *Proc. 2nd Symp. on Coal Mine Drainage Res.*, Pittsburgh Pennyslvania.

Caruccio F.T. (1972) Trace element distribution in reactive and inert pyrite. *Proc. 4th Symp. on Coal Mine Drainage Res.*, Pittsburgh Pennsylvania.

GRAY R.J., SCHAPIRO N. & COE G.D. (1963) Distribution and forms of sulfur in a high volatile Pittsburg seam coal. *Trans. Soc. Min. Eng.* 226, 113–21.

NEAVEL R.C. (1966) Sulfur in coal, its distribution in the seam and mine products. Supplementary Research Report #1966–1, Coal Research Section, Pennsylvania State University.

Plant growth on acid molybdenum mill tailings as influenced by liming, leaching and fertility treatments

W. A. BERG, E. M. BARRAU and L. A. RHODES
Colorado State University, Fort Collins, U.S.A.

Introduction

Tailings, the solid waste remaining after ores have been treated to remove the commercial minerals, can pose aesthetic and air and water pollution problems. In some mill waste disposal systems coarse tailings are used to build dams to hold the finer tailings and the process water. Other systems use earth and rock dams to hold the tailings. In either system, a tailings surface is left exposed that must eventually be stabilized to prevent wind and water erosion when the disposal site is filled to capacity or when the mill is closed down.

Approaches to long term tailings stabilization have been to establish vegetation directly on the tailings (Dean *et al.* 1969; James 1966; Shetron & Duffek 1970; Young 1969) or to cover the tailings with soil and/or geological material (Beverly 1968; U.S.A.E.C. 1963).

A knowledge of the chemical and physical characteristics of the specific tailings in relation to plant growth is needed when seeding and planting directly in tailings. When tailings are covered with soil, this information is still desirable as it should be considered in relation to depth of soil cover and in terms of through-leaching and possible water pollution by chemicals in the tailings.

The tailings used in this study were produced by the Climax Molybdenum Company at Climax, Colorado. Earlier work by Grubb (1965) revealed that the tailings were usually extremely acid and required addition of lime and NPK before they would support plant growth. On-site seeding and planting studies by Grubb (1965) were complicated by windblown tailings, water erosion, and by the very short growing season at Climax where the elevation is 3,400m above sea level (upper subalpine). Grubb (1965) also raised the

Colorado State University Experiment Station Scientific Series Paper No. 1892.

question of possible deficiencies of Mn, Zn, and Mg in plants grown on the tailings.

In this paper we report on greenhouse and laboratory work on (1) the amount of lime required to maintain a pH level suitable for plant growth on the tailings; and (2) the effect of K, Zn, Mn, Cu, and soil treatments and leaching on plant growth on the tailings.

Chemical and physical characteristics of the tailings

The mine and mill at Climax, Colorado is currently the largest producer of Mo in the world. The following brief description of the mineral deposit and milling process is included to familiarize the reader with the composition and treatment of the tailings.

'The molybdenite deposit is in an area of pre-Cambrian granite and schist intruded by dikes of quartz monzonite porphyry. . . . Mineralization is characterized by fine grained quartz replacement on a large scale of the granite, schist, and porphyry . . . the molybdenite is fine-grained and occurs in countless criss-crossing quartz veinlets. . . . In the ore zone considerable secondary orthoclase occurs in addition to fine-grained quartz replacement. . . . The source of the molybdenite is from hydrothermal solutions.' (Vanderwilt 1947.)

'At the mill, the ore undergoes a series of crushing stages, grinding, and sizing in order to reduce it to proper fineness for liberation of values and treatment by froth flotation. The molybdenite is recovered as a relatively pure concentrate . . . beside molybdenite, the ore contains small quantities of pyrite, wolframite (tungsten mineral), and cassiterite (tin mineral) . . . the tailings from the Mo flotation process are treated . . . with gravity concentration devices to separate the remaining minerals from the worthless gangue.' (Wickmann 1960.)

Eight different tailings samples were used in the studies reported here. These samples are identified by a letter and numbers. The letter is for the location (e.g. c for Climax), the first number is the year sampled (e.g. 70 for 1970), and the last number is the sample number taken at that location in the given year. Five samples (c-69-1, c-70-2, c-70-3, c-70-5, c-71-6) were from the outer berm of a tailings pond deposited prior to 1936. Two samples (c-70-1 and c-70-4) were from the upper portion of the Climax Robinson pond tailings dam and were deposited three to five years prior to sampling. The remaining tailing, o-70-1, was from the pound area (now dry) of tailings from lead and zinc ores milled approximately 20 years ago near Ouray, Colorado. A loam soil (A_1 horizon, Platner series) was used as a standard against which to compare plant growth on the tailings.

Table 1. Chemical and physical characteristics of tailings and soil used in greenhouse studies.

Material	pH[1]	Phosphorus ppm		Nitrogen[4]	Extractable Bases[5]				Plant Available Micronutrients ppm[6]				Particle Size[7]			Texture	Field Capacity[8]
		Bray[2]	$NaHCO_3$[3]	ppm	K	Na	Ca	Mg	Mn	Zn	Fe	Cu	Sand %	Silt %	Clay %		%
Tailing c-69-1	4·3	—	—	—	10	5	74	2	·2	·6	44	0·5	89	7	4	Sand	5·0
Tailing c-70-1	3·6	3	1	13	7	3	30	1	·3	·7	46	0·3	88	8	4	Sand	4·5
Tailing c-70-2	4·1	11	6	7	17	3	220	0·5	0·1	0·5	31	0·3	83	13	4	Loamy Sand	4·4
Tailing c-70-3	3·6	8	3	9	16	2	10	1·5	0·3	0·5	70	0·6	88	9	3	Sand	3·9
Tailing c-70-4	4·7	2	1	8	12	3	65	9	3·0	1·7	33	1·1	93	6	1	Sand	4·0
Tailing c-70-5	3·6	7	2	24	11	3	65	1	0·2	0·4	69	0·5	76	20	4	Loamy Sand	4·0
Tailing c-71-6	3·0	—	1	14	12	5	30	2	0·2	1·3	64	0·6	87	10	3	Sand	5·0
Tailing 0-70-1	3·1	6	10	—	8	5	640	94	28·0	17·0	80	10·0	52	33	15	Loam	11·0
Soil	8·1	15	9	1,400	580	11	2,640	284	29·0	1·4	9	0·8	44	36	20	Loam	16·0

1 Determined with a pH meter on a 2:1 by weight distilled water: tailing mixture.

2 Method for acid soils: <3 ppm, very low; 3–7 ppm, low; 7–20 ppm, medium in Black (1965).

3 Method for calcareous soils: <5 ppm, response to fertilizer P; 5–10 ppm, probable response; > 10 ppm, no response likely, pp 1044–7 Black (1965).

4 Kjeldahl method for total N in Black (1965).

5 Extractable in a 10:1 1N ammonium acetate: tailing mixture shaken for 30 minutes. K and Na determined by flame photometer; Ca and Mg determined by atomic absorption spectrophotometry. <60 ppm K is low.

6 Extracted by the DTPA method (Follett & Lindsay 1971). Determination made on tailings after liming and wetting and drying. Mn, Zn, Fe, and Cu determined by atomic absorption spectrophotometry. Mn, 0–1 ppm may respond to Mn; Zn, 0–0·5 ppm low; 0·5–1 ppm marginal; >1 ppm adequate; Fe, 0–2 ppm low, 2–4·5 marginal, >4.5 adequate; Cu, 0–0·2 ppm may respond to Cu.

7 Hydrometer method in Black (1965).

8 The amount of water held at field capacity was estimated by placing a known weight of air-dry tailing or soil into a cylinder that was 7 cm in diameter and 30 cm in height. A measured amount of water was added to the top of the column and the surface covered to prevent evaporation; 24 hours later the dry soil remaining in the bottom of the cylinder was weighed and then the per cent water in the moist soil was calculated.

Chemical and physical information on the tailings are given in Table 1. These analyses indicate that the tailings as a plant growth media are extremely acid, low to extremely low in N, P, and K, low in Ca and Mg, marginal in Mn and Zn, adequate in Cu, and very high in Fe. The tailings are largely sand size and thus hold a rather small amount of water per unit of volume.

The acidity of the tailings is a result of oxidation of sulphide minerals, most likely pyrite. Pyrite can be detected visually in the tailings which are predominantly sand-size quartz. The texture is coarse because these samples were taken from the outside (dam) of the tailings ponds. Tailings in the pond area are finer textured, as is the sample from Ouray.

Study 1—lime requirement

If plants are to grow on these tailings it is necessary to add sufficient lime to obtain and maintain a pH favourable for plant growth. A basic question in liming tailings is whether sulphides will oxidize after the initial neutralization and result in the tailings becoming acid again. To study this, an incubation study was set up to follow the change in pH of limed tailings over a period of time.

METHODS

The lime requirement on the tailings was determined by a titration procedure where 60 ml of 1 N KCl was added to 20 g of tailings and the mixture titrated to a pH of 7 with NaOH. The mixture was allowed to stand for 24 hours and again titrated up to a pH of 7. Results of the lime requirement determinations are shown in Table 2.

Preliminary studies had shown that the titration method underestimated the amount of lime required on the Climax tailings. Thus, in this study lime treatments were added that were zero, two, four, and six times the equivalent of base required to neutralize the tailings by the titration procedure.

Table 2. Neutralization requirements of tailings in parts per million $CaCO_3$.

Tailing	Lime requirement
	(ppm)
c-70-1	250
c-70-2	125
c-70-3	225
c-70-4	100
c-70-5	225
o-70-6	3,500

Six tailings and three replications of each treatment were used in the study. Individual treatments consisted of 2,000 g of tailings into which lime was mixed and then placed into polyethylene-lined 2 l containers. Deionized water was added to bring the treated tailings to field capacity. The tailings were placed in a greenhouse and allowed to dry. At 14-day intervals water was added to bring the treatments to field capacity. Greenhouse temperatures ranged between 15° and 30°C. After reaction periods of 28, 84, 400, and 900 days the treatments were sampled and pH determined on a 1:1 tailing: distilled water mixture.

The lime used in the study was $CaCO_3$ which was a waste product from sugarbeet processing. This lime passed a 100-mesh sieve and contained the equivalent of 86% $CaCO_3$.

RESULTS AND DISCUSSION

The change in tailings pH as a result of the different lime rates and subjection to wetting and drying treatments over a period of time is shown in Fig. 1 for

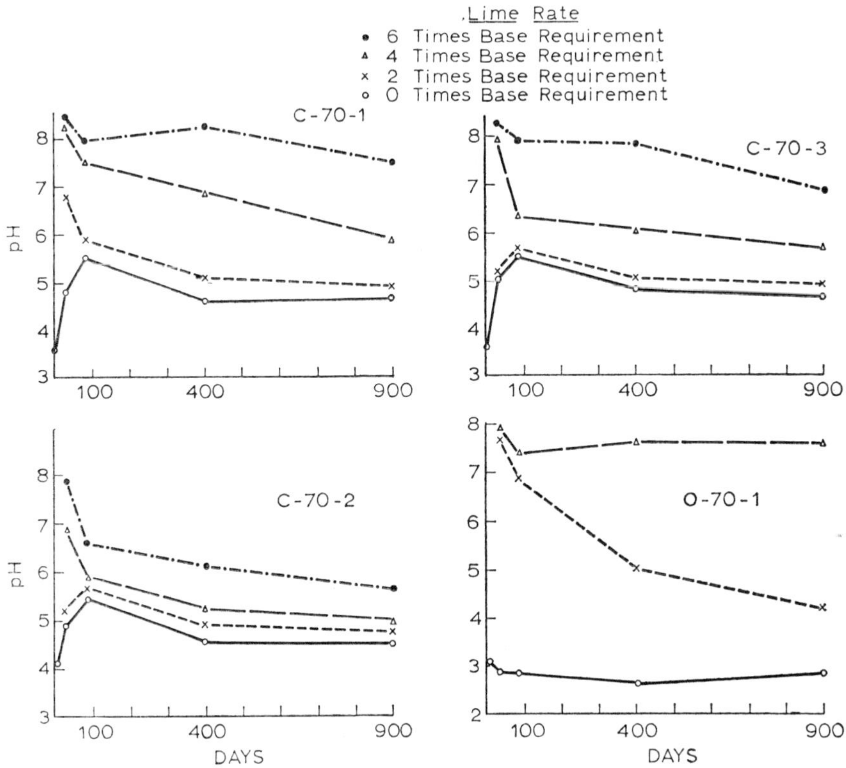

Figure 1. Tailing pH as a function of reaction time after addition of lime at zero, two, four, and six times the equivalent of base required to neutralize the tailing by NaOH titration.

four of the tailings. After 900 days of incubation the pH is still decreasing on most of the limed tailings. This indicates that, although the tailings can be neutralized by liming, there is apparently continuing sulphide oxidation.

The highest lime rate used in these studies was adequate to maintain the tailings at a pH suitable for plant growth over the 900-day reaction period. However, the study indicates that if long-term vegetative stabilization of the tailings is anticipated, higher lime rates or maintenance liming will be required.

The increase in pH of the zero lime rate treatment on the Mo tailings for the first 84 days (Fig. 1) was substantial and unexpected. This change was checked out in terms of titratable acidity by the titration procedure as outlined in the methods section. Only about one-half of the original level of titratable acidity was found in the zero lime rate treatment of the tailings after 400 days of reaction. The reason for the decrease in acidity is unknown; however, Coleman *et al.* (1960) have reported loss of exchangeable acidity in extremely acid soils at high temperatures.

Study 2—fescue growth on a Mo mill tailing as influenced by leaching and additions of K, Mn+Zn, and soil

In this study a number of treatments were investigated for their value in increasing plant growth on a limed and NP-fertilized Mo tailing. Leaching after liming and addition of K were found to be basic requirements for plant growth on the tailing. A plant growth response to the addition of Mn+Zn and soil additions was also found.

METHODS

The tailing c-69-1 was limed by addition of reagent grade CaO and $(MgCO_3)_4.Mg(OH)_2.4H_2O$ equal to four times the base equivalent required to neutralize the tailing by the titration procedure given in the methods section of Study 1. Three-fourths of the base equivalent was added as CaO and one-fourth as $(MgCO_3)_4.Mg(OH)_2.4H_2O$. After liming the tailing was mixed and subjected to three wet and air-dry cycles over a period of 15 days to allow the liming reagents to react. Then half of the limed tailing was leached with seven equal increments of deionized water applied at 12-hour intervals. Total water used was 84 l on 75 kg of tailing.

After air drying, 50 ppm N as $Ca(NO_3)_2$ and 12 ppm P as $Ca(H_2PO_4)_2.H_2O$ were added in solution and mixed into all treatments. An additional 44 ppm P was applied to all treatments with a band placement of triple superphosphate in each pot. An additional 100 ppm N as NH_4NO_3 was added to all treatments during the course of this study. The soil was given the same N and P treatments as the tailing.

A total of 12 treatments and combinations of treatments was applied to the tailing after liming, leaching of one-half, and NP fertilization. The treatments and yields are shown in Table 3. There were three replications on each treatment. Yields reported are the means from the three replications.

Table 3. Treatments and mean yield of fescue grown on limed and NP-fertilized tailing c-69-1 in Study 2.

Treatment	Yield, g/pot			
	Unleached		Leached	
None	·58	a*	·74	a†
100 ppm K as K_2SO_4	1·28	bc	2·37	b
300 ppm K as K_2SO_4 (K_2)	1·00	bc	2·68	b
10 ppm Mn as $MnCl_2.4H_2O$ (Mn_1+3 ppm Zn as $ZnSO_4.7H_2O$ (Zn_1)+K_2	1·45	de	2·99	c
30 ppm Mn as $MnCl_2.4H_2O$ (Mn_2)+9 ppm Zn as $ZnSO_4.7H_2O$ (Zn_2)+K_2	1·60	f	3·28	d
·4 ppm B as H_3BO_3 (B_1)+Mn_1+Zn_1+K_2	1·21	b	3·24	d
1·2 ppm B as H_3BO_3 (B_2)+Mn_2+Zn_2+K_2	1·99	g	3·55	e
2 ppm Fe as Fe-EDDHA+B_1+Mn_1+Zn_1+K_2	1·57	ef	3·32	d
6 ppm Fe as Fe-EDDHA+B_2+Mn_2+Zn_2+K_2	1·89	g	3·73	f
1% soil by weight+K_2	1·38	cd	3·21	d
10% soil by weight+K_2	2·18	h	3·75	f
90% soil by weight+K_2	7·26	j	7·24	h
Soil NP fertilized, not limed, not leached	6·40	i	6·04	g

* Yields within treatments on unleached tailing followed by the same letter are not significantly different ($P < 0{\cdot}05$).

† Yields within treatments on leached tailing followed by the same letter are not significantly different ($P < 0{\cdot}05$).

Each treatment consisted of 2 kg of tailing placed in a 2 l polyethylene-lined container. Tall fescue (*Festuca arundinacea*) was seeded on 7 February 1970 and the tailing brought to field capacity with deionized water. During the 44-day growing period water was added as required and temperatures in the greenhouse were set for 15°C at night and 24°C in the daytime. However, on sunny days the temperature went as high as 30°C. When harvested the plants were cut off 1 cm above the tailing surface, oven-dried at 70°C, and weighed.

The fescue yields on the unleached and leached tailings were subjected to separate analyses of variance. In both cases the treatment effect was highly significant (P < 0·01). Duncan's new multiple range test was then used to test for significant differences among treatments (LeClerq *et al.* 1962).

RESULTS AND DISCUSSION

A wide range in fescue yields was obtained on the tailing as a result of the treatments. The treatments giving some of the larger yield differences are shown in Fig. 2. Other treatments gave statistically significant yield differences (Table 3), but the practical significance of some of the relatively small yield differences is questionable, especially in view of the variability in plant growth among Climax tailings, as will be shown in Studies 3 and 4.

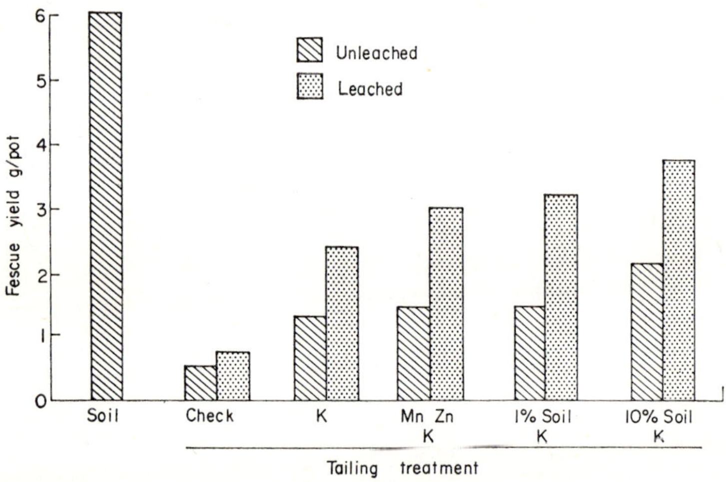

Figure 2. Fescue yields on limed and NP-fertilized tailing c-69-1 as influenced by leaching, potassium, manganese plus zinc, and soil treatments.

The treatment showing the greatest plant growth response was the addition of 100 ppm K. This response was predicted from the very low soil test value for K. A high rate of K addition (300 ppm) gave no additional yield increase.

Leaching the tailing after liming increased yields nearly twofold when K was supplied to the tailing.

The addition of the combination of 10 ppm Mn and 3 ppm Zn gave a slight yield increase. This treatment was suggested by the work of Grubb (1965) which showed that total Mn and Zn were low in the Climax tailings and by the soil test for plant-available minor elements which indicated low Mn and marginal Zn levels on the limed tailing. The response to additions of these elements is explored further in Study 3.

Soil additions to the tailing gave some minor plant growth increases. Soil addition is a method of supplying an array of both chemical and biological factors that may be limiting plant growth on the tailing. The relatively small increase in plant growth due to the soil additions, particularly between the 1% and 10% soil additions, indicates that the absence or low level of one or a combination of factors is probably not limiting plant growth on the tailing.

The mixture of 90% soil and 10% tailing by weight gave a greater yield than the soil (Table 3). In a study not reported here it was found that sulphate was deficient, limiting plant growth on the soil. The tailing served as a source of sulphate. In later studies sulphate was added to the soil to avoid this problem.

Study 3—fescue yields on Mo mill tailings as influenced by Mn, Zn, and Cu additions

In this study plant growth was increased on two of five Mo tailings by addition of Mn or Zn or the combination of Mn and Zn. Yield of fescue seeded in the five tailings was only 35% of the yield on soil during the first growth period of 49 days. In the 24-day regrowth period yield on the tailings averaged 58% of the yield on soil. Leaching after liming increased fescue growth several-fold on four of the five tailings.

METHODS

The five tailings samples used in this study were limed and leached and had P applied as detailed in the methods section of Study 2. A lime–no-leach treatment was also included. 50 ppm N was added as NH_4NO_3 and 200 ppm K as K_2SO_4.

The treatments consisted of 10 ppm Mn as $MnCl_2.4H_2O$, 6 ppm Zn as $ZnSO_4.7H_2O$, the combination of 10 ppm Mn and 6 ppm Zn, and the combination of the above Mn and Zn rates plus 1·5 ppm Cu as $CuSO_4.5H_2O$. All treatments were applied in solution to air-dry tailings, then mixed. An additional 200 ppm N as NH_4NO_3 was applied in 50-ppm increments at approximately 14-day intervals during the course of the study. Three replications were used for each treatment. Yields reported are the means from the three replications. Soil fertilized with N, P, and K at the rates outlined above was included as a standard.

Greenhouse techniques were the same as in Study 2. The treatments were seeded to fescue on 3 February 1971 and harvested 49 days later.

The regrowth was harvested 24 days after the initial harvest. The yields within each tailing were subjected to an analysis of variance. When the F test was significant ($P < 0{\cdot}05$), Duncan's new multiple range test was used to test for differences among treatments (LeClerq *et al.* 1962).

RESULTS AND DISCUSSION

The addition of Mn or Zn increased fescue yields on two of the five tailings (Table 4) during the initial 49-day growth period. The combination of Mn plus Zn significantly increased the yields on these two tailings over yields resulting from separate Mn or Zn additions. The further addition of Cu significantly increased growth on one tailing (c-70-4) over growth on Mn and Zn treatments.

Table 4. First harvest fescue yields in g/pot on limed, leached, and NPK-fertilized tailings as influenced by Mn, Zn, and Cu treatments in study 3.

	Tailing					
Treatment	c-70-1	c-70-2	c-70-3	c-70-4	c-70-5	Soil
None	2·65 b*	1·72 ab	2·52 ab	3·08 a	·89 a	7·73
Mn	1·73 a	1·67 ab	2·01 a	4·91 c	1·48 bc	
Zn	3·20 b	1·74 ab	3·27 b	4·31 b	1·75 c	
Mn+Zn	3·17 b	1·23 a	3·03 b	5·56 d	2·25 d	
Mn+Zn+Cu	3·11 b	2·19 b	2·64 ab	6·09 e	1·18 ab	

* Yields within each tailing followed by the same letter are not significantly different (P —0·05).

During the 24-day regrowth period there were some statistically significant responses to the addition of the minor elements (Table 5); however, the magnitude of the responses was not as great as for the first growth period.

Table 5. Regrowth fescue yields in g/pot on limed, leached, and NPK-fertilized tailings as influenced by Mn, Zn, and Cu treatments in study 3.

	Tailing					
Treatment	c-70-1	c-70-2	c-70-3	c-70-4	c-70-5	Soil
None	6·52 b*	4·95 a	6·00†	7·26 a	2·60 a	10·90
Mn	5·45 a	5·84 b	5·13	7·64 ab	3·01 ab	
Zn	7·34 bc	5·83 b	7·07	7·74 b	3·48 bc	
Mn+Zn	7·61 c	5·05 a	7·22	7·55 ab	3·75 c	
Mn+Zn+Cu	7·02 bc	6·56 b	6·84	8·33 c	2·96 ab	

* Yields within each tailing followed by the same letter are not significantly different ($P < 0{\cdot}05$).

† No significant F test for treatments on this tailing.

There was no consistent relationship between response to addition of the minor elements and the soil test values for Mn, Zn, and Cu on the limed tailings (Table 1).

The average fescue yield for all treatments on the leached tailings was 35% of the yield on soil for the first harvest and increased to 58% of the soil yield for the regrowth. This may indicate that as the plants matured they were less sensitive to a growth-inhibiting factor in the tailings. No differences in plant colour among treatments were noted during this study. However, it was noted that 14-day-old fescue seedlings within the same pot of tailings for any treatment were not as uniform in height as seedlings grown in soil.

Leaching the tailings after liming resulted in considerably greater fescue growth on four of the five tailings used in this study. Fescue yields averaged over all minor element treatments are shown in Table 6. Only on the tailing with the greatest fescue yield (c-70-4) was there no growth response to leaching after liming. It is of interest that this tailing contained more plant available Mn, Zn, and Cu (Table 1) than did the other tailings used in Study 3.

Table 6. Fescue yields in g/pot on limed and fertilized tailings in study 3 as influenced by leaching treatment.

Tailing	Treatment	
	No Leach	Leach
c-70-1	·53	2·65
c-70-2	·10	1·72
c-70-3	·09	2·52
c-70-4	3·01	3·08
c-70-5	·06	·89

Study 4—fescue yield on tailings as influenced by leaching sequence and lime source

In previous studies it was found that leaching after liming acid Mo mill tailings usually resulted in substantially greater plant growth than on unleached tailings. Was this due to a reduction in salt concentration in the tailings after leaching? Or was leaching after liming needed to reduce concentrations of a toxic ion or an ion imbalance resulting from the liming treatment? The ion imbalance investigated was Mg, which had been used in previous liming treatments in a 3:1 Ca:Mg equivalent ratio. The results indicate that total soluble salts probably were not a factor but that excess Mg introduced in the liming reagent probably was.

METHODS

Three lime sources used in this study were:
1. CaO plus $(MgCO_3)_4.Mg(OH)_2{\cdot}4H_2O$ in a 3:1 Ca:Mg equivalent ratio added as powdered reagent-grade chemicals.
2. Same reagents as above except in a 19:1 Ca:Mg ratio.
3. Lime that was a waste product from sugarbeet processing. This lime, also used in Study 1, had a Ca:Mg ratio of 30:1.

The amount of lime added to the tailings was equivalent to four times the base required to neutralize the tailings by the titration method (for method see Study 1).

The three leach-lime sequences were:
1. Lime—no leach. In this treatment lime was mixed with air-dry tailings, subjected to three wetting and air-drying cycles, fertilized, and seeded.
2. Leach—lime. In this treatment the tailings were first leached by placing 14 kg of tailings in a polyethylene bucket which had a diameter of 26 cm and a perforated bottom. Eight equal increments of deionized water were added to the tailings at 12-hour intervals until a total of 84 l of water had been added and leached through the tailings. After leaching, the tailings were air-dried, treated with lime, and subjected to three wetting and drying cycles before fertilizing and seeding.
3. Lime—leach. Tailings in this treatment were limed, subjected to three wet and dry cycles, leached as outlined in Treatment 2 above, dried, fertilized, and seeded.

The tailings used in this study were c-70-3 and c-70-5, both used in Study 1 and 3, and tailing c-71-6 (Table 1). After the lime and leach treatments all the tailings received the following fertility additions in solution: 50 ppm N as NH_4NO_3, 50 ppm P as $Ca(H_2PO_4)_2.H_2O$, 100 ppm K as KCl, 5 ppm Zn as $ZnSO_4.7H_2O$, and 10 ppm Mn as $MnCl_2.4H_2O$. An additional 50 ppm N as $(NH_4)_2SO_4$ and 50 ppm N as NH_4NO_3 were added during the course of the study. The soil received only the N, P, and Zn fertilizers.

Fescue was seeded on 26 January 1972 and harvested 50 days later. Greenhouse procedures as outlined in Study 2 were followed. Three replications were used on all treatments. Yields reported are the means from the three replications. The fescue yields were subjected to analysis of variance and F tests.

RESULTS AND DISCUSSION

Fescue yields were greater on tailings that were limed and then leached than on tailings that were limed and not leached, or that were leached and then limed (Table 7). A possible reason for these treatment differences might be a lower soluble-salt content in the lime-leach treatment. This was checked

Table 7. Fescue yields in g/pot on fertilized tailings as influenced by lime source and leaching treatment.

Treatment	Tailing			
	c-70-3	c-70-5	c-71-6	Soil
Lime source 1—no leach	·85	·46	·15	
Lime source 2—no leach	2·47	1·88	1·39	
Lime source 3—no leach	1·50	·62	·34	
Leach—lime source 1	·48	·32	·25	
Leach—lime source 2	1·29	·54	·52	
Leach—lime source 3	·70	·31	·21	
Lime source 1—leach	3·90	3·15	2·66	
Lime source 2—leach	3·32	3·13	3·59	
Lime source 3—leach	3·98	3·35	2·78	
No treatment				7·54

out by determining the electrical conductivity of extracts from the tailings receiving the various leach-lime sequences (Table 8). Since it was nearly impossible to make a conventional saturated paste from the coarse-textured tailings, a tailings to distilled water ratio was used where the amount of distilled water added was four times the amount of water needed to bring the tailings to field capacity.

Table 8. Electrical conductivity in mmhos/cm of extracts from tailings receiving different leaching treatments.

Treatment	Tailing		
	c-70-3	c-70-5	c-71-6
Lime source 1—no leach	1·77	2·54	3·13
Lime source 2—no leach	1·92	2·39	3·01
Lime source 3—no leach	1·85	2·27	2·67
Leach—lime source 1	1·10	1·28	1·83
Leach—lime source 2	1·31	1·38	2·16
Leach—lime source 3	·99	1·10	1·51
Lime source 1—leach	1·21	·79	1·41
Lime source 2—leach	·90	·69	1·23
Lime source 3—leach	1·09	·76	1·25

The tailings extract conductivity values were evaluated in terms of standards established for soil saturation extracts (U.S. Salinity Laboratory Staff 1954) by doubling the conductivity values for the tailings extracts. (This is because in medium-textured soils a saturation paste is made by adding about twice the water needed to bring the soil to field capacity.) By these standards, soluble salts in the lime—no leach treatment may be restricting plant growth. However, when the differences in conductivity are compared to the differences in fescue yield (Table 7) it is apparent that the factor restricting plant growth is probably not soluble salts *per se*, as yields tend to be greater on the lime-no leach treatment than on the leach-lime treatment despite the higher soluble salt content in the lime—no leach treatment.

The overall yield on each tailing was significantly influenced by lime source (Table 7). However, within the high yielding lime-leach treatment no one lime source consistently produced greater yields than other lime sources. Lime source 2 (19:1 Ca:Mg ratio) produced greater yields than lime source 1 (4:1 Ca:Mg ratio) on the lime-no leach treatment. Thus, it appears that excess Mg is present in unleached treatments where the liming reagent contained a 4:1 equivalent ratio of Ca:Mg. However, this does not account for the generally poor growth on the leach-lime treatments for the three lime sources. Additional study is needed to determine if these effects are due to ion imbalances or toxicities inherent in the tailings or possibly due to artifacts created by the liming reagents.

Field studies on Mo tailings

Limited field studies have been and are being conducted on plant establishment on the Mo tailings at Climax, Colorado. These studies will not be reported here except to note that growth of perennial grass species has been poor to fair when compared to growth of nearby seedings on soil. Cereal rye (*Secale cereale*) has made fair to good growth on the tailings and might be considered in temporary stabilization programmes or as a nurse crop for perennial species. Long-term N fertilization will also be needed if plants are grown directly on the tailings as no adapted N-fixing species are available for seeding at this elevation.

Thus, if vegetative stabilization of the Mo tailings is attempted on a large scale, careful management will be required in view of the apparently continuing oxidation of sulphides in the limed tailings, the need for continuing N fertilization, and the short growing season at this elevation.

Acknowledgments

This co-operative research was funded by Colorado State University Experiment Station, Colorado State Forest Service, Colorado Industrial Mining Development Board and American Metals Climax.

Summary

Coarse-textured acid molybdenum mill tailings were limed and then incubated in a glasshouse for 900 days. The pH of the limed tailings decreased over this time. Lime at the rate of four to six times the base equivalent required to neutralize the tailings was needed to maintain the tailings at a pH suitable for plant growth over the 900-day period.

Fertilization with N, P and K was a basic requirement for plant growth on the tailings in glasshouse studies. A plant growth response to Mn, Zn, and Cu was also found on some of the tailings.

Leaching after liming usually resulted in substantial plant yield increases over liming without leaching. A portion of this yield reduction on unleached tailings may be due to excess Mg introduced by the liming treatment, however, a portion of the beneficial effect of leaching after liming remains unexplained.

In field tests at the 3,400 m elevation tailings disposal site, rye (*Secale cereale*) produced fair cover on limed and fertilized tailings. Perennial grasses did not grow as well as rye.

References

BEVERLY R.G. (1968) Unique disposal methods are required for uranium mill waste. Mining Engineering June 1968, 52–6.

BLACK C.A. (1965) *Methods of Soil Analysis.* American Society of Agronomy, Madison, Wisconsin.

COLEMAN N.T., RAGLAND J.L. & CRAIG D. (1960) An unexpected reaction between Al-clay on Al-soil and $CaCl_2$. *Soil Sci. Soc. Amer. Proc.* **24**, 419–20.

DEAN, K.C., HAVENS R. & HARPER K.T. (1969) Chemical and vegetation stabilization of a Nevada copper porphyry mill tailing. U.S. Dept. of Int., Bur. of Mines Report of Investigations, 7261.

FOLLETT R.H. & LINDSAY W.L. (1971) Changes in DTPA-extractable zinc, iron, manganese, and copper in soils following fertilization. *Soil. Sci. Soc. Amer. Proc.* **35**, 600–2.

GRUBB H.F. (1965) The feasibility of vegetating mine tailings at Climax, Colorado. M.S. thesis, Colorado State University.

JAMES A.L. (1966) Stabilizing mine dumps with vegetation. *Endeavor* **25**, 154–7.

LeClerq E.L., Leonard W.H. & Clark A.G. (1962) *Field Plot Technique*. Burgess, Minneapolis.

Shetron, S.G. & Duffek R. (1970) Establishing vegetation on iron mine tailings. *J. of Soil & Wat. Cons.* **25**, 227–30.

U.S. A.E.C. (1963) A report of the Monticello mill tailing erosion control project, Monticello, Utah. U.S. A.E.C., R.M.O. 3005.

U.S. Salinity Lab. Staff. (1954) *Diagnosis and Improvement of Saline and Alkali Soils*. USDA, Handbook 60.

Vanderwilt J.W. (1947) *Mineral Resources of Colorado*. State of Colorado Mineral Resources Board.

Wickmann A.P. (1960) Minerals processing. In *Mineral Resources of Colorado* (Ed. S.M. Del Rio), State of Colorado Mineral Resources Board.

Young C.A. (1969) The use of vegetation to stabilize mine tailings areas at Copper Cliff. Canad. Mining J. June 1969, 43–6.

Reclamation of uranium-mined areas in the United States

M. MAY *University of Wyoming, Laramie, U.S.A.*

Introduction

At present, over 18 million acres ($7{\cdot}3 \times 10^6$ ha) in the United States are under claim, or lease, for uranium exploration. As a much needed energy source, uranium mining and exploration is receiving unusual interest. Although most of the activity related to uranium is presently confined to further exploration, there are 19 active processing mills supplying 'yellow cake' or uranium oxide (U_3O_8) to over 100 power generating plants within the United States.

The occurrence of uranium ore is almost entirely restricted to sedimentary deposits, about 99·16% of known uranium deposits in the United States are found in sedimentary strata in the western part of the country. As related to geologic time, 53% of the uranium deposits are found in Jurassic deposits and 44% in deposits of Tertiary age, primarily early Tertiary.

Geographical occurrence of uranium-bearing deposits is related to three major areas: the Colorado Plateau, sometimes referred to as the 'Uranium Province', contains approximately 65% of the known uranium deposits and includes parts of the states of New Mexico, Colorado, Utah, and a small portion of Arizona; the Wyoming Tertiary Basins account for about 28% of the uranium deposits and about 4% of the deposits are located in the Texas Gulf Coast area. Although each area of occurrence has its own particular characteristics, the Wyoming Tertiary Basins typify the problems and conditions related to reclamation and will be used as a representative reference throughout this paper.

Contribution S.R.520 of the Wyoming Agricultural Experiment Station, Laramie, Wyoming.

The deposits and methods of mining

To obtain a clearer picture of the problems associated with reclamation of uranium-mined areas, we must first look at the general characteristics of the uranium deposits and methods of mining used in uranium ore removal. There is still debate about the original source of uranium, but being soluble in water, it is assumed that uranium deposits are the result of leaching of granitic or other materials. Uranium has been moved by moving ground water and deposited at geochemical fronts in thin lenses to 'rolls' 15 m or more in depth. These rolls may be found at various depths from surface outcroppings to great depths. Average ore purity in terms of uranium oxide presently being mined varies from 0·15 to 0·25 of 1% and occasionally as high as 0·30% (Beamer *pers. comm.*). Uranium oxide or 'yellow cake' production at these purity levels vary from 1·3 to 2·2 kg/Mt of uranium-bearing ore. Being an extremely active material, uranium combines or is associated with many different compounds—the most common of which are sulphates, phosphates, arsenical compounds, and carbonates. In addition, uranium commonly occurs with selenium and copper. In the past, geobotanists have used botanical prospecting and indicator plant species for uranium and associated elements or compounds in determining surface deposits of uranium ore, which may give us our first clue in plant species selection needed for later revegetation of the areas (Kleinhampl 1962).

About 55% of present uranium production in the United States is from underground mining methods and 45% from open pit or strip mining activities. Many of the underground mines are, however, the result of following the thin deposits and small 'rolls' after open pit mining of the major deposits has been completed. At present (1973), the market price of uranium oxide is depressed at approximately $6.50 per pound as compared to over $9.00 per pound a few years ago, restricting the stripping ratio to 10 cubic yards of overburden removed to obtain one pound of uranium oxide (approximately 16·9 m^3 of overburden per kg of uranium).

Open pit mining in the sedimentary strata usually is conducted by conventional methods and with conventional equipment to remove overburden from the uranium deposits. Most work is accomplished by means of scrapers, bulldozers, and haul trucks and the use of any explosives is rarely required. Once the overburden has been removed to the strata immediately above the uranium deposit, core samples are taken at 60 to 95 cm intervals and the samples analysed immediately for uranium content. The uranium ore is then removed and taken to the processing mill to be dumped into specific bins depending on the quality of the ore. Analyses of the core samples and subsequent quality of ore carried on each haul truck are radioed to the mill prior to the ore arrival so the ore can be separated into the various grades. The ore is then blended, crushed into a fine texture material and the uranium oxide removed by leaching processes, normally employing an acid leaching procedure.

Needless to say, the entire mining and processing activities are monitored constantly for various types of radioactivity. In Wyoming, written reports must be filed with 17 different state and federal agencies on possible radioactive pollution of air, water, overburden, and food, or accumulative toxicity to humans or native fauna at each mine location.

Site characteristics

The sedimentary deposits where uranium is found are almost always at lower elevations or basins in the western U.S. Mining activities are normally in areas which receive from 18 to 25 cm of precipitation annually and high winds are common throughout the year. In addition, these arid to semi-arid areas are subjected to high evaporation and sublimation losses leaving perhaps one half of the precipitation received available for plant growth. Frost-free periods vary from year long in the southern areas to about 70 days growing season in the northern uraniferous basins. Top soils in the mining areas are usually poorly developed, shallow in depth, low in organic mattter, and commonly high in sodium and total soluble salts with pH readings of 7·5 to 9·0. The arid character of the areas of mining activity is one of the most important factors in revegetation of the mine areas.

Characteristics of overburden material, both physical and chemical, vary significantly depending on the geological strata and associated compounds. In most cases, the surface or top soil is neutral to basic but as the overburden layers removed approach the mineral deposit, the overburden becomes quite acid. The lower most layers of overburden material, which usually are deposited on the tops of the spoil piles, may reach an acid pH of 1·9 (paste analysis) and are commonly near a pH of 2·6. The low pH recordings normally associated with uranium bodies usually are attributed to the iron sulphates and sulphites found with uranium ore although much discussion and speculation has been devoted to the extreme acid condition of some overburden piles. Nitrogen content of the overburden as determined by percentage of organic matter is, of course, low and phosphorus content is also low but normally adequate to support plant growth. The sulphur compounds attributing to the acid condition leach readily from the surface of the overburden piles although unvegetated overburden piles exposed for four years have been noted to retain a pH as low as 3·8. With normal climatic conditions, the spoil piles reach a pH of 5·0 to 6·0 when exposed to leaching for a three- to eight-year-period. A limited amount of 'volunteer' seedlings of plant species from contiguous undisturbed rangeland occurs when the pH reaches 4·3 or above.

Revegetation studies

Early attempts to grow plants on uranium ore were made by researchers of several countries in the late 1800s and early 1900s. These early plant studies were initiated as attempts to induce genetic mutations of several grass and forb species by growing them on the radioactive material. Although these initial trials did not consider reclamation as a goal, the results are useful in showing some of the problems that exist. In all cases reviewed, the uranium ore used was extremely acid. When covered with a thin layer of soil, several grass species grew quite well—their roots penetrating into the uranium ore layer. Forbs such as alfalfa, however, usually exhibited restriction of root penetration, the root growth stopping when it reached the uranium layer with root growth often forming a 'club' deformed condition when the radioactive (and low pH) layer was contacted.

Reclamation of uranium-mined areas must consider and adjust to four basic problems: (1) climatic limitations to establishing vegetation on the overburden piles in the arid and semi-arid basins; (2) acid condition of the overburden material; (3) sloping and shaping of the overburden to obtain optimum topographic features; (4) obtaining ecologically adapted plant species that will provide a stable and productive vegetative cover; and, (5) soiling techniques.

CLIMATIC LIMITATIONS

Climatic factors prevailing in the uraniferous basins limit revegetation attempts, directly and indirectly, by limiting moisture available for plant establishment and growth. Surface treatments that enhance water accumulation, infiltration, moisture retention and detention, or reduce moisture loss, show measurable effects on overburden revegetation programmes. Commonly known range improvement practices such as mulching with straw or water soluble asphalt materials, range pitting, jute or excelsior netting may be employed with a high degree of success. Occasionally sprinkler irrigation may be used to obtain an initial establishment of vegetative cover.

In the northern latitudes, two climatic factors are prevalent that may be used in revegetation programmes. These factors, snow and wind, when combined with moveable snow fences can increase water availability in controlled amounts on almost any given area. Standard four foot high snow fences when strategically located and spaced at statistically determined intervals (depending upon sublimation and snowfall rates) can accumulate up to 2·5 cubic metres of water per linear foot of snow fence. The amount of water gained per unit area can be predicted with a high level of confidence. In addition to the moisture gained for vegetative growth, snow fences can also serve to increase acid leaching from the surface layer of the overburden piles as well as reduce surface evaporation during the summer months.

CONDITION OF OVERBURDEN MATERIAL

Reduction of acidity of the overburden material is accomplished in several ways. The addition of lime (calcium carbonate) is used to increase soil pH in some areas. In most uranium mining areas, the amount of lime required and the expenses involved in hauling lime for considerable distances restricts its use. Covering the overburden piles with topsoil, stockpiled during initial stripping activities and required by law seems to be the most practical approach to obtaining a surface pH suitable for plant growth. To use sewage sludge or similar waste materials as a surface treatment is impractical due to low population densities. For example, the state of Wyoming contains 97,914 sq. miles with a population of 300,000 compared to the United Kingdom of approximately the same area (94,214 sq. miles) and over 180 times as many people.

SLOPING AND SHAPING

Initial uranium mining resulted in overburden piles with rough steep slopes, too steep to be prepared or seeded with standard farming machinery. More recently, laws have required that overburden piles be shaped to conform with the surrounding flat to undulating topography making them suitable for top soiling and other surface treatments that require equipment. Shaping of the overburden piles have reduced erosion and increased soil moisture recharge into the piles resulting in more moisture available for plant growth.

SPECIES SELECTION

The approaches available in obtaining ecologically adapted plant species for reclamation use is somewhat unique in uranium-mined areas. First, any species used must be able to withstand the arid climatic conditions of the uraniferous basins. Secondly, species must be adapted to the edaphic factors of low pH values, occasionally high pH values from the topsoil or salt-bearing sedimentary deposits and to the effects, if any, resulting from radioactive material. Thirdly, the plants must fit the reclamation objectives of cover, productivity, aesthetic values, etc.

Ecotypes for spoil revegetation may be selected from areas of similar climatic conditions by measuring surface radiation to correspond with radiation readings from the overburden material to be reseeded. In a similar approach, plants from similar climatic zones may be sampled for tissue analysis of uranium and selected when uranium quantities in the plant tissue approach the amount absorbed by plants on the overburden material. Similar selections can be made by selecting plants that are established and apparently stable on areas bearing minerals that are commonly associated with uranium deposits. The best example of the latter is probably selenium indicator plants as selenium has a high frequency of occurrence with uranium ore deposits. A simple and very effective approach is to select plants that have actually encroached onto the overburden piles from the surrounding plant communities and maintain these selections for seed increase and eventually a total revegetation programme.

Plant species that have commonly invaded uranium overburden piles include the following:

Grasses

Agropyron smithii (western wheatgrass)
Agropyron spicatum (bluebunch wheatgrass)
Hordeum jubatum (foxtail barley)
Koeleria cristata (prairie junegrass)
Oryzopsis hymenoides (Indian ricegrass)
Poa secunda (Sandberg bluegrass)
Sitanion hystrix (bottlebrush squirreltail)
Stipa comata (needleandthread)
Stipa viridula (green needlegrass)

Forbs

Astragalus spp. (milkvetch)
Cryptantha affinis (common cryptantha)
Grindelia squarrosa (curlycup gumweed)
Salsola kali (common Russian thistle)

Woody or semi-woody plants

Artemisia nova (black sagebrush)
Artemisia tridentata (big sagebrush)
Atriplex nuttalli (Nuttall saltbush)
Chrysothamnus viscidiflorus (Douglas rabbitbrush)
Sarcobatus vermiculatus (black greasewood).

Many of the above species have established on salty spots occurring in the otherwise acid profile of the overburden deposits. Most of the forbs and shrubs have invaded the neutral to basic material that occurs, but many of the grasses as well as the milkvetches (*Astragalus* spp.) have established on acid overburden material, or on both acid and salt spots. For example, foxtail barley (*Hordeum jubatum*) has a high tolerance to pH ranges and has successfully established on overburden with a pH value of 4·3 to 9·0.

These species and their selected ecotypes should be given strong consideration in any reclamation attempt.

SOILING

Perhaps an easier solution to obtaining ecologically stable plant species for uranium overburden revegetation lies in the soil rather than the plants on the surface. Initial soil samples from the uraniferous basins have shown that over 600 viable seeds may be found in the surface 25 mm of soil on 9·5 dm^2 area and over 800 viable seeds if the 25–50 mm depth is included for the same area. These determinations were made by scattering a thin layer of the soil samples over sterile sand and germinating the included seed in greenhouse studies. As might be expected, the amount of viable seed contained in the surface soil varies considerably between the major vegetative types, but the resulting 'native' species are certainly adapted to the existing ecological parameters when topsoil is used in reclaiming overburden piles. By careful selection of topsoil to cover the overburden piles, excluding those with high salt concentrations and selecting others with high potentials for viable seed, it is possible that reseeding the uranium-mined areas with topsoil is a reality without further treatment.

Conclusions

In conclusion, it should be noted that despite all existing conditions, adverse or detrimental to plant establishment and growth in the uranium-mined areas, successful revegetation can be accomplished. Selection of plant ecotypes that have invaded overburden piles and the use of these ecotypes as a seed source for revegetation shows much promise. The concept of selecting topsoil to be used and 'seeding with topsoil' can be applied to many reclamation areas, and the use of treatments such as snow fencing or other moisture retaining methods can be useful tools in uranium spoil reclamation and similar reclamation programmes.

Acknowledgments

Much of the information reported here is contained in unpublished reports to the Wyoming Uranium Producers to whom grateful acknowledgment is made.

Summary

An account is presented of the areas of known uranium deposits and present mining activities together with the associated reclamation problems. Initial reclamation attempts are described and the search for ecologically stable plant species for revegetation outlined. The possible environmental benefits from uranium strip-mined areas are discussed.

Reference

KLEINHAMPL F.J. (1962) Botanical Prospecting for Uranium on South Elk Ridge, San Juan County, Utah. *S.S. Geol. Surv. Bull.* 1085-D.

Physiological mechanisms of heavy metal tolerance in plants

S. J. WAINWRIGHT and H. W. WOOLHOUSE
University College of Swansea, Swansea, Wales, U.K. and University of Leeds, Leeds, U.K.

Introduction

Bradshaw (1952) demonstrated that populations of *Agrostis tenuis* from lead- and zinc-contaminated soils were tolerant to high concentrations of these metals; plants from normal pasture soils were not able to survive on metal-contaminated soils.

Since that time many studies have been made on metal tolerance and species which have evolved tolerant strains to one or more metals have been found in all the major groups of the plant kingdom. A review of the ecological, genetical and physiological aspects of metal tolerance in plants has been published by Antonovics *et al.* (1971).

Any discussion of the mechanisms of heavy metal tolerance in plants must take into account three fundamental observations. Firstly, metal tolerance is specific. If an organism is tolerant to a given metal, it is not automatically, or indeed usually, tolerant to another metal. There are occasional exceptions to this; for example a population of *Agrostis tenuis* collected from an area contaminated with zinc showed a range of zinc tolerances which were linearly related to its range of nickel tolerances (Gregory & Bradshaw 1965), (Fig. 1). Secondly, metal tolerance is not an all-or-none phenomenon; different degrees of tolerance can be recognized within a population of a given species (Jowett 1958; Gregory & Bradshaw 1965). It is generally found that there is a positive correlation between the amount of metal in the soil and the degree of tolerance to the metal, of plants growing on the soil (Gregory & Bradshaw 1965). Thirdly, metal tolerance is inherited (Jowett 1959; Wilkins 1960; Bröker 1963; Antonovics 1966; Allen 1971; Urquart 1971).

In this paper we shall consider what are the physiological mechanisms which enable some plants to tolerate high levels of heavy metals?

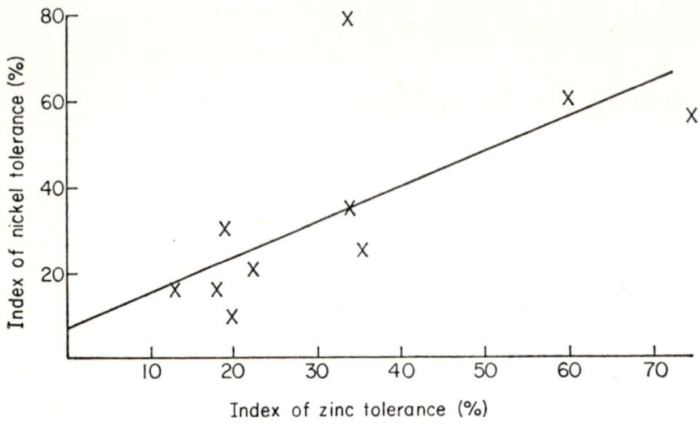

Figure 1. The index of tolerance to nickel of several populations of *Agrostis tenuis* plotted against their index of tolerance to zinc. Fitted regression: $y = 7{\cdot}4 + 0{\cdot}815x$. (From Gregory & Bradshaw 1965.)

Tolerance mechanisms in micro-organisms

Some copper-tolerant varieties of yeast have been found to produce and release large amounts of H_2S, resulting in the precipitation of copper sulphide in and around their cell-walls. This would seem a feasible mechanism for inactivating the copper, but there is little correlation between the degree of copper tolerance and the amounts of H_2S produced and mutants have been found which have high copper tolerance and yet only produce small quantities of H_2S (Ashida 1965). Moreover, in the production of a simple molecule such as H_2S, there can be no basis for the observed copper specificity. A number of other metals could be precipitated as insoluble sulphides.

Mercury resistant strains of bacteria complex the metal to form a volatile compound which is then lost from the medium (Tonomura *et al.* 1968a, 1968b, 1968c). In the case of arsenite resistance in *Pseudomonas pseudomallei*, there is a decreased permeability of the cells to arsenite (Beppu & Arima 1964). Similarly cadmium tolerance in *Staphylococcus aureus* is associated with a decreased uptake of the metal (Chopra 1970), (Fig. 2).

Tolerance mechanisms in angiosperms

THE CELL WALL HYPOTHESIS

Copper and zinc-tolerant strains of *Agrostis tenuis* take up more copper or zinc respectively into the roots than do non-tolerant plants. In this case the metals are accumulated predominantly by the cell walls as revealed by fractionation and analysis of root homogenates; the greater the degree of

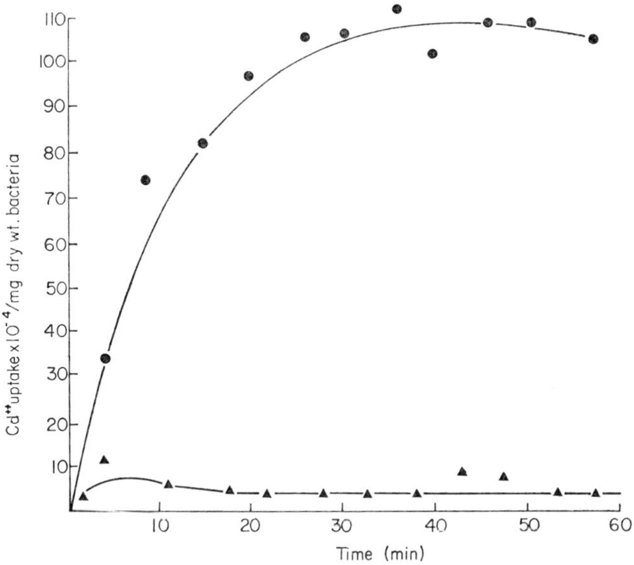

Figure 2. Uptake of Cd^{++} ions, on a dry weight basis, by cadmium-sensitive (●) and by cadmium-resistant (▲) staphylococci. Initial specific activity of $CdCl_2$, 2·5 μCi per μ Mole. $CdCl_2$ concentration, $10^{-4}M$. All values are corrected for non-specific binding of tracer by the growth medium. (From Chopra 1971.)

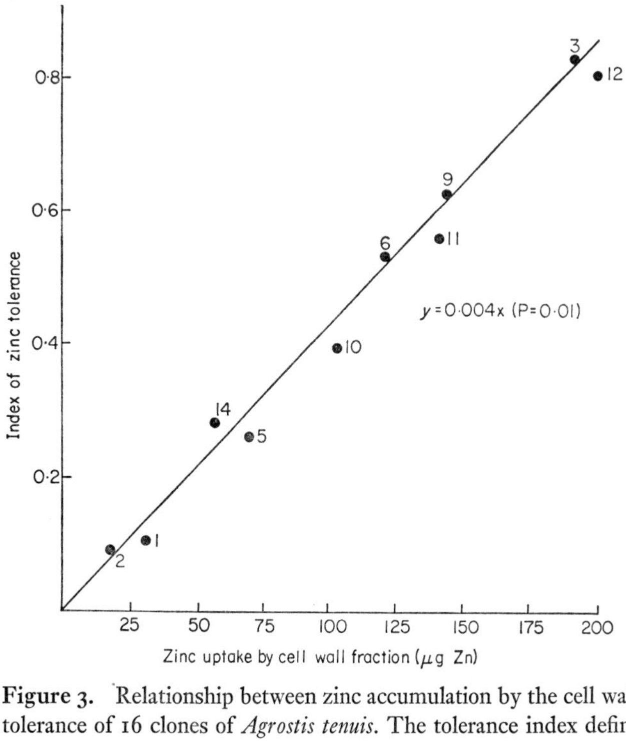

Figure 3. Relationship between zinc accumulation by the cell wall and the index of zinc tolerance of 16 clones of *Agrostis tenuis*. The tolerance index defined as: Mean length of longest root in Zn solution divided by mean length of longest root in distilled water. The numbers against points on the graph are reference numbers to the clones. (From Turner & Marshall 1972.)

metal tolerance, the greater is the metal binding capacity of the cell-wall fraction (Turner & Marshall 1971, 1972) (Fig. 3). The binding of zinc by these cell-wall fractions is independent of temperature, increases with increasing concentrations of the metal and is virtually complete after 5 minutes (Fig. 4).

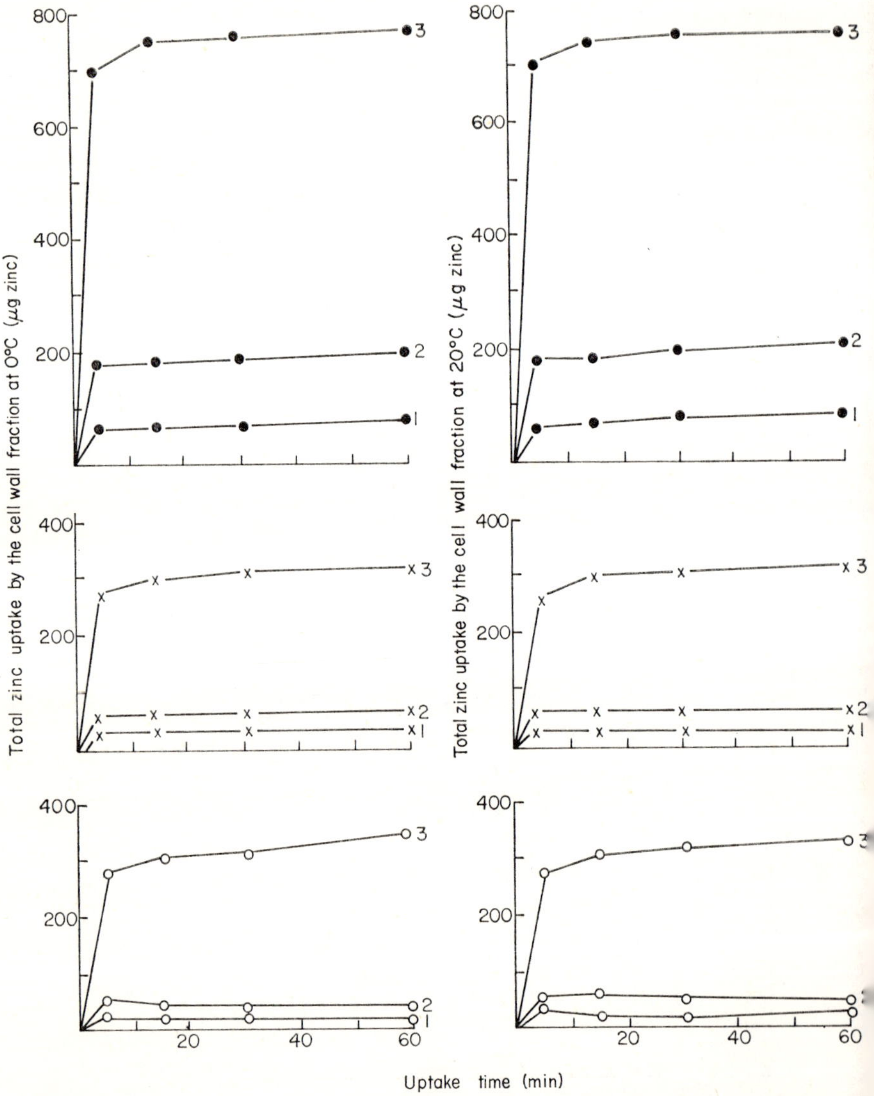

Figure 4. Zinc uptake by the cell wall fraction of roots of *Agrostis tenuis* as a function of time, temperature and zinc concentration. (●) zinc-tolerant clone; (x) non-tolerant clone; (○) copper-tolerant clone. 1, 10μg zinc per ml; 2, 25 μg zinc per ml; 3, 100 μg zinc per ml. (From Turner & Marshall 1971.)

There is uncertainty concerning the molecular structures in these cell-walls which can confer the observed specificity for the binding of metal ions. Turner & Marshall (1972) showed that there is no correlation between the nitrogen content of the cell wall and the capacity to bind zinc (Fig. 5); moreover the zinc, once bound, could not be released from the walls by protease treatment. From these observations it was concluded that cell-wall proteins could not be involved in the binding process. However, a number

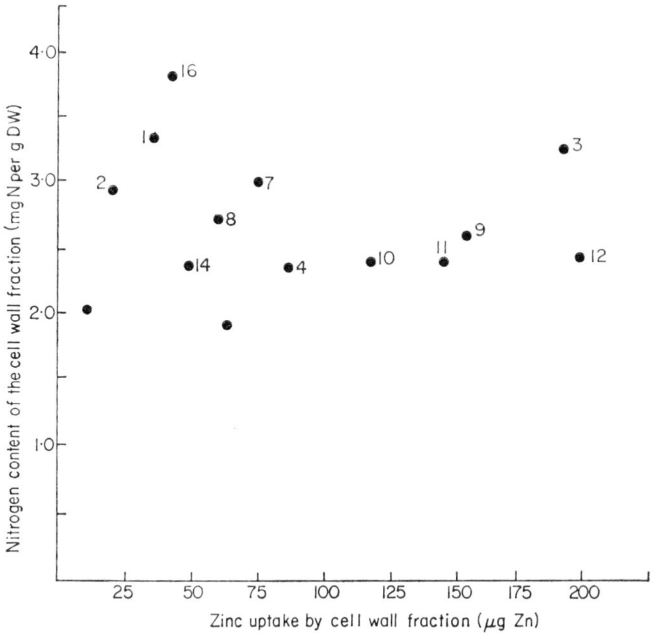

Figure 5. Relationship between zinc accumulation and the nitrogen content of the cell wall fraction of roots of *Agrostis tenuis*. The numbers against points on the graph are reference numbers to different clones. (From Turner & Marshall 1972.)

of objections can be raised against this interpretation. Firstly, if one assumes that all of the nitrogen in the cell walls is present as amino acids, then it can be calculated that there is a molar ratio of 35 amino acid residues for every zinc ion bound and 120 amino acid residues for every copper ion bound when the cell walls bind metal to the levels reported by Turner (1970). It is clear therefore, that there is more than enough nitrogen present in the cell walls to account for all of the observed metal binding. Moreover, if one was going to invoke components of the cell walls containing nitrogenous compounds in the binding process, there would be no *a priori* reasons for expecting a correlation between the extent of metal binding and the total nitrogen content of the cell walls.

Failure to release zinc by digestion with proteolytic enzymes does not prove that proteins are not involved in binding of the metals. Woolhouse (1970) has pointed out that the binding of proteins to metals or to other inert matrices such as cellulose frequently renders them inaccessible to hydrolytic attack. Lamport (1965) used 10 different proteolytic enzymes in an attempt to digest cell wall proteins. In no case was he able to digest more than 40% of the wall protein, and in most cases considerably less. Wyn Jones *et al.* (1971) report that trypsin does release significant quantities of zinc from cell-walls of *A. tenuis*. They found however, that 66% of the bound zinc was released by cellulase digestion, as a complex which could be partially purified by Sephadex G25 filtration (Fig. 6). The complex contained both amino acid and sugar residues. These observations led the authors to the hypothesis that in metal tolerant plants the metal is bound as a stable complex in the walls of the root cells and is thus prevented from reaching the metabolic sites where it could do damage.

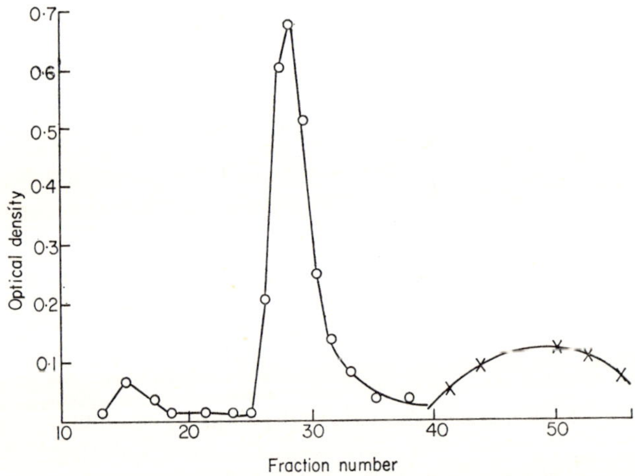

Figure 6. Sephadex G25 fractionation of cellulase hydrolysates of cell walls of roots of *Agrostis tenuis*. The major peak (●) shows a fraction containing an organo-zinc complex; the minor peak (x) shows free Zn^{++} ions. (From Wyn Jones *et al.* 1971.)

Although one cannot dispute the presence of this metal-binding phenomenon, there is evidence which shows that this binding does not constitute the whole of the tolerance mechanism. If *A. tenuis* is grown in water culture in the presence of 10^{-4} M zinc, after 9 days zinc-susceptible plants have lost 50% of their chlorophyll whereas tolerant plants have not suffered any loss of chlorophyll. On analysis it is found that both susceptible and tolerant plants have accumulated similar quantities of zinc in the leaves (Fig. 7). *Becium homblei* is found to contain more copper in the leaves when it is grown on soils containing increased amounts of copper (Reilly 1969) (Fig. 8).

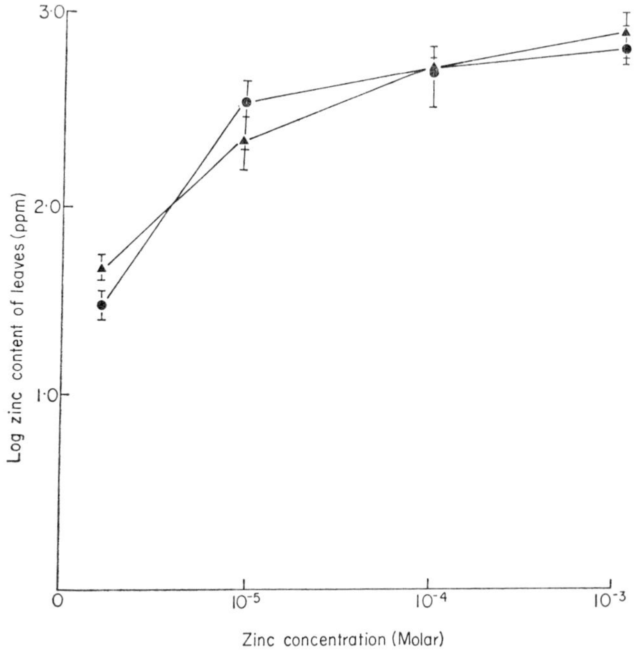

Figure 7. Zinc content of leaves of *Agrostis tenuis* grown in culture solutions containing a range of concentrations of zinc. ▲-zinc-tolerant strain; ●-zinc-susceptible strain. (Note: Fe was supplied as ferric nitrate, there was no chelating agent in the culture solution.)

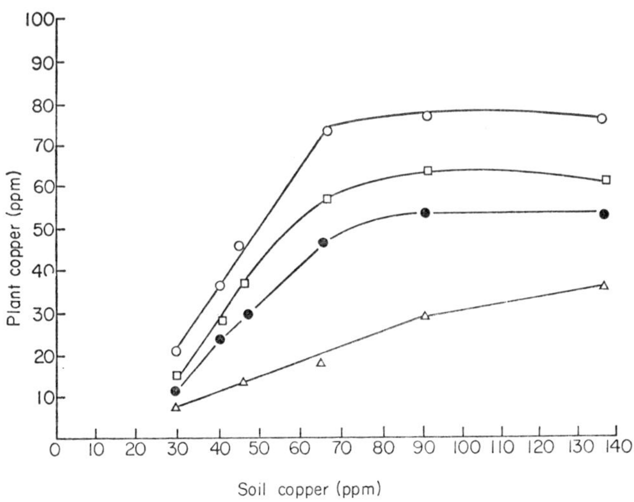

Figure 8. Relationship of leaf to soil copper at different stages of the growing season of *Becuim homblei*. Samples collected from marked plants in the months indicated were dry-ashed and analysed directly for Cu. Soil samples were taken from around the roots of the plants and Cu extracted with 1·22 M sodium acetate buffer at pH 6·0. Samples collected in (○) December, (□) January, (●) March and (△) May. (From Reilly 1969.)

In other species the relationships between the metal content of the leaves and amounts of that metal in the soil are not always linear and they differ for different metals (Nicholls *et al.* 1965). However, when studies such as these are made it must be remembered that they do not constitute controlled uptake experiments. It is known that the degree of metal tolerance of plants found growing on metal-contaminated soils depends on the concentration of metal in the soil. It follows that in many of the publications on this subject which give graphs showing the amount of metal in the leaves plotted against the amount of metal in the soil, each point on the graph is derived from a plant of different tolerance. In no case has this difference in tolerance been measured and then taken into account in *in situ* uptake studies.

The burden of all these studies is that even though the cell-walls of metal-tolerant strains of many species do bind appreciable quantities of metal, this binding does not significantly prevent the metal from being taken up by the roots, translocated and accumulated in the aerial parts. Moreover, accumulation by the cell walls is not a universal feature of metal tolerance. Rathore *et al.* (1972) studied a zinc-tolerant cultivar of *Phaseolus vulgaris*; there was no enhanced accumulation of zinc in the cell walls and the cytoplasmic fraction contained the greatest quantities of zinc.

METABOLIC ADAPTATIONS

Figure 9 shows a remarkable correlation between the copper content and total nitrogen content of the leaves of *Becium homblei* (Reilly 1969). It is not known whether this relationship has any physiological significance, but we may hazard the speculation that some of the nitrogen-containing material may be associated with a mechanism for the chelation of the copper. Further evidence that a part of the tolerance mechanism may reside in the leaves as well as in the roots comes from observations on excised leaves of *A. tenuis*. Leaf segments were floated on solutions containing zinc or copper ions in the dark and the rate of loss of chlorophyll was measured (Figs 10, 11 & 12). It is seen that after 4 days there was a greater loss of chlorophyll in leaf segments of susceptible ecotypes but the loss was not so marked in tolerant ecotypes.

There is a positive linear relationship between the degree of zinc tolerance in *A. tenuis* and the capacity of mitochondrial preparations to bind zinc (Turner & Marshall 1972). Oxygen uptake by mitochondria extracted from zinc tolerant *A. tenuis* was not inhibited by zinc to the same extent as it was in mitochondria prepared from susceptible ecotypes (Wyn-Jones *et al.* 1971).

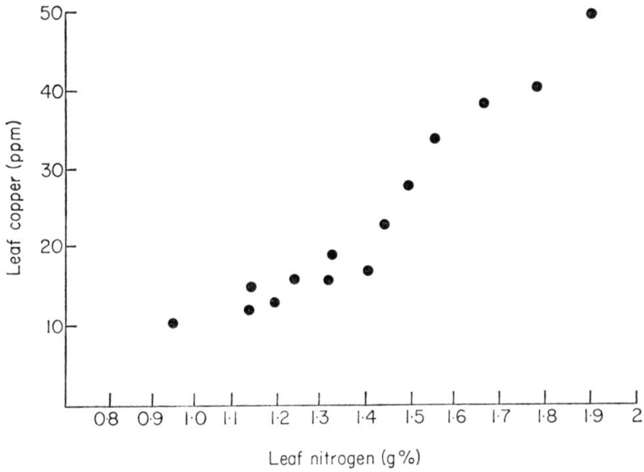

Figure 9. Relationship between nitrogen and copper in the leaf tissue of *Becium homblei*. Dried samples of leaf were analysed for Cu by dry ashing. The ash was dissolved in 1N HCl and analysed with the atomic absorption spectrophotometer for Cu. Nitrogen was estimated by Kjeldahl digestion and microdiffusion. The correlation coefficient for the two variables is 0·84. (From Reilly 1969.)

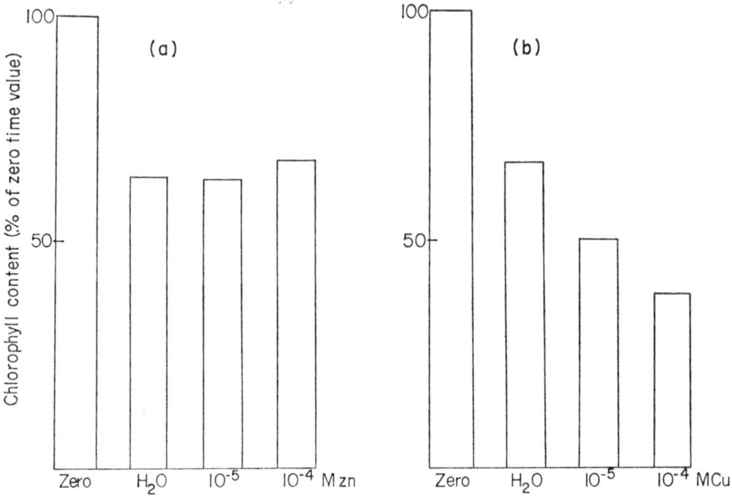

Figure 10. The effect of Zn^{++} and Cu^{++} ions on the chlorophyll content of leaf segments of a zinc-tolerant race of *Agrostis tenuis*. The left hand column marked zero shows the chlorophyll content of the segments at the time of excision of the segments, set as 100%. The values for the water controls and the range of metal concentrations represent the chlorophyll content of the segments after incubation for 4 days in darkness at 25°C under these treatments, expressed as percentages of the zero time value.

Chlorosis is a commonly observed symptom of heavy metal toxicity; it is probably a secondary characteristic because there is usually a lag of several days between the onset of toxic conditions and the chlorosis. The inhibition of root growth on the other hand is one of the most rapid responses to toxic concentrations of a heavy metal which has been observed and indeed it is this which forms the basis of many of the tests of tolerance.

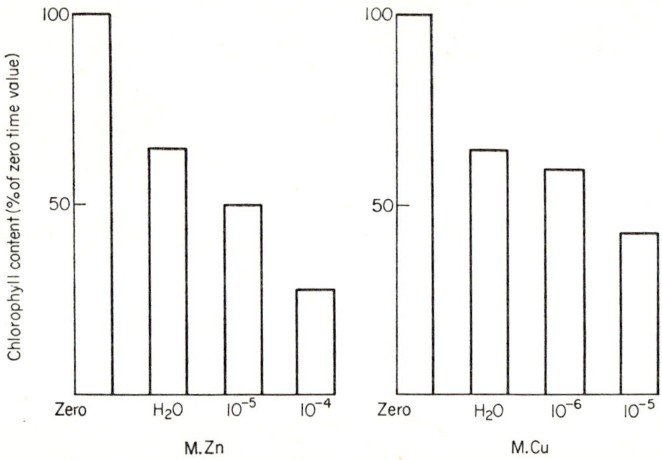

Figure 11. The effect of Zn^{++} and Cu^{++} ions on the chlorophyll content of leaf segments of a copper-tolerant race of *Agrostis tenuis*. Experimental details as in legend to Fig. 10.

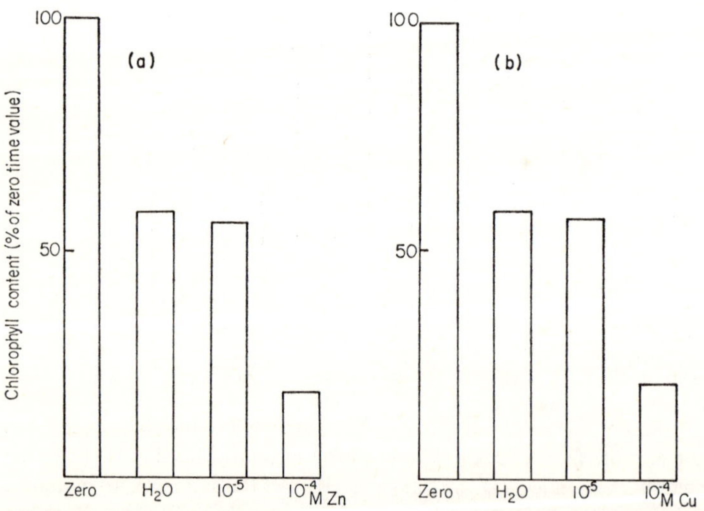

Figure 12. The effect of Zn^{++} and Cu^{++} ions on the chlorophyll content of leaf segments of a non-tolerant race of *Agrostis tenuis*. Experimental details as in legend to Fig. 10.

Metal tolerance and cell elongation

Cell division in onion roots is inhibited by aluminium ions (Clarkson 1969) and this metal is also found to inhibit the incorporation of carbohydrate residues into the cell walls of developing roots. This latter observation is consistent with our finding that there is a rapid inhibition of elongation of root cells following applications of heavy metal ions. The meristem of roots of *A. tenuis* is situated in the first 1 mm of the root tip. Almost all of the cell elongation takes place within the second 1 mm segment behind the tip. Segments were cut from this elongating zone and allowed to elongate in a 2% sucrose solution on a shaking incubator at 25°C for 24 hours in the dark. In a series of treatments a range of concentrations of Zn^{++} and Cu^{++} were added and the volume of medium was controlled so that uptake of metal by the segments had no significant effect on the concentration of metal in the medium. It was found that cell elongation was more severely inhibited in susceptible than in metal-tolerant ecotypes (Figs 13 & 14). In these experiments the volume of incubation medium was such that no significant lowering of toxic metal concentration could be detected so that the observed differences in elongation could not be due to binding of the metal ions by the cell walls.

Metal tolerance and cell permeability

The variety and complexity of the observed relationships between tolerance, toxicity and metal accumulation, noted earlier, suggests that there may be a number of different physiological mechanisms responsible for the specificity of the growth responses. This leads to some degree of conceptual difficulty with mechanisms for selective tolerance if one takes the view that at the cell surface there are likely to be a number of 'exposed' metabolic sites such as membrane-bound ion pumps, which might be affected by all metals. All adaptations to heavy metals may therefore have of necessity to include some method of protection of the membranes and other exposed sites on the surface of the root cells.

It is a feature of most normal roots that if they are placed in distilled water, they exhibit a slow leakage of potassium ions across the plasmalemma into the surrounding fluid (Briggs *et al.* 1958; Nassery 1971, 1972). Aluminium ions will damage the cell membranes, probably the plasmalemma in the first instance, resulting in an increased leakage of K^+ ions from the roots of *Agrostis* (Woolhouse 1969). A series of experiments were performed to investigate the effects of copper and zinc ions on K^+ leakage. The roots were pre-incubated in aerated 30 mM KCl and 0·5 mM $CaSO_4$ for 4 hours at 20°C to obtain equal high levels of K^+ in all of the roots. The roots were then

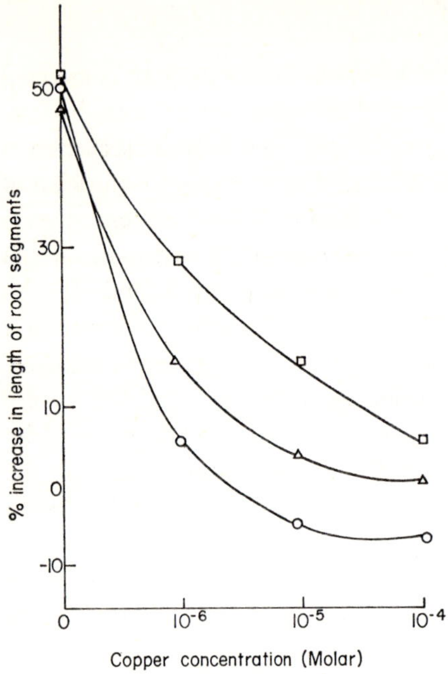

Figure 13. Effect of Cu^{++} ions on the elongation of excised segments of roots of *Agrostis tenuis*. Segments 0·9 mm in length were cut from the elongating zone of the roots and incubated in darkness for 24 hr. at 25°C in a medium containing 2% sucrose and the appropriate concentration of copper. The elongation is expressed as a percentage of the initial length (0·9 mm). For details see text. (○) Non-tolerant race. (□) Copper-tolerant race (△) Zinc-tolerant race.

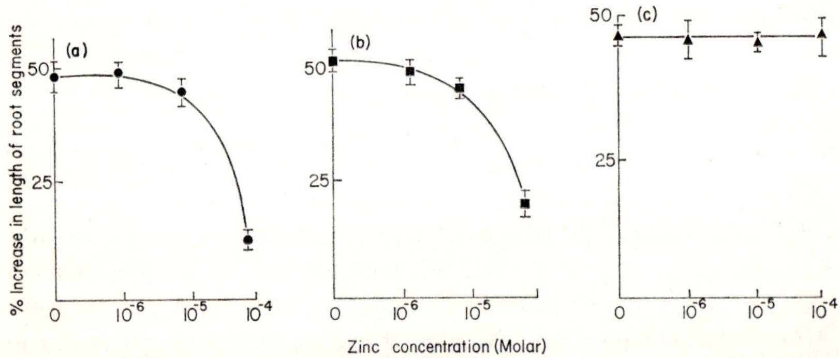

Figure 14. Effect of Zn^{++} ions on the elongation of excised segments of roots of *Agrostis tenuis*. For details see legend to Fig. 13 and text. (a) (●) Non-tolerant race. (b) (■) copper-tolerant race. (c) (▲) zinc-tolerant race

given three five-minute washes in distilled water to wash out the free space K^+ before metal treatment was started. The roots were pre-incubated for 5 minutes in the metal solution and then transferred to the treatment solution for measurement of K^+ leakage. Figures 15 and 16 show that zinc ions had no effect on the leakage of potassium in any of the races of *A. tenuis*. Copper on the other hand, caused K^+ leakage from the roots in all three races but caused much less leakage with the copper-tolerant race. These results may be interpreted to suggest that the plasmalemma is damaged by toxic concentrations of copper, leading to increased leakage of potassium ions from the cells. Clearly, however, this is not a universal feature of tolerance, since in the case of zinc the metal is without effect on the permeability of the membrane to K^+ in any of the races. It would seem therefore, that although copper and zinc ions are both toxic to *Agrostis*, they are toxic in different ways. Zinc ions do not damage the plasmalemma whereas copper ions do (this may be related to the different stereochemistry of the hydrated ions of the two metals). This suggests that plasmalemma modifications are involved in copper tolerance but not in zinc tolerance. It may be that zinc ions are inactivated by formation of complexes which are then compartmented in the cell, whereas some part of Cu and Al tolerance may operate through an exclusion mechanism at the plasmalemma.

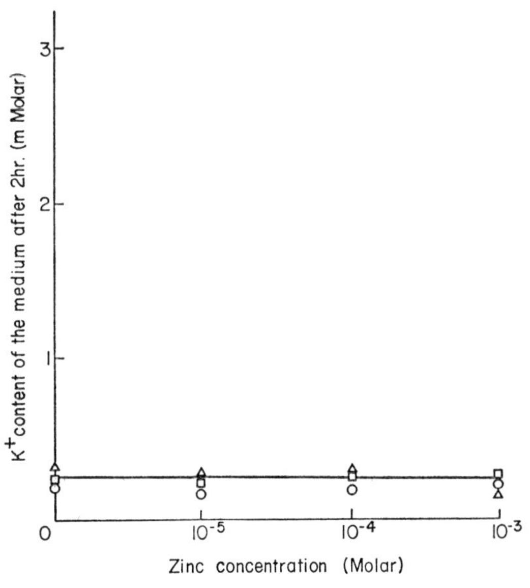

Figure 15. Effect of Zn^{++} ions on the leakage of K^+ from excised root tips of *Agrostis tenuis*. (○) Non-tolerant race. (□) Copper-tolerant race. (△) Zinc-tolerant race. The results are expressed as increase in μ Molarity of the external medium per mm^2 of root surface per 2 hours. For details see text.

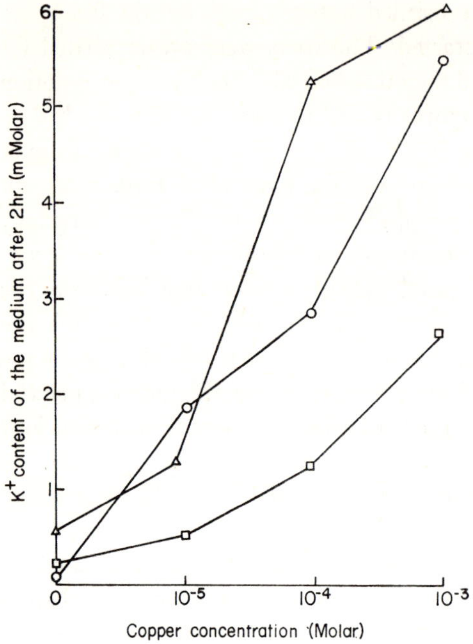

Figure 16. Effect of Cu^{++} ions on the leakage of K^+ from excised root tips of *Agrostis tenuis*. (○) Non-tolerant race (□) Copper-tolerant race (△) Zinc-tolerant race. The results are expressed as increase in μ Molarity of the external medium per mm^2 of root surface per 2 hours. For details see text.

Metal tolerance and enzyme adaptation

Recently it has become generally recognized that there are a number of enzyme systems which are localized in the cell-wall *in vivo* (Lamport 1965). In the case of root cells, the cell-wall enzymes will come under the influence of any metal ions which are present in the soil solution. Vacuum extraction of the soil solution into porous cups in the field shows that the soil solution can contain up to 10^{-3} M zinc at Trelogan mine in Flintshire and up to $2{\cdot}5 \times 10^{-5}$ M copper at Parys Mountain copper mine in Anglesey. Cell-wall enzymes fall into two classes, those which are ionically bound to the cell-wall matrix and can be eluted by salt gradients, and those which are covalently bound (Lamport 1965). A differential susceptibility of cell-wall bound ATPase activity towards aluminium ions has been demonstrated in edaphic ecotypes of *A. tenuis* (Woolhouse 1969) (Fig. 17).

The effects of copper ions on covalently bound cell-wall acid phosphatase from copper-tolerant and copper-susceptible ecotypes of *Agrostis* have been studied, on the hypothesis that the metal-tolerant ecotype could have evolved

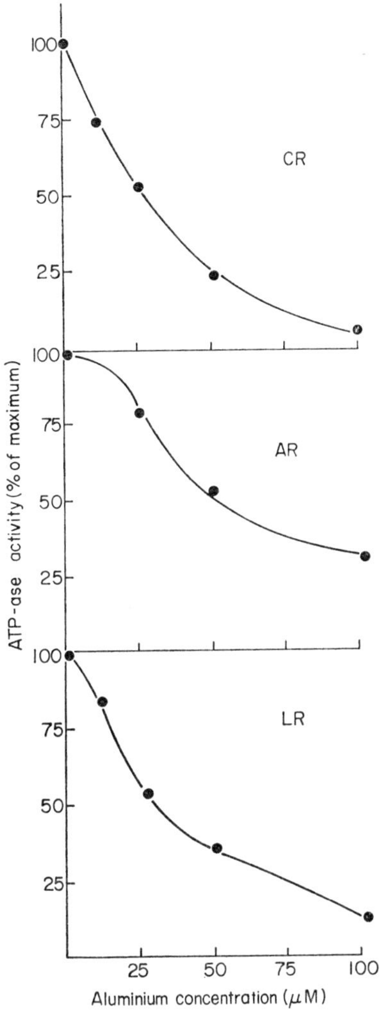

Figure 17. Effect of Al^{+++} ions on the cell wall-bound ATPase activity of roots of edaphic ecotypes of *Agrostis tenuis* (CR: calcareous soil ecotype; AR: acid soil ecotype; LR: lead soil ecotype). Extraction: known weights of roots were ground in a pestle and mortar with 2 volumes of a solution containing 0·25 M sucrose, 3 mM Tris-HCl buffer, ph 7·4. The extract was strained through cheese cloth and then centrifuged at 1,400 × g for 10 min. The supernatant was discarded and the cell wall pellet washed and recentrifuged with 2 further aliquots of extraction medium before finally suspending in 10 mM Tris-HCl buffer, pH 7·4. Assay conditions: the reaction was started by adding 0·5 ml of enzyme extract containing 0·2 mg protein to a solution containing 3 mM ATP, 50 mM citrate buffer pH 4·8 and 1·5 mM $MgCl_2$, in a final volume of 3·5 ml. The preparation was incubated for 60 min at 30°C and the reaction then stopped by addition of 2·0 ml of 20% trichloroacetic acid. Phosphate released was estimated by the method of Fiske & Subbarow (1924) and protein by the method of Lowry *et al.* (1951).

a cell-wall enzyme system which is better able to function in the presence of copper ions than the enzyme from a non-tolerant ecotype. Enzyme adaptation conferring resistance to an inhibitor has been demonstrated in the case of the local anaesthetic dibucaine on normal and variant forms of the enzyme cholinesterase from human serum (Kalow & Davies 1958). The genetically variant cholinesterase was less strongly inhibited by dibucaine than the normal enzyme (Fig. 18).

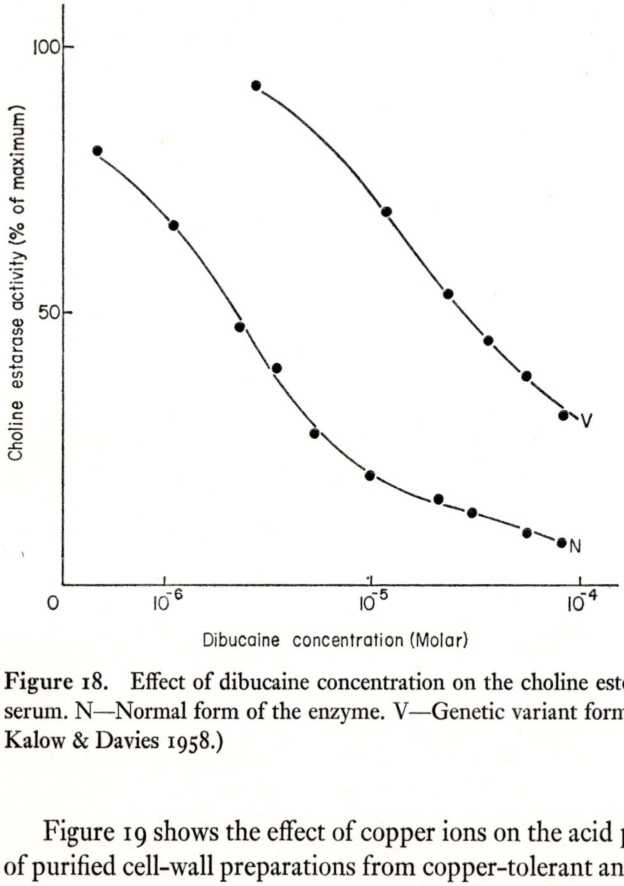

Figure 18. Effect of dibucaine concentration on the choline esterase activity of human serum. N—Normal form of the enzyme. V—Genetic variant form of the enzyme. (From Kalow & Davies 1958.)

Figure 19 shows the effect of copper ions on the acid phosphatase activity of purified cell-wall preparations from copper-tolerant and non-tolerant roots of *A. tenuis*. The figures are corrected for cell-wall binding of copper, so that concentration values on the horizontal axis represent total concentrations of copper in the solution after cell-wall binding has taken place, and are thus available for enzyme inhibition. It is evident that the concentration of copper required to bring about serious inhibition of enzyme activity is orders of magnitude greater than the $2{\cdot}5 \times 10^{-5}$ Molar copper found in the soil solution at Parys Mountain. However, in an enzyme assay system it is usual to include a buffering agent. This maintains the pH at the optimum for the enzyme but also, by preventing pH fluctuations from occurring, precludes the possibility of the ionic state of any of the constituents of the extract being altered and thus changing their reactivity. When measuring the effect of copper ions on enzyme activity however, a problem arises because at a pH of 5·6 which is optimal for this acid phosphatase, the ligands present in all buffer systems react with copper ions to form either chelation complexes or partly-dissociated salts. This leads to a reduction in the concentration of free copper ions

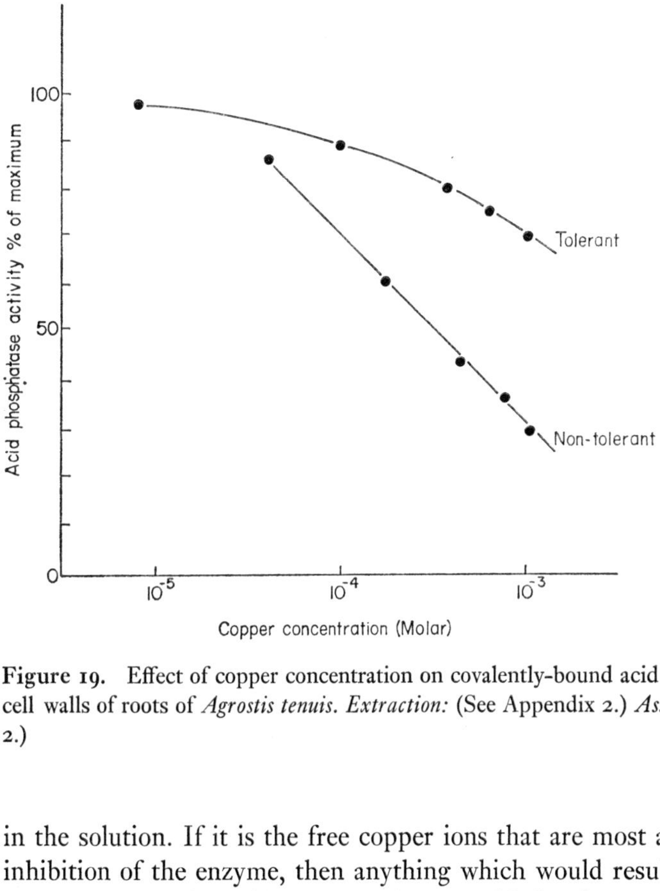

Figure 19. Effect of copper concentration on covalently-bound acid phosphatase of the cell walls of roots of *Agrostis tenuis*. *Extraction:* (See Appendix 2.) *Assay:* (See Appendix 2.)

in the solution. If it is the free copper ions that are most active in causing inhibition of the enzyme, then anything which would result in a reduction of the concentration of free copper ions would also decrease enzyme inhibition. It has been found that a given concentration of copper ions produces less inhibition in the presence of a citrate buffer than in the presence of an acetate buffer. This is because citrate forms more stable complexes with copper than does acetate (O'Sullivan 1969). It is therefore necessary to calculate the concentration of free copper ions in an enzyme assay at any given total copper concentration.

O'Sullivan (1969) gives expressions from which the relationship between ligand concentration, total metal concentration and free metal ion concentration can be derived. This relationship takes the form of a quadratic:

$$[M]_F = \frac{-\left(1+\frac{K[L]_T}{\alpha}-\frac{K[M]_T}{\alpha}\right)+\sqrt{\left(1+\frac{K[L]_T}{\alpha}-\frac{K[M]_T}{\alpha}\right)^2+\frac{4K[M]_T}{\alpha}}}{\frac{2K}{\alpha}}$$

where M_F = Concentration of free metal ions

M_T = Total concentration of metal

L_T = Total concentration of ligand

K = Stability constant of metal–ligand complex

α is defined as: $\left(1 + \frac{[H^+]}{Ka}\right)$

$[H^+]$ = Hydrogen ion concentrations.

Ka is the dissociation constant of the protonated ligand.

Figure 20 shows the computed relationship between total copper concentration in a 40 mM acetate buffer and the concentration of free copper ions. An acetate buffer system was used because it buffers at pH 5·6 and produces a simple partly-dissociated salt rather than chelation complexes. Also acetic acid has only one ionizable hydrogen. These considerations make the use of cumulative stability constants unnecessary and thus considerably facilitate computation. Whilst recognizing that different methods of measuring stability constants yield different values and that one must accept an inherent error in computed free copper ion concentrations from this source, it can be seen that the abnormally high concentrations of copper apparently needed to bring about the inhibition shown in Fig. 19 are reduced to the order of concentration found in soil solutions by salt formation with the buffer acetate.

Whilst one cannot say that the copper-buffer complex is not in itself somewhat inhibitory to the enzyme activity it is possible to show that free copper ions are the major inhibitor. A commercial soluble acid phosphatase was used to check whether any complications arise in interpretation of the results when the cell-wall enzyme is used. In this experiment the total concentration of copper was maintained constant at 10^{-3} M, and the acetate concentration was varied to produce different concentrations of free copper ions; the pH was maintained constant. Figure 21a shows the effect of increasing acetate concentration on the activity of the enzyme in the presence of 10^{-3} M copper. Over this range of concentration, the enzyme activity was inversely linearly related to the concentration of free copper ions (Fig. 21b). Figure 20 shows a linear relationship between log total copper concentration and log free copper ion concentration and it is demonstrable that because the slope of this log—log plot is equal to 1, the concentration of free metal ions is a linear function of total metal concentration (Appendix 1). In view of this linear relationship it is legitimate to plot total copper concentration on these dose-response curves without calculating free copper concentrations in every instance.

Double reciprocal plots (Figs 22 & 23) show that the inhibition of the cell-wall phosphatase by copper is non-competitive and that the Kms for the enzyme from the copper-tolerant plants and from the copper-susceptible plants is the same. Using Dixon's (1953) method of plotting $1/V$ against inhibitor concentrations in order to obtain the inhibitor constant K_i (Figs 24 & 25), shows that the enzyme from the copper-tolerant plants has a higher K_i than the enzyme from the copper-susceptible plants.

Equilibria of the form:

$$E+I \underset{K6}{\overset{K5}{\rightleftarrows}} EI$$

$$ES+I \underset{K6}{\overset{K5}{\rightleftarrows}} ESI$$

can establish in a non-competitive situation, between enzyme (E), inhibitor (I) and substrate (S).

$$Ki = \frac{K6}{K5}$$

Therefore, a large K_i indicates an unstable EI or ESI complex.

This evidence indicates therefore that the cell-wall acid phosphatase from roots of copper-tolerant *Agrostis* produces a more unstable complex with copper than does the enzyme from copper-susceptible plants. An analogous situation to this has recently been found in *E. coli* where Silver *et al.* (1972) have shown that manganese acts as a competitive inhibitor of the magnesium-uptake carrier system. In this case the Mg carrier system of a manganese-tolerant strain shows an altered K_i with manganese such that it has a lower affinity for manganese than does the carrier system of a normal strain.

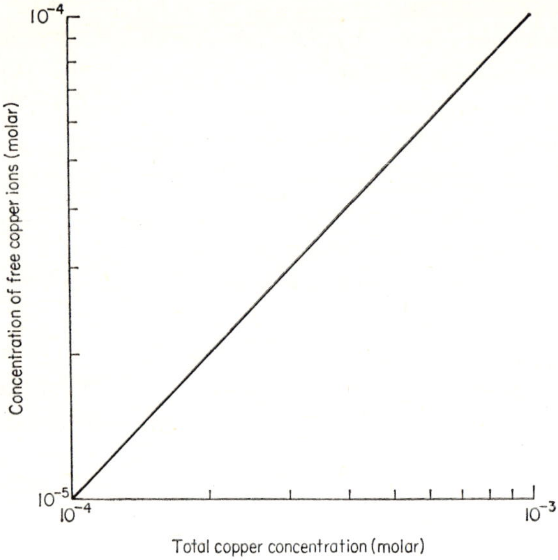

Figure 20. Relationship between the concentration of free Cu^{++} ions and the total copper concentration in a 40 mM solution of acetate buffer. For details of the calculation see text.

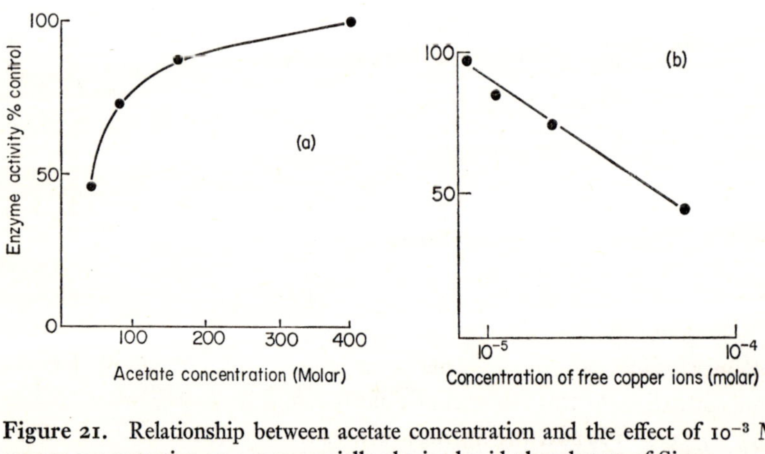

Figure 21. Relationship between acetate concentration and the effect of 10^{-3} M total copper concentration on a commercially obtained acid phosphatase of Sigma.
(a) the enzyme activity plotted against acetate concentration.
(b) the enzyme activity plotted against the concentrations of free Cu^{++} ions which correspond to the acetate concentrations shown in (a). Assay conditions otherwise as in Appendix 2.

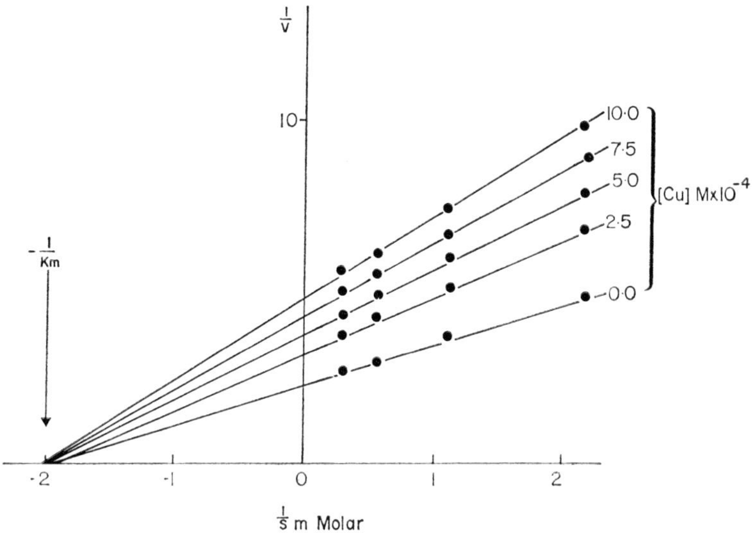

Figure 22. Double reciprocal plot showing the non-competitive inhibition by copper of the activity of covalently-bound acid phosphatase of cell walls from roots of a copper-tolerant race of *Agrostis tenuis*. Assay conditions as in Appendix 2.

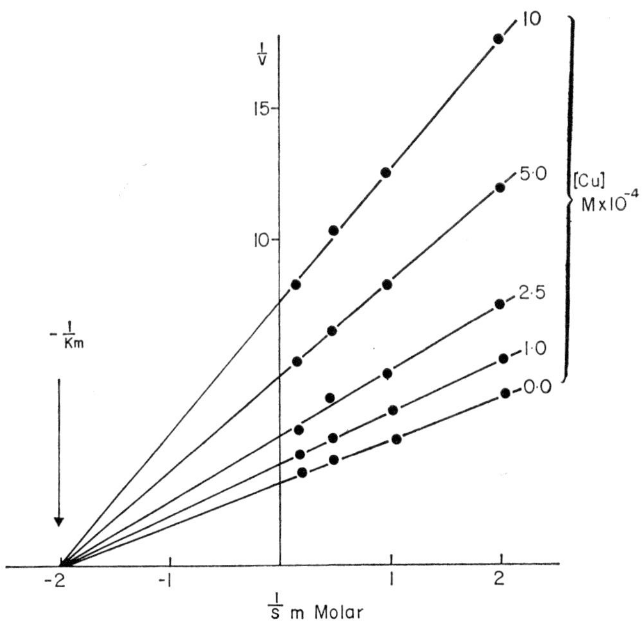

Figure 23. Double reciprocal plot showing the non-competitive inhibition by copper of the activity of covalently-bound acid phosphatase of cell walls from roots of a copper-susceptible race of *Agrostis tenuis*. Assay conditions as in Appendix 2.

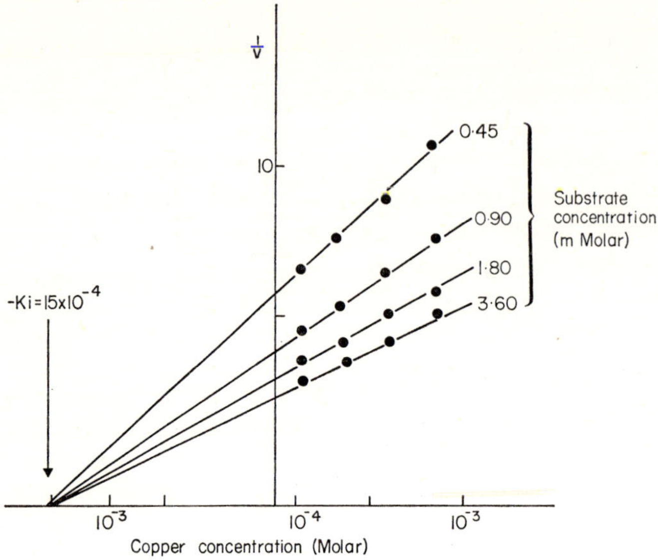

Figure 24. A plot of the reciprocal of the rate of enzyme reaction (l/v) against inhibitor (Cu) concentration following Dixon (1953), to obtain the inhibitor constant (Ki) for the covalently-bound acid phosphatase of cell walls from roots of a copper-tolerant race of *Agrostis tenuis*. Assay conditions as in Appendix 2.

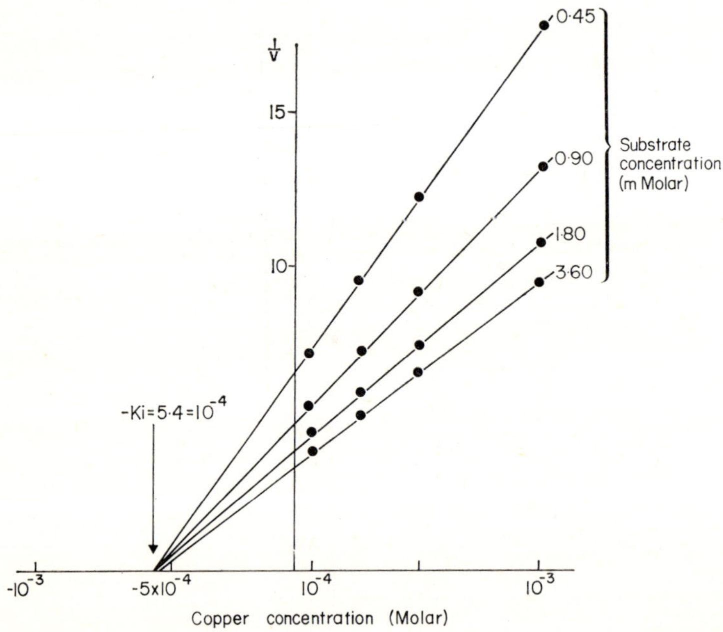

Figure 25. A plot of the reciprocal of the rate of enzyme reaction (l/v) against inhibitor concentration following Dixon (1953), to obtain the inhibitor constant (Ki) for the covalently-bound acid phosphatase of cell walls from the roots of a copper-susceptible race of *Agrostis tenuis*. Assay conditions as in Appendix 2.

Conclusions

We may conclude therefore, that plants have evolved a variety of mechanisms for achieving tolerance to heavy metals. It is also clear that the mechanisms of tolerance in any individual may involve a variety pf physiological processes each of which contribute to the total metal tolerance. This is consistent with the fact that populations show various degrees of tolerance to a given metal. It also consistent with such genetical studies as have been made on the metal tolerance of higher plants, which show that the inheritance of tolerance is a complex phenomenon under the influence of a number of genes with modifiers for dominance which are themselves probably affected by the genome as a whole (Urquhart 1971).

Summary

The available information on the mechanisms of metal tolerance makes it increasingly clear that these are complex and involve a number of metabolic systems. It also seems that the mechanisms of tolerance differ between species.

References

ALLEN W.R. (1971) Copper tolerance in some Californian populations of the monkey flower, *Mimulus guttatus*. *Proc. R. Soc. B* 177, 177–96.

ANTONOVICS J. (1966) The Genetics and Evolution of Differences Between Closely Adjacent Plant Populations with Special References to Heavy metal Tolerance. Ph.D. thesis, Univ. of Wales.

ANTONOVICS J., BRADSHAW A.D. & TURNER R.G. (1971) Heavy metal tolerance in plants. *Adv. in Ecol. Res.* 7, 1–85.

ASHIDA J. (1965) Adaptation of Fungi to metal toxicants. *Ann. Rev. Phytopathol.* 3, 153–74.

BRADSHAW A.D. (1952) Populations of *Agrostis tenuis* resistant to lead and zinc poisoning. *Nature, Lond.* 169, 1098.

BRIGGS G.E., HOPE A.B. & PITMAN M.G. (1958) Measurement of ionic fluxes in red beet tissue using radioisotopes. *J. Radioisotopes in Sci. Res.* 4, 391.

BRÖKER W. (1963) Genetisch—Physiologische Untersuchungen uber die Zinkuertvaglichkeit von *Silene inflata* Sm. *Flora, Jena, B.* 153, 122–56.

BEPPU M.L. & ARIMA K. (1964) Introduction and mechanisms of arsenite resistance in *Pseudomonas pseudomallei*. *J. Bacteriol.* 88, 151–7.

CHOPRA I. (1971) Decreased uptake of Cadmium by a resistant strain of *Staphylococcus aureus*. *J. Gen. Microbiol.* 63, 265–7.

CLARKSON D.T. (1969) Metabolic aspects of aluminium toxicity and some possible mechanisms for resistance. In *Ecological Aspects of the Mineral Nutrition of Plants* (Ed. I.H. Rorison), Blackwell, Oxford.

DIXON M. (1953) Determination of enzyme inhibitor constants. *Biochem. J.* 55, 170–1.

FISKE C.H. & SUBBAROW Y. (1925) The colorimetric determination of phosphorus. *J. biol. Chem.* 66, 375–400.

GREGORY R.P.G. & BRADSHAW A.D. (1965) Heavy metal tolerance in populations of *Agrostis tenuis* sibth. and other grasses. *New Phytol.* 64, 131–43.

HARLEY J.L. & WAID J.S. (1955) A method of studying active mycelia on living roots and other surfaces in the soil. *Trans. Br. mycol. Soc.* 38, 104–18.

JOWETT D. (1958) Population of *Agrostis* sp. tolerant to heavy metals. *Nature. Lond.* 182, 816–17.

JOWETT D. (1959) Genecology of Heavy Metal Tolerance in *Agrostis*; Ph.D. thesis, Univ. of Wales.

KALOW W. & DAVIES R.O. (1958) The activity of various esterase inhibitors towards a typical human serum cholinesterase. *Biochem. Pharmac.* 1, 183–92.

LAMPORT D.T.A. (1965) The protein component of primary cell walls. *Adv. in Bot. Res.* 2, 151–218.

LOWRY O.H., ROSEBROUGH N.J., FARR A.L. & RANDALL R.J. (1951) Protein measurement with the Folin reagent. *J. biol. Chem.* 193, 265–75.

NASSERY H. (1971) Some Aspects of potassium loss from excised barley roots. *New Phytol.* 70, 113–17.

NASSERY H. (1972) The loss of potassium and sodium from excised barley and bean roots. *New Phytol.* 71, 269–74.

NICHOLLS D.W., PROVAN D.M.J., COLE M.M. & TOOMS J.S. (1965) Geobotany and Geochemistry in mineral exploration in the Dugald River Area. Cloncurry District Australia. *Trans. Inst. Min. Metal.* 74, 165–799.

O'SULLIVAN W.J. (1969) Stability Constants of Metal Complexes. In *Data for Biochemical Research* (2nd Edition, Ed. by R.M.C. Dawson, D.C. Elliott, W.H. Elliott & K.H. Jones), Clarendon Press, Oxford.

RATHORE V.S., BAJAJ Y.P.S. & WITTWER S.H. (1972) Sub cellular localisation of zinc and calcium in Bean (*Phaseolus vulgaris L.*) tissues. *Plant Physiol.* Lancaster 49,207–11.

REILLY C. (1969) The uptake and accumulation of copper by *Becium homblei* Duvig. et Plancte. *New Phytol.* 68, 1081–7.

SILVER S., JOHNSEINE P., WHITNEY E.L. & CLARK D. (1972) Manganese—resistant mutants of *Escherichia coli*: physiological and genetic studies. *J. Bacteriol.* 110, 186–95.

TONOMURA K., MAEDA F. & FUTAI F. (1968a) Studies on the action of mercury resistant micro-organisms on mercurials. *J. Ferment. Technol.*, Osaka 46, 685–92.

TONOMURA K., NAKAGAMI T., FUTAI F. & MAEDA K. (1968b) Studies on the action of mercury resistant micro-organisms on mercurials. *J. Ferment. Technol.*, Osaka 46, 506–12.

TONOMURA K., MAEDA K., FUTAI F., NAKAGAMI T. & YAMADA M. (1968c) Stimulative vaporisation of phenylmercuric acetate by mercury resistant bacteria. *Nature, Lond.* 217, 644.

TURNER R.G. (1970) The subcellular distribution of zinc and copper within roots of metal-tolerant clones of *Agrostis tenuis*. Sibth. *New Phytol.* 69, 725–31.

TURNER R.G. & MARSHALL C. (1971) The accumulation of ^{65}Zn by root homogenates of Zn—tolerant and non-tolerant clones of *Agrostis tenuis*. Sibth. *New Phytol.* 70, 539–545.

TURNER R.G. & MARSHALL C. (1972) The accumulation of zinc by subcellular fractions of roots of *Agrostis tenuis*. Sibth in relation to zinc tolerance. *New Phytol.* 71, 671–6.

URQUHART C. (1971) Genetics of lead tolerance in *Festuca ovina*. *Heredity* 26, 19–33.

WILKINS D.A. (1960) The measurement and genetic analysis of lead-tolerance in *Festuca ovina*. *Rep. Scott. Pl. Breed. Stn.* 85–98.

WOOLHOUSE H.W. (1969) Differences in the properties of the acid phosphatases of plant roots and their significance in the evolution of edaphic ecotypes. In *Ecological Aspects of the Mineral Nutrition of Plants* (Ed. by I.H. Rorison), Blackwell, Oxford.

WOOLHOUSE H.W. (1970) Environment and enzyme evolution in plants. In *Phytochemical Phylogeny* (Ed. by J.B. Haborne), Academic Press, London.

WYN JONES R.G., SUTCLIFFE M. & MARSHALL C. (1971) Physiological and biochemical basis for heavy metal tolerance in clones of *Agrostis tenuis*. In *Recent Advances in Plant Nutrition* (Ed. by R.M. Samish), Gordon & Breach, New York.

Appendix 1

The relationship between free copper ions and total copper concentration in the enzyme assay system.

Let M_F be the concentration of free metal ions and M_T the total concentration of that metal in the solution.

If a graph is plotted of log M_F against log M_T and is found to be linear then:

$$\log_{10} M_F = \log_{10} M_T + C$$

or

$$\log_e M_F = m \log_e M_T + K$$

and

$$\log_e M_F = \log_e M_T{}^m + K$$

and so

$$e^{\log_e M_F} = e^{\log_e M_T{}^m} e^K$$

$$M_F = M_T{}^m . e^K$$

This can only be a linear relationship if $m = 1$.

Examination of Fig. 20 shows that the slope of the line does $= 1$. Note that $10^{-3}M$ total copper corresponds to concentrations of approximately $10^{-4}M$ free copper ions.

Appendix 2 The preparation of purified cell walls

100 g of roots from water cultured plants were vigorously washed with 10 changes of 250 ml distilled water before homogenization. Harley and Waid (1955) have shown that when roots are cleaned by washing with water, most of the removable fungal contaminants of the root surface are removed by 5–10 washings.

The procedure for producing purified cell-wall preparations suitable as a source of covalently bound cell-wall acid phosphatase is described by the flow diagram. The initial homogenization by the Atomix blender was carried out in Tris/HCl buffer at pH 7·2 to prevent damage to the cell-wall enzymes by drastic changes in pH caused by the release of organic acids during homogenization. Sucrose was included to osmotically protect the organelles preventing unnecessary contamination of the cell-wall preparation by osmotically ruptured organelles. 4 mM mercaptoethanol was included as a reducing agent. Low speed centrifugation on a bench centrifuge was used so that small particles such as organelles and bacteria would tend to remain in suspension. Indeed, the smallest fragments of cell-wall also remained in suspension. 2% Triton × 100 was used to remove membrane contaminants. 0·5 M NaCl was used to extract ionically bound enzymes from the cell wall, both those which are present *in vivo*, and those which adsorbed onto the wall during the

Flow diagram summarising the procedure for making purified cell-wall preparations from roots of Agrostis tenuis.

100g wet weight of roots homogenised in ice-chilled Atomix with 100 ml 10mM Tris/H 1 buffer pH 7.2 containing 4mM mercaptoethanol and 0.5M sucrose, for 3 mins.

↓

Grind to a fine brei with a pestle and mortar, using a little acid washed sand.

↓

Centrifuge at low speed on M.S.E. bench centrifuge for 2 mins.

Supernatant discarded. ← → Pellet resuspended in 50ml ice-cold 4mM mercaptoethanol and centrifuged at low speed for 1 min.

Pellet. ← → Supernatant discarded.

Repeated 15 times.

↓

Pellet suspended in 50ml ice-cold, 2% Triton × 100 in 4mM mercaptoethanol, and stirred for 30 mins. Then centrifuged at low speed for 1 min.

Supernatant discarded ← → Pellet suspended in 100ml ice-cold 0.5M NaCl in 4mM mercaptoethanol, stirred for 30 min, and centrifuged at low speed for 1 min.

Pellet. ← → Supernatant discarded.

Repeated 5 times.

↓

Pellet suspended in 50ml ice-cold 2.5×10^{-3}M EDTA and 4mM mercaptoethanol, stirred for 10 min, then centrifuged at low speed for 1 min.

Pellet. ← → Supernatant discarded.

Repeated twice.

↓

Pellet suspended in 50ml ice-cold 4mM mercaptoethanol and centrifuged at low speed for 1 min.

Pellet. ← → Supernatant discarded.

Repeated 5 times.

↓

Pellet suspended in 100ml ice-cold 4mM mercaptoethanol and strained through a fine tea strainer to removed long fibers.

↓

1.0 ml aliquot of wall suspension removed, dried and weighed.

↓

Suspension of walls diluted to give a concentration of 3.5g wall/100ml suspension. This suspension was then stored at 1.0°C until used.

initial homogenization. The cell-wall preparation was washed with EDTA solution to remove metallic contaminants. This was important because the effects of metal ions on the phosphatase activity of the cell-walls was to be studied. The enzyme was found to be stable for periods of up to a month if stored at 1 °C. Freezing however, substantially reduced the activity of the enzyme. Cell-wall preparations more than 7 days old were not used for enzyme assays. The use of a fine tea strainer to filter the wall preparation greatly facilitated the micropipetting of the cell-wall preparation by removing long fibres from the suspension.

THE ENZYME ASSAY

The enzyme assay depended on the release of p-nitrophenol by the hydrolysis of p-nitrophenyl phosphate under acid conditions at pH 5·6. Each assay tube contained p-nitrophenyl phosphate and metal ions of known concentration, and 7·0 mg cell-wall material in 2·4 ml 40 mM acetate buffer. The digests were incubated at 25°C for 30 min in a shaking incubator. The reaction was stopped by adding 3 ml N NaOH. The cell-walls were pelleted by centrifuging for 1 min at low speed in an M.S.E. bench centrifuge. The absorption at 410 nm of the supernatant was measured with a Unicam S.P. 600 spectrophotometer. A blank was made up from 7 mg cell-walls, 2·4 ml water, and 3 ml N NaOH. This blank was used because the NaOH extracted a very slight straw-coloured coloration from the cell-walls which absorbed at 410 nm, and in the concentrations at which it was present, was just detectable by the spectrophotometer. The metal ions had no effect on the absorption at 410 nm and so were not included in the blank.

Investigations into the nitrogen cycle in colliery spoil

P. J. WILLIAMS *University of York, York, U.K.*

Introduction

Colliery spoils can supply little or no nitrogen for plant growth despite the fact that they may contain appreciable amounts of fossilized organic nitrogen. The nitrogen requirements of non-nitrogen fixing species of vegetation introduced on to colliery spoil heaps in revegetation programmes must initially be supplied in the form of fertilizers. Since the cost of fertilization is high, it is economically desirable that the spoil ecosystem becomes self-supporting as quickly as possible. This situation will only be achieved when the applied nutrients in general, and nitrogen in particular become incorporated into efficient recycling systems. This paper reports the results of investigations undertaken to evaluate the extent, importance, and factors affecting the establishment of nitrogen cycling in ameliorated and untreated colliery spoils at two sites in the West Riding of Yorkshire. These sites are Mitchells Main and Upton.

Site description

(a) MITCHELLS MAIN (Grid. Ref. SE392043)

A field trial was set up on this site in 1967 to investigate the failure of an initially successful revegetation scheme. A randomized block design was employed and the following treatments were included: Shoddy at 4 tons per acre; Sewage Sludge at 20 tons per acre; Limestone at 8 tons per acre; a Control. Each treatment was replicated three times and divided into six sub-plots. These were sown with a *Festuca-Agrostis* seed mixture (80% *Festuca rubra* and 20% *Agrostis tenuis* by weight), left unseeded or planted with Silver Birch, Alder, *Robinia* or Poplar saplings. The differentiation of each main plot into distinct sub-plots soon became difficult because trees died and their sub-plots became grassed over by the sown species. For the purposes of the present investigations the division of treatments into sub-plots was ignored, pooled samples being prepared from individual samples taken at random within the whole of the treatment plot.

(b) UPTON (Grid. Ref. SE484143)

This spoil heap was regraded in 1970 and a field trial set up immediately. The trial design was similar to that described for Mitchells Main except that each treatment was sown with the *Festuca–Agrostis* mixture and was not sub-divided.

Experimental work

The experimental work falls into three sections, each will be described separately before the results of all three are considered together in a general discussion.

PERIODIC ANALYSIS OF MINERAL NITROGEN, pH AND ESTIMATES OF THE NUMBERS OF NITROSOMONAS AND NITROBACTER

The initial investigations were concerned with the levels of mineral nitrogen in the spoils and the relationships between these and the size of the nitrifyer population.

Materials and methods

Spoil samples were taken at approximately monthly intervals from the trial plots. A bulked sample was prepared for each treatment by pooling six randomly selected spoil cores. Individual cores were approximately 7·5 cm in diameter and 15 cm deep. All determinations were made on the fresh, < 1 cm fraction of the spoil. Mineral nitrogen was determined using the magnesium oxide/Devarda's alloy technique described by Bremner (1965). pH was measured electrometrically in a 1:2 spoil to 0·01 M. calcium chloride solution. The numbers of *Nitrosomonas* and *Nitrobacter* were estimated using the technique recommended by the N.A.A.S. (undated publication) and the most probable number technique of Finney (1951).

Results

The results of these determinations are presented graphically in Figs 1–6. Each value plotted represents the mean for the three field replicates.

Interpretation and discussion

(a) *Mitchells Main.* The results of the periodic mineral nitrogen determinations (Fig. 1) show that the concentration of ammonium is persistently higher than that of nitrate or nitrite in the Shoddy, Sewage and Control treatments. The ammonium concentration is highest in the Shoddy and lowest in the Sewage treatments. In the Lime treatment all three forms of mineral nitrogen occur at approximately equal concentrations.

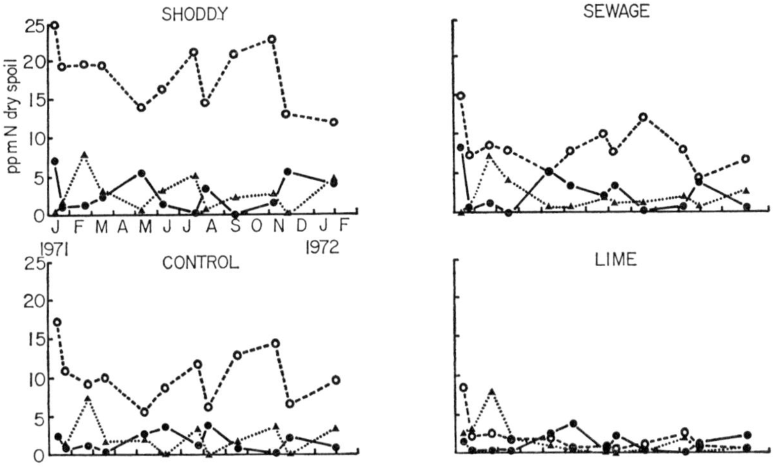

Figure 1. Mitchells Main. Mineral nitrogen levels.
○ - - - - Ammonium ▲ · · · · · · Nitrite ● ——— Nitrate

The levels of nitrite and nitrate are of the same magnitude in all treatments and vary between 0–8 ppm N. No seasonal variation in any form of mineral nitrogen is apparent.

The probable explanation for the higher levels of ammonium than nitrate or nitrite can be found by examining the counts of *Nitrosomonas* and *Nitrobacter* (Fig. 2). For all treatments the numbers, especially of *Nitrobacter*, are very low. Since *Nitrosomonas* and *Nitrobacter* together are responsible for the great majority, if not all of the conversion of ammonium to nitrate, it is not surprising to find a predominance of ammonium in spoils where these organisms are present in such low numbers.

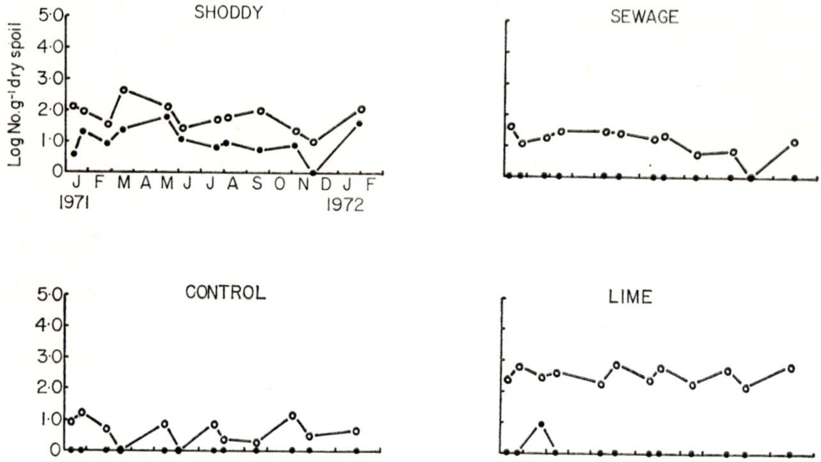

Figure 2. Mitchells Main. Nitrifying microorganisms.
○ ——— Nitrosomonas ● ——— Nitrobacter

It is well known that both *Nitrosomonas* and *Nitrobacter* are very intolerant of acid conditions and the results of the pH determinations (Fig. 3) show that the mean pH for the Shoddy, Sewage and Control treatment is approximately pH 4·0 whilst the lime treatment is higher at pH 5·5.

These results suggest that the lower numbers of nitrifying bacteria are due, at least to some extent, to the acid nature of the spoils. The pH value cannot, however, be the only important factor because the highest numbers

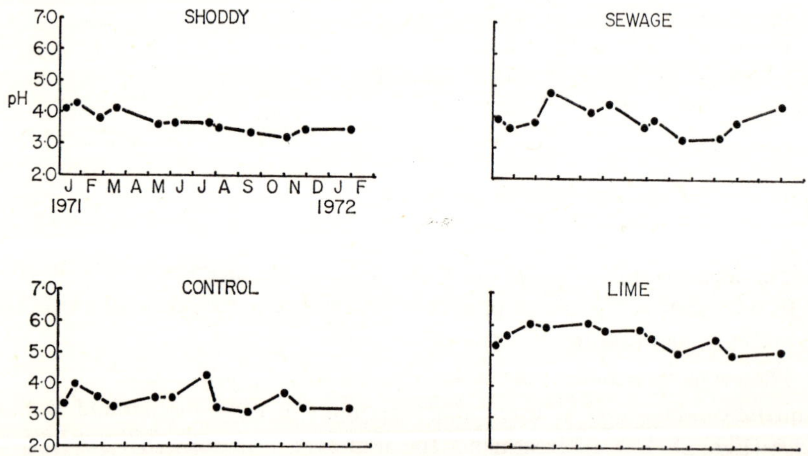

Figure 3. Mitchells Main. pH determinations.

of *Nitrobacter* occur not on the Lime treatment but on the Shoddy. This might indicate that substrate availability is also of importance.

Despite the low numbers of nitrifying organisms some nitrate is found even in the most acid spoil. Weber & Gainey (1962) have reported similar findings for acid soils where nitrate was produced at pH values as low as 4·0.

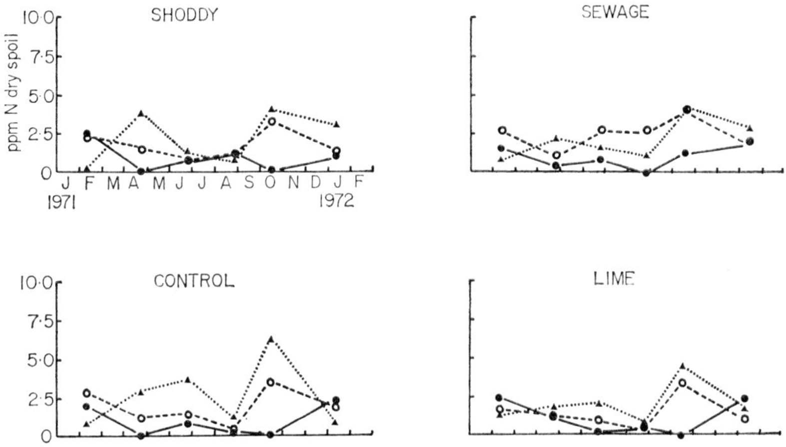

Figure 4. Upton. Mineral nitrogen levels.
○ - - - Ammonium ▲ · · · Nitrite ● ——— Nitrate

(b) *Upton.* The level of all three forms of mineral nitrogen (Fig. 4) is approximately the same in all four treatments and closely resembles that found in the Lime treatment at Mitchells Main. The pH of the spoil is relatively high (Fig. 6) and the numbers of *Nitrosomonas* and *Nitrobacter* (Fig. 5), are very

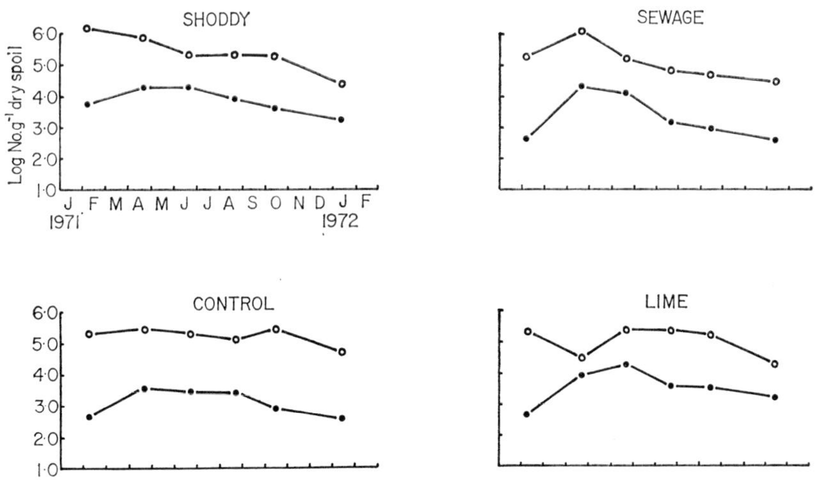

Figure 5. Upton. Nitrifying microorganisms.
○ ——— Nitrosomonas ● ——— Nitrobacter

high, being comparable to the highest levels found in unfertilized soil (Alexander 1965). Under these circumstances no preponderance of ammonium over nitrate would be expected, and none is observed.

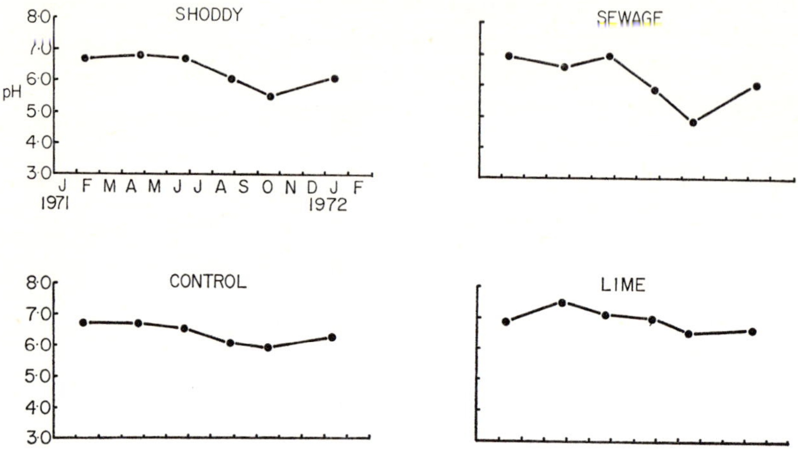

Figure 6. Upton. pH determinations.

The concentrations of ammonium, nitrite and nitrate range between 0–7 ppm N. No seasonal variation occurs.

The mineral nitrogen status of a soil merely represents the net result of all the interacting factors of nitrogen production and consumption. No information is provided by such data on the rate or magnitude of nitrogen transformations in the ecosystem and hence the ability of a soil to provide nitrogen for plant growth cannot easily be assessed from mineral nitrogen data alone. Some indirect information on these points can sometimes be obtained when, as in the present investigation, microbial enumerations are performed as an adjunct to mineral nitrogen determinations. Although the interpretations of numbers of microorganisms in terms of activity presents its own problems, numbers of *Nitrosomonas* and *Nitrobacter* are generally accepted to be a reliable index of their activity. Thus, in the situation which exists at Upton where low levels of mineral nitrogen are accompanied by large populations of nitrifying bacteria, the conclusion reached is that rapid mineralization, nitrification and nitrate removal is taking place. A rapid cycle is in operation. The situation at Mitchells Main is complicated by the inhibition of the nitrifying micro-organisms and little can be deduced about the nitrogen cycle in this spoil.

MINERALIZABLE NITROGEN DETERMINATIONS

It is generally accepted that of the various methods regularly used to assess the potential ability of a soil to provide nitrogen for plant growth, those involving estimation of the mineral nitrogen formed when soil is incubated under conditions which promote mineralization of soil organic nitrogen are the most satisfactory. Many workers have found that such incubation techniques provide a fairly accurate index of the availability of soil nitrogen to plants (Black *et al.* 1947; Allison & Sterling 1949; Gasser & Williams 1963). Incubation techniques have therefore been used in the present investigations.

Material and methods

A single pooled sample was prepared for each treatment. The fresh, < 1 cm fraction was obtained and divided into a number of sub-samples. Replicate 10 g quantities of sub-sample were incubated in the dark at 15% moisture at 27°C in a loosely capped 250 ml glass container. The moisture lost through evaporation was replaced every two days. Certain sub-samples were incubated with either 1.0% $CaCO_3$ and/or 50 ppm of nitrogen, supplied as an ammonium sulphate solution. These additions were made in order to resolve the effect of substrate availability and pH on the nitrifying bacteria. After 0, 10, 20 and 40 days of incubation the mineral nitrogen content of replicate samples of the incubated spoil was estimated using the method previously described.

In order to make meaningful comparisons between the mineralization and nitrification processes in spoils and soils, two soils were treated in the same way as the spoils. One soil was of nearly neutral reaction (pH 6·3), obtained from an unfertilized Beech copse, whilst the other, of acid reaction (pH 4·5) was obtained from a grassy area on an upland moorland at Osmotherly in North Yorkshire.

Samples from all four locations were collected within a few days of each other, at the beginning of the growing season in March 1972.

Results

The results are shown graphically in Figs 7 to 9. The pH at the beginning of the incubation period is shown on each graph. The values plotted represent the mean of two replicates examined at each time interval. Nitrite concentrations were always very low (< 2 ppm) and have therefore been omitted.

Interpretation and discussion

The results of this incubation experiment provide information on the processes of nitrification and mineralization. Nitrification will be discussed first.

The experimental results tend to confirm the conclusions drawn from the periodic study of spoil mineral nitrogen levels and nitrifying populations. Thus at Mitchells Main (Fig. 7), nitrification of added or mineralized ammonium only occurs in the Shoddy, Sewage and Control treatments when the pH is raised by the addition of calcium carbonate. This shows that low pH is the factor limiting the population size of the nitrifying bacteria. The Lime treatment at Mitchells Main accumulates nitrate even in the absence of calcium carbonate and this indicates that the pH value of this spoil is high enough for nitrifying activity and suggests that the low numbers usually found are due to a deficiency of substrate.

At Upton (Fig. 8) ammonium does not accumulate under any condition; applied ammonium is rapidly converted to nitrate. This confirms the results presented in the previous experimental section.

The similarity between the nitrification processes in the acid soil and spoils and neutral soil and spoils demonstrates that there are no fundamental differences between the soils and spoils in this process.

The quantity of mineralizable nitrogen released after 40 days incubation for all the soil and spoil samples is given in Table 1. Mineralizable nitrogen

Table 1. Mineralizable nitrogen ($gN.m^{-2}$ to depth of 18 cm).

	Mitchells Main		Upton	
	$-CaCO_3$	$+CaCO_3$	$-CaCO_3$	$+CaCO_3$
Shoddy	3·0	11·8	0·1	0·7
Sewage	6·2	8·4	0·0	1·2
Control	2·9	4·8	1·0	1·2
Lime	1·4	2·0	0·5	1·0
	Moorland		Woodland	
Soil	12·7	24·9	9·3	8·3

is taken here to mean the total quantity present at the end of the incubation period less that present initially. The values for the samples incubated both with and without calcium carbonate are shown. The values for incubations with ammonium sulphate are not shown but generally, the addition of this nitrogen source decreased the quantity of mineralizable nitrogen recorded. This is probably due to the increase in acidity that results from the oxidation of ammonium sulphate.

The representation of each treatment by a single bulked sample was

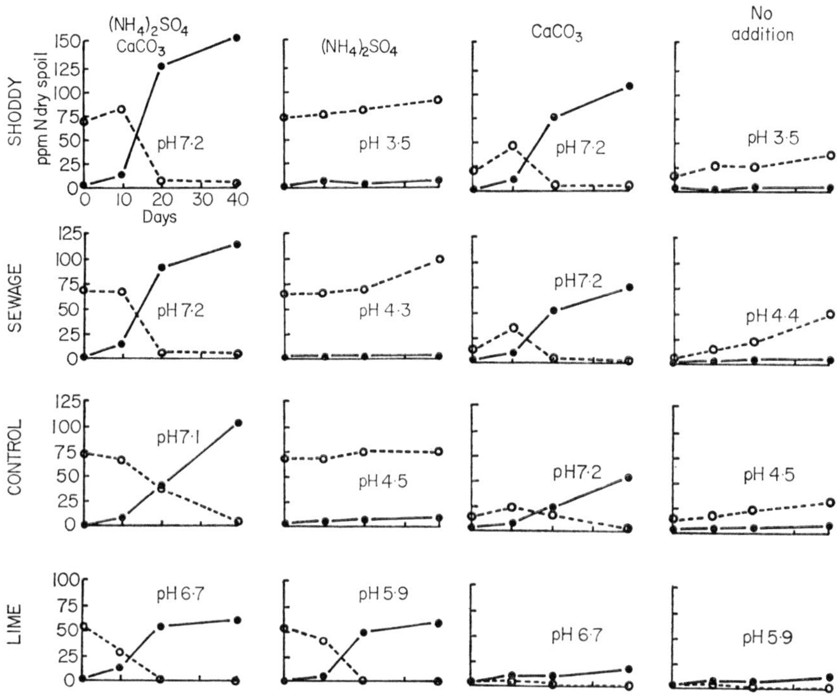

Figure 7. Mitchells Main. Incubation of spoil samples.
○ - - - Ammonium ● ——— Nitrate.

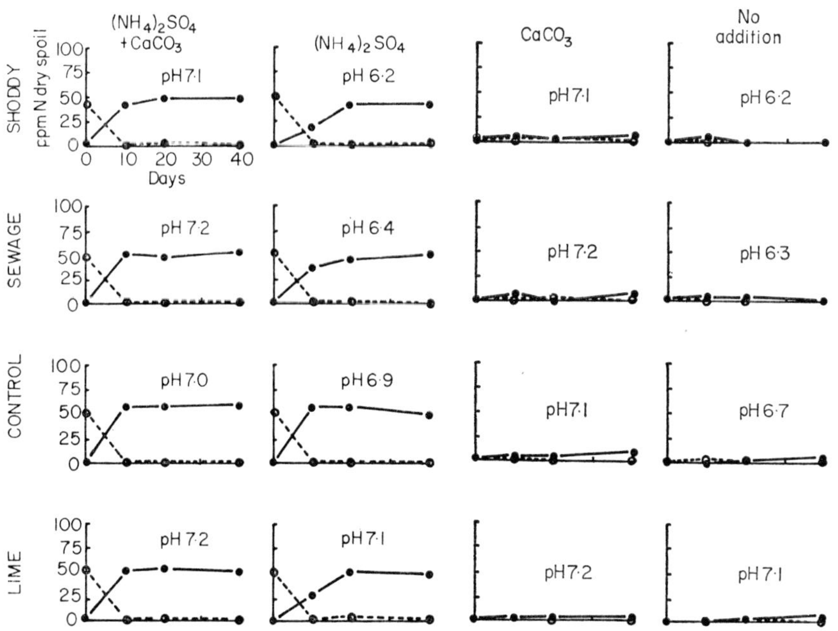

Figure 8. Upton. Incubation of spoil samples.
○ - - - - Ammonium ● ——— Nitrate

necessary because of the large number of determinations that had to be made. However, this procedure precluded a statistical interpretation of the results. Nevertheless several interesting points emerge from the data obtained and these will be considered.

Incubating the acid Mitchells Main and Moorland samples with calcium carbonate not only influences the conversion of ammonium to nitrate, but also increases the quantity of mineralizable nitrogen formed. This result suggests that when the pH is raised, many groups of heterotrophic micro-organisms, previously inhibited by the acid conditions, proliferate and promote the mineralization of materials that previously were unattacked, or

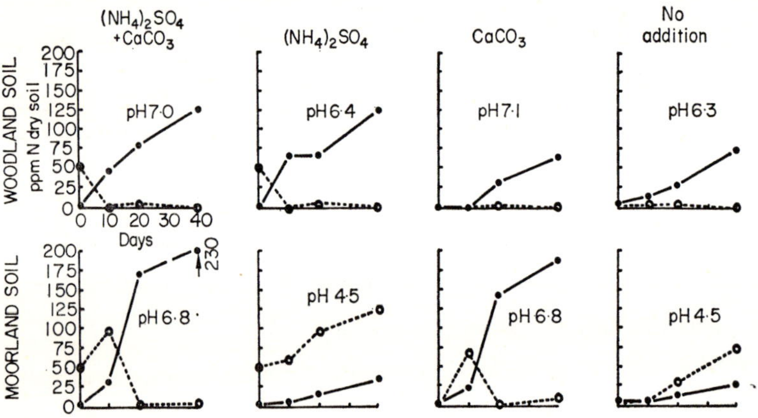

Figure 9. Soils. Incubation of soil samples.
○ - - - - Ammonium ● ——— Nitrate

attacked only slowly. Calcium carbonate additions also increased the quantity of mineralizable nitrogen in the Lime treatment at Mitchells Main and all the treatments at Upton. However, the actual quantities of nitrogen involved are smaller than in the more acid spoils and soil.

The difference between mineralizable nitrogen produced on incubation of samples with and without added calcium carbonate demonstrates the difference between potential mineralizable nitrogen reserves and the actual quantities that may be released under field conditions. It is these latter values that are most likely to be indicative of the nitrogen supply situation under field conditions and future discussion will be limited to these.

The lack of statistical information makes comparisons of the effect of treatments on the quantity of mineralizable nitrogen present in the spoils difficult. At Mitchells Main the values are variable but might suggest that the Shoddy, Sewage and Control treatments can supply more mineral nitrogen than the Lime treatment. No obvious treatment effect is however discernible. Similarly, the treatments at Upton have not produced any marked effect.

In this case all values are very similar and are much lower than those recorded for Mitchells Main spoils.

Direct comparisons between the mineralizable nitrogen produced on incubation of the soils and control spoils shows that the soils contain a very much greater mineralizable nitrogen reserve than the spoils.

THE NITROGEN STATUS OF VEGETATION GROWING ON THE FIELD TRIALS

The investigations so far described have provided some information on the form and availability of nitrogen in the different spoils. The most sensitive indicator of nitrogen availability to plants must be the plants themselves. Data on the nitrogen status of the vegetation is therefore desirable. In order to obtain this information and also to establish whether any relationship exists between the nitrogen status of the vegetation and the predicted availability of nitrogen provided by the mineralization study, sample cuts were taken. These could not be made at the time when the spoil samples for the mineralization experiment were taken because mineralization studies indicate the future availability of mineral nitrogen whilst the nitrogen status of vegetation indicates past availability. Spoil samples were therefore taken at the very beginning of the growing season in March, 1972 when mineralizable nitrogen is at a maximum (Richardson 1937) and the vegetation samples in late July, towards the end of the growing season.

Materials and methods

Three random 0·0625 M^2 quadrats were taken from each treatment plot and the vegetation sorted into living and dead components. The dry-weight of each component was measured by drying at 105°C for 24 hours before cooling and weighing. The nitrogen determinations were performed using a Kjeldahl wet-digestion method.

At both sites the vegetation was very largely composed of the sown *Agrostis* and *Festuca* grasses.

Results

The results are presented in Table 2. Presentation of the data allows direct comparison with that for the incubation study presented in Table 1. Each value in Table 2 is the mean for the three field replicates. The data was statistically analysed using a split plot analysis of variance technique. Variance ratios, and where appropriate, least significant difference values are given (Table 3).

Table 2. Nitrogen in above ground vegetation (gN. m^{-2}).

	Mitchells Main		Upton	
	Living	Dead	Living	Dead
Shoddy	0·7	4·4	1·5	1·4
Sewage	0·5	2·2	0·8	0·3
Control	0·2	1·6	0·6	0·4
Lime	0·5	4·4	0·5	0·2

Table 3. Statistical analysis.

	Mitchells Main		Upton	
Source	Variance ratio	5% L.S.D.	Variance ratio	5% L.S.D.
Treatments	1·58 n.s.		5·16*	0·77
Subtreatments (living v. dead)	16·62**	1·51	12·21**	0·18
Interaction	1·57 n.s.		0·39 n.s.	
Replicates	0·99 n.s.		0·81 n.s.	

Interpretation and discussion

The statistical procedure demonstrates that for Mitchells Main, whilst treatment effects (Shoddy, Sewage, etc.) are not significant, the dead vegetation represents significantly more nitrogen than the living. At Upton a significant treatment effect does occur and this is found to be a difference between the vegetation on the Shoddy treatment and the others. Here significantly more nitrogen is represented by the living, rather than the dead vegetation, the reverse of the Mitchells Main situation. No significant replicate or interaction effects are observed.

It could be argued that no treatment effects are observed at Mitchells Main because the ameliorants were applied five years before the present sampling was undertaken. In this time any initial effects could have been lost. This is probably not the case however because Cooper (*pers. comm.*) examined this field trial eighteen months after establishment and found that significant treatment effects could not be observed. It therefore seems likely that the treatments were ineffective at this site. The finding that the dead vegetation represented significantly more nitrogen than the living at Mitchells Main shows that there has been an accumulation of undecomposed material over the years. This commonly occurs on reclamation sites. The failure of decomposition can result in the formation of a dense mat of vegetation that can physically inhibit new growth. This situation has not yet been reached at Mitchells Main and the mineralization study indicates that some breakdown of dead material must be occurring.

At Upton the vegetation on the Shoddy treatment contains more nitrogen than that on the other treatments and indicates that the Shoddy must have been broken down by the spoil microflora. It is interesting to note at this point that when the spoil samples were collected from the sites for the incubation experiment, remnants of Shoddy were visible in the spoil at Mitchells Main, but not Upton. This confirms the view expressed above that the Shoddy had been broken down at Upton and suggests that the lack of a Shoddy treatment effect at Mitchells Main might have been due to a partial failure of decomposition processes.

The presence of greater quantities of nitrogen in the living vegetation at Upton, as compared with the opposite situation at Mitchells Main, could be a reflection of the different ages of the trials. The situation cannot be interpreted only in these terms however, because the overall pH is much higher at Upton than Mitchell. This would suggest that the breakdown rate at Upton would be greater than at Mitchell and both factors are therefore probably important.

General discussion

The inhibition of nitrification that results in ammonium being the predominant form of mineral nitrogen in the acid colliery spoil at Mitchells Main is not unique, occurring widely in acid soils (Cornfield 1952). Such inhibition is not in itself very important because it has been established for many years that grasses can utilize ammonium as a nitrogen source (Richardson 1938). The uptake of ammonium by roots can, however, affect the uptake of other mineral elements by the plant. This is important under some conditions and will be discussed later. On neutral spoils this situation does not arise because the large population of nitrifying bacteria suggests that ammonium is converted to nitrate before being taken up by the vegetation. The large population of nitrifying bacteria at Upton is very interesting because in normal soils the numbers occurring below a grass sward are usually very small. Various explanations of this phenomenon have been put forward. These include the inhibition of the micro-organisms by plant root exudates (Theron 1951) and the effect of adverse pH conditions or substrate availability (Robinson 1963). Possibly the numbers are high at the Upton site because the grass sward is newly established and may decrease with time. The conversion of ammonium to nitrate increases the possibility of nitrate loss through leaching. Walker (1956) however has emphasized the very high capacity of grasses to utilize nitrogen and accumulation of mineral nitrogen in the nitrate form should not occur during periods of active growth of the sward. At other times, especially in the winter, although the activity of the mineralizing and nitrifying populations of micro-organism would be low, some loss of nitrate could take place.

The relatively high levels of mineral nitrogen throughout the year on the Shoddy, Sewage and Control treatments at Mitchells Main suggests that the grass cover is unable to utilize the available nitrogen. Similarly, comparisons between the quantity of mineralizable nitrogen present in these spoils at the beginning of the growing season and that present in the above ground vegetation towards the end of the growing season again indicates that considerably more nitrogen was available than actually used. This holds true even if the quantity of nitrogen present in the vegetation tops is doubled to account approximately for that present in the root systems. Thus nitrogen would not appear to be a limiting factor for plant growth on these spoils. Unpublished data on the levels of other mineral nutrients in the acid spoils at Mitchells Main shows that the level of manganese and more especially aluminium may reach toxic concentrations during the summer months, water soluble aluminium concentration of over 100 ppm, often being recorded. Such toxicities may explain why the levels of mineral nitrogen remained high in the acid spoils.

The form in which the plant has to take up its nitrogen is important in this respect because uptake of ammonium leads to a fall in the pH of the medium immediately surrounding the roots (Jackson 1967). The solubility of aluminium can be greatly increased by relatively small changes in acidity and hence absorption of ammonium may accentuate the development of aluminium toxicity. The pH of the Limed spoil at Mitchells Main is high enough to prevent the solution of aluminium and manganese salts and the results of the mineralization study show that ammonium may be converted to nitrate before being taken up by the vegetation. Toxicity factors are not important in this spoil and it is therefore interesting to note that if the quantity of nitrogen in the vegetation is doubled, to account for root nitrogen, the value obtained is only just smaller than that suggested from the mineralization study. On spoil amended in this way, nitrogen availability probably is a growth limiting factor.

For all the Upton treatments the levels of mineral nitrogen remain low throughout the year and the mineralization study indicates that very little nitrogen is available for plant growth. The living vegetation however, usually contains more nitrogen than at Mitchells Main (except the lime treatment). This suggests that whilst nitrogen availability is almost certainly a growth limiting factor, the high pH allows the development of an active microflora that promotes the rapid decomposition of dead material and results in the rapid recycling of the small quantities of nitrogen present in the ecosystem. Breakdown of the applied Shoddy had occurred before the spoil samples for the mineralization study were taken and this again indiates the existence of an active microbial population at this site.

Since the vegetation on the control treatments at both sites contained some nitrogen, the proposition made in the first sentence of this paper, that colliery spoil can supply little or no nitrogen for plant growth, must be justified. The probability is that most, if not all, of this nitrogen has reached the spoil in the rainfall and in animal- and wind-borne organic debris. Dennington (*pers. comm.*) has measured the quantity of plant available nitrogen arriving at the surface of colliery spoil heaps in rainfall in the West Riding of Yorkshire. The mean value obtained is equivalent to 0·9 g N. m^{-2}. $year^{-1}$. This quantity of nitrogen must be of significance to the vegetation growing on these spoils and would appear to be large enough to account for a high proportion of the nitrogen in the vegetation growing in the control treatments.

The claim made by Cornwell & Stone (1968), that acid colliery spoil can supply some nitrogen for plant growth as a result of the release of previously fixed ammonium that occurs when lattice structures in the clay minerals are broken down, has not been substantiated by experimental evidence. Further ammonium released in this way may not be of any practical significance because the degree of acidity needed to effect its release in any significant amounts would also prevent the growth of most species of vegetation.

In conclusion, the results of the present investigation show that a nitrogen cycle has become established on the Upton spoils, but the availability of nitrogen in untreated spoil is low and limits plant growth. At Mitchells Main the availability of nitrogen is higher (but is still low when compared with a normal soil) but factors other than nitrogen availability are more likely to be limiting plant growth. Some breakdown of organic matter is occurring and, despite the low pH, some nitrogen recycling is taking place. At both sites, nitrogen deficiencies are slowly being remedied by natural agencies.

Summary

Analysis of ameliorated and untreated colliery spoil for ammonium- nitrite- and nitrate-nitrogen, together with estimates of the numbers of *Nitrosomonas* and *Nitrobacter* have been performed on acid and neutral colliery spoil. The relationship between mineralizable nitrogen and the nitrogen status and yield of plants growing on colliery spoil is described.

References

ALEXANDER M. (1965) Nitrification. In *Soil Nitrogen* (Ed. by W.V. Bartholomew & F.E. Clark) *Agronomy* No. 10, The American Society of Agronomy, Madison, Wisconsin.

ALLISON F.E. & STERLING L.D. (1949) Nitrate formation from soil organic matter in relation to total nitrogen and cropping practices. *Soil Sci.* 67, 239–52.

BLACK C. A., NELSON L.B. & PRITCHETT W.L. (1947) Nitrogen utilization by wheat as affected by rate of fertilization. *Soil Sci. Soc. Am. Proc.* 11, 393–6.

BREMNER J.M. (1965) Inorganic forms of nitrogen. In *Methods of Soil Analysis* (Ed. by C.A. Black) *Agronomy* No. 9, The American Society of Agronomy, Madison, Wisconsin.

CORNFIELD A.H. (1952) The mineralization of the nitrogen of soils during incubation: influence of pH, total nitrogen, and organic carbon contents. *J. Sci. Food Agric.* 3, 343–9.

CORNWELL S.M. & STONE E.L. (1968) Availability of nitrogen to plants in acid coal mine spoil. *Nature, Lond.* 217, 768–9.

FINNEY D.J. (1951) The estimation of bacterial densities from dilution series. *J. Hyg.* 49, 26–35.

GASSER J.K.R. & WILLIAMS R.J.B. (1963) Soil nitrogen. VII. Correlations between measurements of nitrogen status of soils and nitrogen percentage and nitrogen content of crops. *J. Sci. Food Agric.* 14, 268–77.

JACKSON W.A. (1967) Physiological effects of soil acidity. In *Soil Acidity and Liming* (Ed by R.W. Pearson & F. Adams), *Agronomy* No. 12, American Society of Agronomy, Madison, Wisconsin.

National Agricultural Advisory Service (undated) *Techniques in Soil Microbiology* 2.

RICHARDSON H.L. (1938) The nitrogen cycle in grassland soils with especial reference to the Rothamsted Park Grass Experiment. *J. agric. Sci.* 28, 73–121.

ROBINSON J.B. (1963) Nitrification in a New Zealand grassland soil. *Pl. Soil* 19, 173–83.

THERON J.J. (1951) The influence of plants on the mineralization of nitrogen and the maintenance of organic matter in the soil. *J. agric. Sci.* 41, 89–96.

WALKER T.W. (1967) The nitrogen cycle in grassland soils. *J. Sci. Food Agric.* 7, 66–72.

WEBER D.F. & GAINEY D.L. (1962) Relative sensitivity of nitrifying organisms to hydrogen ions in soils and in solutions. *Soil Sci.* 94, 138–45.

Physical problems of growing plants on colliery waste

J. A. RICHARDSON *University of Newcastle upon Tyne, Newcastle upon Tyne, U.K.*

Introduction

A handful of builders' sand is not a soil, and neither is a lump of sticky clay from the brickworks nor the surface layers of a colliery spoil heap. In this material there is no organic matter, no microorganisms, and no soil structure exists. Most of the work described here was done in north-east England where, 15 years ago, it was a great step forward when a few pit heaps were planted, as they stood, with trees and grass in order to improve the scene and to cut down air pollution by dust particles. At the present time money is available for fuller reclamation schemes in which heaps are reshaped and blended into the surrounding land; this no doubt will be the pattern in the future until all the old pit heaps disappear. In Durham, which is currently reclaiming more derelict land than any other English county, the County Council aims to reclaim about 6,600 acres (2,666 ha) between 1973 and 1981. It is proper for some enthusiasts to talk in terms of returning the land to full agricultural use, provided it is realized that there is much to be done towards improving the fertility and structure of the soil before this ideal solution can be approached. It takes many years before a stable soil structure is developed and therefore the first task of the ecologist or engineer in charge of a reclamation scheme must be to determine, as far as he can, what the future land use is to be. Clearly it would be a waste of time and money if a site shown as grassland on the plan is treated so as to build up soil structure and fertility and then in a few years is used for factories, houses or roads.

Whereas this paper is concerned with the long term management of planted spoil material leading to the formation of a fertile soil, it must be stated that in the short term raw pit waste can become the instant green field. Spectacular growth has been achieved in north-east England, where the rainfall is over 25 in/yr (0·64 m/yr), using heavy applications of high nitrogen fertilizers e.g. 6 cwt/acre (756 kg/ha) and a ryegrass mixture sown at 75 lb/acre (84 kg/ha). This form of camouflage may serve the useful purpose of temporarily disguising derelict and unsightly land and as an exercise in hydroponics it does have some interest; but this treatment fails when the high annual application of nitrogen is withdrawn and little improvement to soil structure and fertility accrues.

Soil structure depends on the size and arrangement of the different soil particles and on the size of the pores (or capillaries) between the particles. For healthy root growth it is necessary that the size and packing of the particles give a wide range of pore sizes. In a fertile soil, clay particles are bound together in groups, currently called domains, which are up to 0·005 mm in size; domains may be grouped with silt particles to form microaggregates (0·005 to 1·0 mm) or with sand particles to form aggregates (1·0 to 5·0 mm). The small pores between domains and aggregates (0·002 to 0·02 mm) are the capillaries which hold the available water used for plant growth, and the larger pores ($>$ 0·02 mm) are mainly concerned with drainage of water and gaseous exchanges. Provided the structural units are maintained, the pores and the soil will be in a favourable physical condition for root growth. Most soils depend heavily on organic stabilizing agents to retain their structure and their physical condition varies with any changes which may take place in the level of soil organic matter. The most active organic substances in this respect are thought to be polysaccharide gums which are able to bond and cross link soil particles (Swincer *et al.* 1969). These gums are produced by microorganisms that live mainly in the small pores between domains and microaggregates and which feed on soil organic matter. It can be demonstrated that soil crumbs from pasture soils tend to be much more resistant to dispersion in water because of their higher organic content than crumbs from arable soils. Raw shale material, because it has no organic matter and no stabilizing components, is either rapidly dispersed or it is so compacted as to be brick-like. It follows that until much organic matter is added to pit heap soils they will be easily deformed by rainfall and by tillage implements. Some soils are naturally well-structured and they need little effort to keep them in good heart. In other soils, because of high clay content, there is a tendency to become waterlogged, and, where there is a high silt content, soils often compact to form crusts which restrict seedling growth and restrict the entry of water and air (Richardson *et al.* 1971). The evidence is that many soils derived from colliery waste will fall into the latter categories unless they receive careful management from the beginning.

In the sections which follow an account is given of some of the physical properties of pit waste as it is found (a) on the original heaps; (b) on reshaped heaps, and a method is outlined by which changes in soil fertility can be monitored.

Untreated pit heaps

In Durham the colliery waste was tipped in the most convenient place near to the pit head and this resulted in a wide variety of mounds, ridges and cones of all sizes, aspects and slopes. Weathering commenced as soon as the waste was brought out of the pit, and, within a few years after tipping ceased on a particular heap, the surface material contained sufficient small particles for the processes of soil formation and plant colonization to commence. Richardson (1957, 1958) and Richardson & Greenwood (1967) give an account of the effect on colonization of pit heaps of (i) weathering processes, (ii) high surface temperatures, (iii) soil moisture tension.

Table 1. Particle sizes in the 3 in (7·62 cm) surface layer of spoil heaps A (aged 4 yr), B (10 yr), C (18 yr).

Particle size	A	B	C
2 mm	100	172	269
2·0– 6·4 mm	172	250	200
6·4–12·7 mm	131	156	140
12·7 mm	145	125	94
Available water	9%	12%	14%

In Table 1 the relative amounts of certain particle sizes are listed for three spoil heap soils of different ages but comparable parent material. The amounts are calculated with reference to the volume of particles of less than 2 mm occurring in uncolonized soil A (4-year-old). This amount is called 100. After 18 years heap C carried an open plant community; there was a high proportion of fine particles, and the surface soil was beginning to show some structure. This was promoted by increased organic matter being returned to the soil by dead roots and shoots. The water holding capacity of the soil also showed a steady improvement with no loss of porosity. The available water in pit heap soils was determined from a series of pF curves constructed for heaps of different ages but formed from comparable shales. The available water, as per cent dry weight, increased from 9% in a 4-year-old uncolonized soil A, to 14% in the 18-year-old soil C on which characteristic pit heap pioneer species were established. For soil from a 100-year-old heap, supporting a vigorous vegetation of grasses, herbs, shrubs and trees, the value was 23%. Increasing quantities of fine particles lead to a receptive state in the surface layers (soil B in Table 1). At this stage the seeds of pioneer species can germinate and the seedlings survive, and thereafter the interaction of plants and soil produces increased fertility and speeds colonization. The key to the whole process is the production, by weathering in the raw shale, of an appropriate mixture of small hard shale and clay particles of the correct size sufficient to retain water and still remain porous. Clearly, the nature of the spoil material, its rate of disintegration and the rate of removal of the fine particles by wind and by rain water are controlling factors in soil formation and colonization.

Immediately after tipping has stopped the infiltration capacity of a spoil heap is high. On a flat surface there is never any evidence of standing water, and the surface dries out rapidly after rain ceases. As weathering proceeds and the pore size in the surface layers becomes less, the rapid rate at which water enters the surface shows signs of decreasing until a point is reached when, for a given rain, the infiltration rate is just less than the rate of rainfall. At this point run-off occurs, and in the stream is carried the soil-making particles. Depending on the properties of the material forming the heap, the rainfall intensity may only exceed the infiltration capacity during heavy rains associated with thunderstorms. In this case the amount of soil particle accumulation may exceed the amount lost in the run-off and a fine textured surface layer can be built up. On the other hand, if the permeability of the surface is low, it is likely that even slight rain will produce some losses due to run-off. Often it is the flat top of a heap that becomes impermeable and water runs off in minor torrents down the sides. It is from the sides that the fine particles are washed away. Greenwood (1963) examined the erosion of the surface of a conical pit heap at Ouston, Co. Durham and he carried out weekly sampling from March until December 1961. The south-west side of the heap supported only a few pioneer plants whereas on the north-east side colonization was well advanced and there was 60% cover by herbs and grasses. For the south-west side the average weekly solids collected in the erosion gauges weighed 2·77 g with a range from 11·39 to 0·04 g. Predictably there was less erosion from the north-east side due to the ameliorating effect of the plant cover. Here the average weekly figure was 0·5 g and the range was 3·39 g to nil. Tests with sticky plate dust collectors suggested that, especially on the south-west side of the heap, the weight of solids carried off in rainwater was only a fraction of the whole, and that the largest part of any eroded material was blown off by the wind. Certainly there was no evidence from deep holes dug into the heap that surface material had been washed down in great quantities into the lower layers where it accumulated to form an impervious centre. Indeed, the material towards the centre appeared to be little changed, even in old heaps, from when it was first tipped.

Experiments on erosion and infiltration, reported by Pickersgill (1971), were carried out at Pittington, Co. Durham, where there was considerable variation in the composition of the surface material from place to place on the heap. Observations were made of adjoining sites which had the same climate, slope and aspect, but different texture of parent material. The heap ran north-west to south-east, it was a flat-topped ridge 760 yd (0·7 km) long by about 40 yd (36·6 m) wide and 50 ft (15·24 m) high. The slopes were cut in places by wide gullies down which large quantities of water poured, from places on the flat top, during rainstorms. At sites 1 to 5 and at site 10 along the south-west slope there was a hard crust at the surface and there were no plants growing. In contrast, sites 6, 7 and 8 had open textured, free draining surfaces and the vegetation provided up to 80% cover. The characteristics of the sites were investigated by the run-off plot method using a rain simulator and the results from two of them are given in Table 2.

Table 2. Infiltration and erosion measurements at a barren site (4) and a colonized site (7) on a spoil heap at Pittington, Co. Durham.

Site	4	7
Time for first ml run-off	10 sec	40 sec
% run-off	94·4	26·0
% in filtration	3·6	74·0
Infiltration capacity	0·0038 in/min (0·010 cm/min)	0·0854 in/min (0·23 cm/min)
Rate of erosion	8·4 g/min	0·25 g/min

The physical properties of the parent material have controlled plant growth on these sites. Where there was a high silt and clay fraction, as at site 4, the surface became compacted and waterlogged in wet weather and in dry conditions a hard crust was formed through which seedlings could not penetrate. Site 7 on the other hand possessed less extreme properties and colonization was initiated, and at the time of the experiments the vegetation almost formed a complete surface cover. On site 4 it was found that the rate of erosion was as high as 15·1 g/min at the end of 3 min, but the rate fell to 0·9 g/min after 22 min. This suggests that the rain first swept off loose material from the hard crust at the surface, and then the rain decreasingly eroded the surface until such time as the forces binding the clay particles were able to retain the surface with only small losses. Under the extremely difficult conditions of site 4, most seeds arriving at the surface would either be blown off or be washed off. Root penetration of any germinating seedlings would be poor. Furthermore, the small amount of water entering the surface would evaporate quickly in summer and fatally high soil moisutre tensions would develop.

Regraded pit heaps

A loam soil possessing a well-developed structure can resist compaction by tillage machinery at moisture contents up to around the field capacity;

but raw shale, which has no soil structure, suffers compaction even at low moisture contents. In the process of reshaping (regrading) a pit spoil heap the surface is crossed and recrossed many times by the machines used for moving and levelling and therefore the material becomes compacted, sheared and smeared in a way that does not occur on arable land or on unmodified pit heaps. Crawler (or track-type) tractors, such as the Caterpillar machines, which are used for bulldozing, levelling and hauling other machines, range from 13,700 lb (6,200 kg) to 68,000 lb (30,800 kg) weight. A feature of the lightest machine is that it can be fitted with extra-broad tracks to reduce the ground contact pressure from 7·3 lb/in² (0·51 kg/cm²) to 3·8 lb/in² (0·26 kg/cm²), but even the heaviest machines do not exceed 10·7 lb/in² (0·75 kg/cm²) because of the broad tracks. In contrast, the wheeled scrapers, which are extensively used in earth-moving operations, have large rubber tyres and a much smaller ground contact area than tracked machines. Consequently, the ground contact pressure in scrapers is high and ranges from 56 lb/in² (3·93 kg/cm²) to 70 lb/in³ (4·92 kg/cm²) or up to nearly seven times that of crawler tractors. Clearly, when pit heaps are reshaped, there could be serious deformation and compaction throughout the whole volume of the disturbed spoil, and this could act to retard or inhibit root growth. The extent of the deformation will increase as the moisture content of the spoil increases, and, as in agricultural practice, it will be an advantage if heavy machinery can be kept off the land until the shale is in a condition when it can be worked without too much compaction occurring.

It has long been known that plant roots can exert considerable force if the pore size in soils enables them to penetrate into pores. When compaction occurs in a normal seed bed, the porosity is adequate to allow the roots to grow through the spaces between the particles. If the intergranular spaces in a soil are too small to accommodate the fully grown roots, then during its growth the root exerts force to displace the soil particles elastically. The mechanical constraint on growth under these conditions is low. However, where the soil particles are cemented together, as in a compacted soil, the cohesive strength and hence the mechanical constraint is high and the growing root may not generate sufficient force to overcome the intergranular cohesion and allow turgor extension to take place. In Newcastle, Dr Hettiaratchi (see Abdalla *et al.* 1969) and his team examined the effect of soil pressures on root growth, and Table 3 is based on results from experiments on root growth in a sandy soil, and in a system of small glass beads (ballotini)

Table 3. Effect of confining pressures on root growth. Assessed over a ten-day growth period.

Pressure lbf. in^{-2}	0	2	4	6	8	10	12	14
Elongation (% of control)	100	97	75	50	30	22	20	15

1 mm diameter giving a pore size of 0·15 mm. Root growth was retarded as the mechanical impedance increased in the seed bed. Elongation was halved at 6 lb/in^2 (0·42 kg/cm^2) and was reduced to only a fifth of the control at 12 lb/in^2 (0·84 kg/cm^2). It seems that even in lightly compacted soils root growth will be a fraction of the control. The ground contact pressures of tractors and scrapers would therefore appear to present a serious obstacle to healthy plant growth. In the field there will be the additional adverse effects of restricted aeration and water supply which could further interfere with growth. The form of the root system is also modified by pressure and pore size. A greater number of lateral roots developed per unit length of seminal axes and they often emerged within 0·5 cm of the root tip. In control plants grown in media with larger pores (0·4 mm) the laterals emerged 5 to 8 cm behind the root tip. This proliferation of lateral branches in roots makes the plant vulnerable to drought. Although the laterals have been shown to be as effective as seminal roots in salt and water absorption, there is often insufficient moisture in the surface layers to maintain a positive water balance in the plant and hence allow turgor enlargement of root cells. The smaller volume of soil exploited may require larger amounts of water and fertilizer to sustain optimum yields.

In a reclamation scheme at Whitehills, Felling, Co. Durham, the compaction of certain areas in April 1973 was so great that the mean penetrometer readings were the same as for a nearby tarmacadam road. So firm was the surface that holes for tree planting could not be dug manually by spade or with a tractor-driven auger. Furthermore, a mole-plough blade attached to a tractor also failed, and to open the ground for planting it was necessary to employ a three-shank parallelogram ripper (or chisel plough) attachment on a track-type tractor. It is by no means certain that the cheapest and most effective methods of cultivation are employed on regraded colliery spoil heaps. Only one aspect has been touched on here, namely, the weight of machinery used, and there are others that would repay investigation. A number of large and expensive full reclamation schemes have been partially completed. Perhaps this is the appropriate time to examine what has been done and, if necessary, to intensify the present monitoring procedures and to modify the existing cultivation techniques.

Soil microorganisms and fertility

In many reclamation schemes it is not possible to provide supplies of well-structured top soil for use either as a blanket layer or for mixing with the shale surface. Indeed it is questionable whether such luxury treatment is necessary on more than a few sites, because there are many sites in Co.

Durham where a grass sward, or a young plantation, has been established with only a low fertilizer treatment for two years. In a successful scheme plants are able to grow to maturity so that organic material is returned to the ground and a process of natural soil formation initiated. Organic matter and associated humus improves the texture and structure of the young soil; a source of major nutrients is provided and also a store for moisture and added fertilizers. Organic matter is the primary source of food for microorganisms, one activity of which is the release of inorganic ions into the soil; another is the production of polysaccharide gums that keep the soil particles in place and aids the development of soil structure.

The activity of soil microbes is a measure of soil fertility. High activity shows that processes concerned with the improvement and the maintenance of soil structure are taking place. The improved physical conditions not only help root growth but also provide a better environment for the soil flora and fauna. At several reclaimed colliery tips in Co. Durham the effects of treatment given to the shales before, during and after planting are being monitored. The activity of soil microorganisms is one of the soil properties measured. The method devised for this purpose, which is described below, could provide a rapid assessment of the effect of any chemical additives (fertilizers, herbicides, fungicides, pesticides) and hence lead to greater confidence in prescribing crop or land treatment.

A wide range of techniques has been used to study the growth of soil microorganisms. For example, Weston *et al.* (1965) examined the activity of microorganisms occurring naturally in waste tips at Swansea by observing the rates of colonization of cellulose film buried in the ground. These same workers also obtained another measure of the activity of soil microbes in waste tips by measuring the CO_2 evolved from incubated samples of tip material either untreated or amended with sewage sludge. Many methods consist essentially of extracting the microbes from the soil in solution and then measuring their numbers as they grow on culture media. Many nutrients may be included in the media used to make counts. A disadvantage of this method is that some microbes, that are not active in the soil, may grow extremely well on the culture medium. On the other hand, others, very active in the soil, may not be suited to the type of culture material currently in use.

In an actively dividing population of soil microorganisms there is a synthesis of protoplasm which is complementary to cell division. Prior to cell division synthesis of DNA (deoxyribonucleic acid) takes place and this process requires a supply of deoxynucleosides. Thymidine is the deoxynucleoside which is utilized in DNA synthesis but not normally in that of RNA. Consequently, the incorporation of thymidine into DNA may be used to indicate the activity of the microbial population of the soil sample. In the method described below (being developed at Newcastle by D. R. Thomas, J. A. Richardson and R. J. Dicker), a known weight of soil was placed in contact with radioactive thymidine of known activity for a given time. The amount of labelled thymidine that was incorporated into microbial DNA was determined by extracting the DNA and then precipitating it; finally the radioactivity was measured by means of a liquid scintillation spectrometer. From the results a measure of the growth activity of at least part of the soil microbes was obtained.

The soil sample (5 g) was placed into a Pyrex weighing bottle and then 2 ml sterile ^{3}H-Tdr (99·76 × 10^4 counts/min per ml) was pipetted on to the soil. The sample was incubated at room temperature for 3 to 5 hours and then 50 ml 5% trichloroacetic acid (TCA) was used to kill the microorganisms and to wash the soil into a 150 ml conical flask. The 5% TCA and soil were heated to 90°C and maintained at this temperature for 20 min to break open the microorganisms and dissolve the DNA (Schneider 1945). After cooling, the mixture was centrifuged at 8,000 g for 10 min. The supernatant was retained. The soil pellet was resuspended in 5% (w/v) TCA and centrifuged at 8,000 g for 10 min. The supernatants were bulked and to the resulting solution was added one-seventh of its volume of saturated aqueous copper sulphate. The solution was allowed to stand overnight at room temperature when the DNA was precipitated as copper nucleate (Deluca *et al.* 1953). The final solution was passed through a Millipore filter (pore size 0·8 $\mu \pm 0{\cdot}05$ μm) which retained the DNA. The copper nucleate was washed under gentle suction (4 ml per min) with 20 ml distilled water and five successive 20 ml aliquots 20% (w/v) TCA to remove free ^{3}H-Tdr in the filter. The filter was air dried under suction and transferred to a 56 mm × 25 mm diameter scintillation vial. Then 2 ml 2-methoxyethanol and 3 ml of PPO-POPOP scintillation fluid were added to the vial. The PPO-POPOP scintillator consisted of a solution of 5 g 2,5-diphenyloxazole (PPO) and 0·3 g 1,4-di-2 (5 phenyloxasolyl)-benzene (POPOP) dissolved in 1 l Analar toluene. The methoxyethanol was added to dissolve the Millipore filters and to absorb any water. The addition of methoxyethanol greatly improved the count rate, presumably because it reduced quenching by residual water and by the filter itself. Also it would aid dispersal of ^{3}H-DNA impregnated on the filter.

Using this method an experiment was performed with a loam soil from a permanent pasture and a young soil developing on a grassed area of a reclaimed colliery site. 5 g samples were incubated for 0, 3 and 24 hours and the results are presented in Table 4. The counts obtained at 0 hours, between 30 and 50 counts/min per g soil, indicate that traces of unchanged ^{3}H-TDR remained on the filters. However, these quantities of ^{3}H-TDR were small and constant and did not affect the overall picture which emerged from the results. It may be concluded that ^{3}H-Tdr was incorporated into the replicating DNA of microorganisms and that the maximum rate of incorporation was in the first three hours of incubation. The loam soils incorporated larger amounts of ^{3}H-Tdr into DNA than did the shale, demonstrating that microorganisms were dividing more rapidly in the loam soil.

Table 4. Activity incorporated from ^{3}H-Tdr with time in a young shale soil and a loam soil.

Soil	Incubation time (h)	Counts/min per 5 g soil
Shale	0	53·3
	3	226·0
	24	346·5
Loam	0	57·9
	3	725·2
	24	812·7

Following on from these experiments a possible future extension might be to separate labelled DNA of different molecular weights in the samples. The DNA with the highest specific activity would come from the most actively dividing microorganisms in that particular soil sample. By isolating DNA from pure cultures of soil microorganisms and comparing its properties with labelled DNA from soil samples, it may be possible to distinguish between those organisms which are actively dividing and those which are dormant in the soil. However, a prior task in the development of this method is to obtain information about soil microorganisms which may not incorporate Tdr at all, and about the possibility that ^{3}H-Tdr might be absorbed by some microbes but then degraded by Tdr-phosphorylase to ^{3}H-thymine and deoxyribose-l-phosphate. In this event the labelled material may be incorporated into other metabolic products as well as into DNA.

Summary

There is often a wide variation in the physical properties of colliery waste material and an account is given of experiments to determine these properties. Attempts are also made to find ways of obtaining amelioration in order to provide conditions which encourage an improvement in soil fertility and hence plant growth.

References

ABDALLA A.M., HETTIARATCHI D.R.P. & REECE A.R. (1969) The mechanics of root growth in granular media. *J. Agric. Engng. Res.* 14, 236–48.

DELUCA H.A., ROSSITER R.J. & STRICKLAND K.P. (1953) Incorporation of radioactive phosphate into the nucleic acids of brain slices. *Biochem. J.* 55, 193–200.

GREENWOOD E.F. (1963). *Studies in the Spread of Plants on to Pit Heaps.* M.Sc. thesis, Univ. Newcastle upon Tyne.

PICKERSGILL B. (1971) *Studies in the Growth of Plants on Pit Heaps.* M.Sc. thesis, Univ. Newcastle upon Tyne.

RICHARDSON J.A. (1957) Derelict pit heaps and their vegetation. *Planning Outlook* 4, 15–22.

RICHARDSON J.A. (1958) The effect of temperature on the growth of plants on pit heaps. *J. Ecol.* 46, 537–46.

RICHARDSON J.A. & GREENWOOD E.F. (1967) Soil moisture tension in relation to plant colonization of pit heaps. *Proc. Univ. Newcastle Phil. Soc.* 1, 129–36.

RICHARDSON J.A., SHENTON B.K. & DICKER R.J. (1971) Botanical studies of natural and planted vegetation on colliery spoil heaps. *Landscape Reclamation* 1, 84–99, I.P.C. Press, Guildford.

SCHNEIDER W.C. (1945) Phosphorus compounds in animal tissues. I. Extraction and estimation of desoxypentose nucleic acid and of pentose nucleic acid. *J. Biol. Chem.* 161, 293–303.

SWINCER G.D., OADES J.M. & GREENLAND D.J. (1969) The extraction, characterisation and significance of soil polysaccharides. *Advanc. Agron.* 21, 195–235.

WESTON R.L., GADGIL P.D., SALTER B.R. & GOODMAN G.T. (1965). Problems of vegetation in the Lower Swansea Valley, an area of extensive industrial dereliction. In *Ecology and the Industrial Society* (Ed. by G.T. Goodman, R.W. Edwards & J.M. Lambert), Blackwell, Oxford.

Pulverized fuel ash as a medium for plant growth

W. N. TOWNSEND and E. W. F. GILLHAM *Department of Plant Sciences, University of Leeds and the Central Electricity Generating Board, U.K.*

Introduction

In the 1950s, the British Electricity Authority (now the Central Electricity Generating Board) recognized the magnitude of the ash disposal problem with which it would be faced in the ensuing decades. For example, a modern generating station of 2,000 mW capacity (of which about 10 are now operating) produces ash at the rate of about 2 tons per minute. This is normally discharged to worked-out gravel pits and similar holes in the ground but occasionally may be mounded above the original land surface. This creates a need for environmental rehabilitation calling for the development of some form of plant cover, which may range from simple amenity restoration to the establishment of sites capable of the agricultural production of economically valuable crops.

Research work was accordingly sponsored at the Universities of Birmingham and Leeds to find acceptable solutions to this problem; this work has subsequently been continued and expanded on the farm scale by the C.E.G.B.

The nature of ash

The ash produced from coal contains the inorganic materials present in the plants from which the coal was formed (representing no more than 1 or 2% of the coal), together with extraneous mineral matter, sometimes representing as much as 20% of the bulk. This mineral matter approximates closely to the soil in which the coal measure plants grew, and is similar in general composition to many present-day mineral soils.

In a boiler using pulverized coal, the coal is ground to the consistency of face powder and combustion takes place in suspension in air. Individual mineral particles such as silica, clay and magnetite remain physically separate and a small proportion of incompletely burned fuel may remain in the ash. This is produced in the form of a light- to dark-grey powder largely composed of spherical glassy particles. The particle size distribution of ash is akin to that of a light silty soil with about 60% of the particles within the 'fine sand' category (0·2 to 0·02 mm) with about 40% in the 'silt' (0·02 to 0·002 mm) fraction. Clay size particles (below 0·002 mm) are present in negligible proportions. This size range is of considerable interest agriculturally, since it has been shown (Cope 1961) that the available water capacity of a soil is positively correlated with the fine sand fraction and negatively correlated with the proportion of coarse sand size particles present.

As a result of the combustion process by which it is produced, the ash is completely sterile and virtually devoid of nitrogen. Apart from this however plant nutrients are present in acceptable quantities.

ASH DISPOSAL SITES

The structural characteristics of ash sites vary according to whether the ash has been deposited in situ by hydraulic transport or whether the material has been deposited mechanically, either after excavation from a lagoon or by a completely dry system involving merely the addition of sufficient water to control the potential dust nuisance. Disposal to clay pits (as at Peterborough) can present mechanical problems due to the difficulty of removing the excess water, but in gravel pits the ash is normally free-draining and even when the water table is high a stable surface is formed. Nevertheless, layers of low permeability can be formed in lagoons due to spontaneous cementation in the ash particles, probably resulting from the presence of free lime and its subsequent reaction with the siliceous particles present; such layers may give rise to adverse drainage situations and can impede root development and penetration. Where excavation and transportation to new sites has occurred, these problems are minimized and ash drainage conditions are usually good.

A further physical problem associated with bare ash deposits is found in the susceptibility to wind erosion, and 'blowing' can be initiated in certain cases when wind velocities reach 16 to 18 km per hour. This problem is not, however, peculiar to ash sites. A survey of wind erosion on sandy soils in the East Midlands (Salter *et al.* 1966) found that the soils liable to erosion had a total sand content of 80 to 90% and that the size fraction between 100 and 200 microns was particularly active in initiating blowing. Studies in the Vale of York (Holliday & Townsend 1959) suggested that soils with less than 3% of clay size particles were those most susceptible to wind erosion. Only a small proportion of ash particles fall within the critical range of 100 to 200 microns; most are finer and the ash thus appears capable of stabilizing 'blowing' soils.

The effects of weathering

The ash of the original coal measure plants is largely present in the form of water soluble salts and may appear in the final product as discrete salt particles retaining their solubility in water. Fractional distillation processes occurring during combustion can lead to the deposition of soluble matter by condensation on to the surface of the glassy spheres, whilst other material may be released more slowly by hydrolytic reactions occurring within the spheres. These effects give rise in young deposits to salinity problems which diminish with exposure and aging. During combustion, any calcium carbonate present in the ash is converted to calcium oxide and this results in aqueous extracts of the ash having pH values rising to 11 or above. Both pH and salinity diminish with age and the pH often stabilizes at about 8.

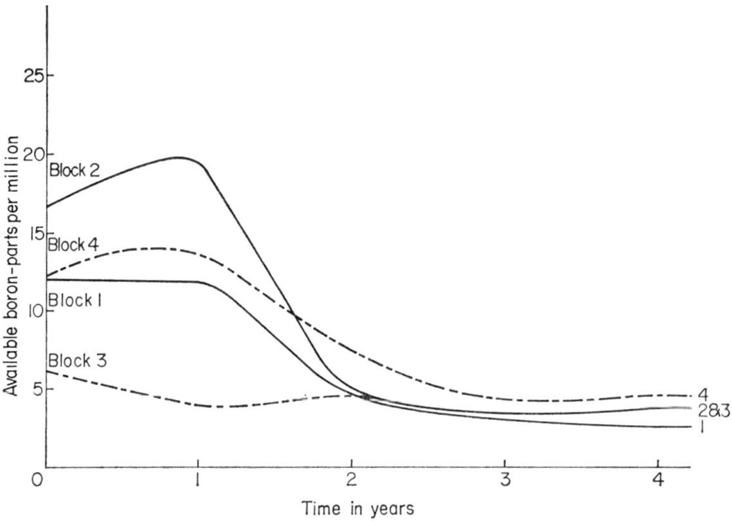

Figure 1. Variation in boron with time—Ironbridge—blocks 1 to 4.

The weathering of ash deposits has been studied and is illustrated graphically in Figs 1, 2 & 3. Figure 1 represents the variation in concentration of boron (the principal growth inhibitor present in fresh ash) with time; the two solid lines represent new ash and the two broken lines ash which had weathered for 20 years in a stockpile before being formed into trial plots. Figures 2 and 3 illustrate the variation in concentration of four elements with depth and a comparison of these two diagrams illustrates the restriction on water movement imposed by the presence of a hard pan in the ash at a depth of 20 to 50 cm.

These curves illustrate the beneficial effects of weathering on ash to be used as a plant growth medium. It may be inferred that, with an ash whose boron content is moderately high (and it must be noted that the boron content of some ashes is considerably higher than that used in this experiment), weathering for a period of three years produces an acceptably low

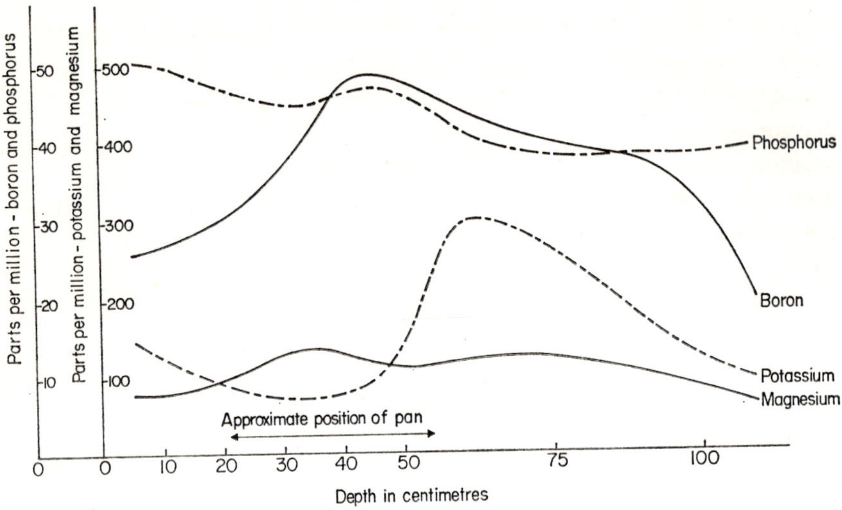

Figure 2. Variation of selected elements with depth—pan present.

boron content. The fresh ash used in these experiments had not been subjected to a lagooning process but was merely conditioned by the addition of sufficient water to control dust. Lagooning, in which excess water is used to transport the ash can, affect a significant reduction in the soluble salt concentration remaining in the ash. This is of material benefit in establishing plant growth on the final ash surface.

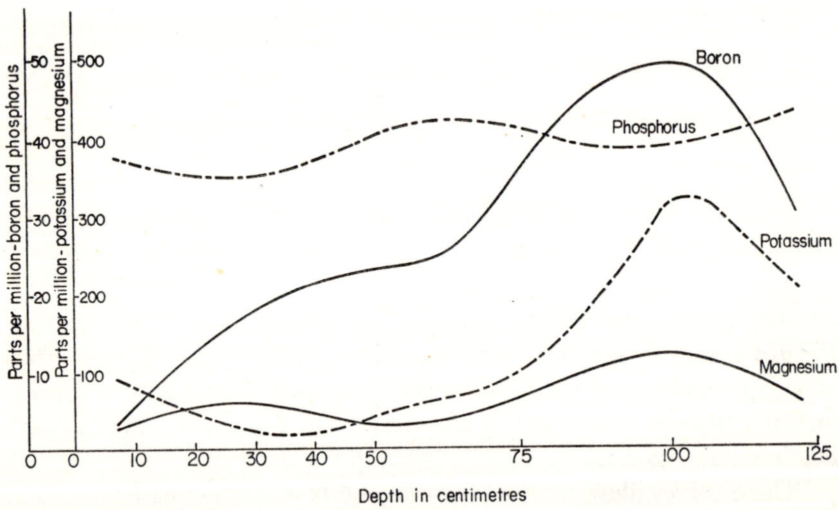

Figure 3. Variation of selected elements with depth—pan absent.

Plant nutrient status

Assessment of plant nutrient status, in terms of macro- and micro-elements, shows considerable differences between ash samples and typical fertile soils. Table 1 gives comparative figures obtained from analyses of many samples. The data suggest that p.f.a. is amply supplied with all mineral nutrients required for plant growth with the exception of nitrogen, but field trials have demonstrated that the high level of available phosphate indicated by laboratory analyses is not borne out in practice as crops respond well to phosphate

Table 1. Nutrients and trace elements in ash and soil (from Cope 1961).

Nutrient	Ash			Typical Soil
	Total %	'Available' %	range	'Available' %
P_2O_5	0·144	0·033	(0·013–0·054)	0·022
K_2O	2·68	0·042	(0·008–0·115)	0·027
Na	0·77	0·055	(0·012–0·123)	0·020
Ca	4·69	0·090	(0·28 –1·83)	0·080
Mg	0·71	0·150	(0·008–0·516)	0·024
Fe	6·09	0·057	(0·001–0·229)	0·013
S	0·48	0·390	(0·07 –0·55)	0·060
N	0·035	—	— —	(Total) 0·180
	ppm	ppm		ppm
Mn	848	99	(12–347)	4·8
B	236	43	(3–150)	2·5
Zn	283	2	(0–4)	2·5
Cr	248	25	(10–50)	2·5
Mo	42	5	(0·7–13)	0·2

applications and may show phosphorus deficiency symptoms when grown without fertilizers. This may be associated with high levels of soluble aluminium in the ash deposits which may interfere with the uptake of phosphorus by the plant.

Even when amply supplied with nitrogen and phosphate, many plants grown on ash show poor growth and leaf symptoms indicative of nutritional disorders. The syndromes exhibited indicate nutritional disorders and the presence of plant growth inhibitors in the ash. The high pH values associated with ash deposits, whilst not necessarily a direct cause of growth inhibition, may adversely influence the availability of normal mineral nutrients, causing increased or decreased uptake by plants. Natural or artificial reduction of the pH is accompanied by a commensurate improvement in growth although toxicity symptoms are not necessarily obliterated.

High salinity, as referred to earlier, can also depress growth through the high osmotic pressures of the soil solution restricting transpiration. Salinity effects are usually assessed from electrical conductivity measurements made upon saturation extracts of soil. Where values exceed 4·0 millimhos/cm, crop yield reductions may be expected. Electrical conductivities of fresh ash samples ranged from 8·4 to 13·1 millimhos/cm but natural weathering in the field for four years has reduced figures to less than 3. In this connection, it is of interest to note that one of the early colonizers of ash surfaces left to weathering naturally is the typical salt-marsh plant, *Atriplex hastata*.

Several growth inhibiting elements are also present in p.f.a. and these may give rise to rehabilitation problems. Aluminium, manganese and boron in particular may cause plant nutritional problems although in practice only boron appears to present a major hazard. Analyses of ash samples for water soluble boron suggests that figures indicating more than 20 ppm are likely to prove toxic to most agricultural crops, the more sensitive crops such as barley, peas and beans showing symptoms at values as low as 7 ppm. Available boron levels do reduce with time and weathering, as illustrated in Fig. 1, but the rate of reduction may be slow due to the progressive decomposition of what are believed to be borosilicate complexes within the ash, maintaining levels against leaching losses. No rapid means of ameliorating this toxicity has been found and prolonged weathering or covering the ash with a soil layer appears the only solution. Nevertheless, encouraging results have been obtained with the growth of a number of agriculturally important crops on experimental plots and, more recently, on the field scale.

Plant growth

Following the earlier experimental work at Birmingham and Leeds Universities, a preliminary plot study was made in 1962 at Ironbridge Power Station. Fresh ash from the power station dust collectors was formed into a plot approximately 10 m by 20 m, extending to a depth of approximately 1·25 m. Attempts to grow various crops on this plot, half of which was treated with a balanced fertilizer and half left untreated, indicated that fertilizers were essential for successful crop production.

In the following year (1963) two pairs of plots were established, one pair consisting of fresh ash from the power station and the other pair of ash weathered on a stockpile for approximately 20 years. One plot of each of these pairs received a top dressing of 3 inches of silt from the River Severn which was intimately mixed with the ash to a depth of 6 inches. The object of this admixture was to improve the texture of the surface ash and to act as a diluent for the boron known to be present. In the first year, the eastern half of each plot received no fertilizer but the western half received fertilizer at the rate of approximately 250 kg per hectare of nitrogen and phosphate and 125 kg per hectare of potash on crops other than legumes; the legumes received approximately 60 kg per hectare of nitrogen, 200 kg per hectare of phosphate and 150 kg per hectare of potash.

The plots were sown with cereals, grasses, legumes and other crops. Some crops grew well, particularly the beet and spinach (both members of the Chenopodiaceae) whilst others failed partially or even completely. The sensitivity to ash conditions closely followed that recorded by Cope (1961) in his experimental work at Leeds. The fresh ash inhibited the growth of many crops with the exception of those already mentioned (the beet family and a few brassicas). Many plants died early in the season. The results obtained from these trials, although incomplete in themselves, gave a useful basis for the planning of the continuing experimental work on these plots. The growth experiments were paralleled by an investigation of the chemical nature of the two ashes and of the silt and by chemical analyses of plant material for trace element uptake.

In later years, the fertilizer treatment was varied to permit a study of the effects of individual plant nutrients and following this pattern it was apparent by the end of 1964 that although the principal fertilizer response was (as might be expected) to nitrogen, responses were also being shown by potatoes and beans to potassium. The principal growth retarding factors still appeared, however, generally to be shortage of nitrogen and excess boron. In 1964, nitrogen was applied to all crops at the rates of 500 kg per hectare and 125 kg per hectare but it was noted that the legumes failed on the plots receiving the higher rate of nitrogen addition; this was attributable either to the salinity of the ash or to ammonia poisoning due to the release of ammonia from its salts by the alkalinity of the ash. This failure could however also be attributed to the failure of the legumes to form nodules in the presence of excessive nitrogen additions. In subsequent years, changes were made both in the amount and type of nitrogenous fertilizer added; the two forms of nitrogen used were nitro chalk and calcium nitrate and the rates of addition were reduced to 62 kg per hectare on the legumes and 250 kg per hectare on other crops.

The legumes generally grew well on Blocks 3 and 4 (of old ash, with and without silt respectively) but initially the growth was less good on Blocks 1 and 2 (of new ash), the plants on these blocks showing symptoms of boron excess. Because of the initial sterility of the ash, the legume seeds were inoculated with the appropriate strain of bacteria.

Red clover (S123) white clover (S100) and lucerne have been grown regularly for a number of years on the plots and apart from failures in the early years on the new ash plots, attributable to excess boron in the ash, all have grown well. Lucerne appears to be somewhat more tolerant of the boron than the clovers. In the later years, the clovers particularly responded well to additions of potash. Peas failed (except on Block 3) in 1963 but in 1968, after five years additional weathering, growth was good but again a marked response to potash was noted.

Of the grasses, timothy, cocksfoot and ryegrass have been grown in each year of the experiment and, apart from Block 2 (new ash with no silt) have grown reasonably well under all conditions. Symptoms of excess boron were noted in the early years of the experiment but these were reduced with the progress of time.

Attempts were made to grow wheat, barley, oats and rye but because of the depredations of jackdaws, the growing of cereals was discontinued in 1969. In the early years, the growth of wheat and barley was poor but there was a progressive improvement as weathering of the ash progressed. In 1968, all plots showed themselves capable of growing barley but there were some symptoms of manganese deficiency on the old ash Block No. 4 (no silt). All cereals have shown a marked response to additions of nitrogen.

Various root crops have been grown but, apart from potatoes and beet, other crops have been grown sporadically. In 1963, potatoes grew well on Block 3 (old ash with silt) but failed completely on Block 2 (new ash alone); potatoes have proved generally to be sensitive to boron but, as with other crops, their performance has improved progressively with the years as the boron content of the ash has been reduced by weathering. Red beet and sugar beet have proved to be useful pioneer crops; even on Block 2, sugar beet survived the first growing season. Progressive improvement has accompanied the reduction in boron content by weathering. Some manganese deficiency was noted on Block 4 in 1967. In 1963, turnips grew well except on Block 2. Again, after weathering, it has proved possible to grow turnips and carrots on all plots. Spinach has usually grown well and has proved reasonably tolerant of ash conditions except on Block 2 in the early years. Kale has behaved similarly.

Plates 1 & 2 illustrate two typical blocks in 1963 whilst Plates 3 & 4 illustrate the same pair of blocks in 1969 and demonstrate clearly the effects of weathering. Table 2 shows that the ash has weathered to an acceptable degree.

From time to time, attempts were made to measure crop yields but because of the small scale of the experiment, the results cannot be regarded as highly accurate; nevertheless, useful comparisons can be drawn of the effects of added nutrients. The results show that the principal fertilizer response has been uniformly to nitrogen, with smaller responses in some cases to phosphate and potash (Tables 3 & 4).

Plate 1. Block 2 at Ironbridge—September 1963 (New ash alone).

Plate 2. Block 1 at Ironbridge—September 1963 (New ash with river silt).

Plate 3. Block 2 at Ironbridge—September 1969.

Plate 4. Block 1 at Ironbridge—September 1969.

Table 2. 'Soil' analyses—1969–70 (Ironbridge Experiment).

Block Plot Ref.	1969 ppm P	K	Mg	Mn	B	pH	Specific cond. μmhos. cm^{-1}	1970 ppm P	K	Mg	B	pH
1B	25	95	30	10·8	2·80	7·0	2150	40	165	34	2·30	7·7
1J	18	55	32	11·5	2·84	7·7	2100	17	75	38	3·57	8·0
1 Gen.	25	85	30	9·0	2·76	7·4	2130	33	110	29	2·48	7·6
2B	37	80	25	9·5	3·72	7·6	2080	32	115	20	3·32	8·0
2J	17	78	25	10·8	3·00	7·8	2130	16	110	24	2·85	8·1
2 Gen.	45	57	26	9·8	3·28	7·9	2080	31	125	21	3·13	8·1
3B	22	123	26	14·0	3·20	7·8	2060	27	125	45	3·47	7·9
3J	11	70	42	15·0	2·64	7·7	2060	11	90	46	2·80	7·9
3 Gen.	16	130	40	15·0	2·70	7·6	2120	19	110	50	2·67	7·7
4B	32	135	54	12·0	4·28	7·9	2080	31	180	49	3·90	8·1
4J	19	88	28	14·0	4·40	8·3	2060	18	125	46	3·60	8·2
4 Gen.	29	106	32	11·6	4·68	8·0	2050	28	130	47	3·80	8·1

Note

Block References
1 New ash with river silt
2 New ash alone
3 Old ash with river silt
4 Old ash alone

Plot References
B All nutrients added
J No fertilizer

Table 3. Yield of kale in 1965, averaged to show effects of nutrients (Ironbridge Experiment).

Block	Nitrogen Plots	Yield	Phosphate Plots	Yield	Potash Plots	Yield
1	All nutrients	7·2	All nutrients	7·2	All nutrients	7·2
	+N*	7·0	+P†	0·3	+K‡	2·1
	−N*	0·4	−P†	4·4	−K‡	4·1
2	All nutrients	5·7	All nutrients	5·7	All nutrients	5·7
	+N*	3·2	+P†	0·2	+K‡	1·2
	−N*	0·1	−P†	1·9	−K‡	1·6
3	All nutrients	10·9	All nutrients	10·9	All nutrients	10·9
	+N*	7·6	+P†	0·4	+K‡	2·1
	−N*	0·6	−P†	4·8	−K‡	4·7
4	All nutrients	12·8	All nutrients	12·8	All nutrients	12·8
	+N*	7·8	+P†	0·0	+K‡	2·7
	−N*	0·0	−P†	3·8	−K‡	3·8

* P and/or K present or absent
† N and/or K present or absent
‡ N and/or P present or absent

Table 4. Yields of root crops averaged to show effect of nitrogen (Ironbridge experiment).

Block	Plots	Potatoes				Sugar Beet			
		1967	1968	1969	1970	1967	1968	1969	1070
1	All nutrients	18·1	20·2	21·1	23·0	6·5	14·3	11·8	—
	+N*	11·9	14·7	8·2	9·1	13·4	9·1	12·3	6·5
	−N†	6·8	8·2	5·2	8·0	1·1	4·0	6·2	5·6
2	All nutrients	9·4	13·5	13·2	11·9	16·1	10·3	15·5	3·1
	+N*	5·0	9·8	9·4	5·2	8·1	9·0	6·0	6·8
	−N†	1·9	1·9	2·7	2·6	0·9	1·0	0·9	4·2
3	All nutrients	12·2	14·2	15·4	21·0	13·9	11·0	25·6	17·5
	+N*	3·6	10·7	10·3	14·2	5·2	12·5	14·7	18·6
	−N†	4·7	6·2	3·8	10·1	1·3	4·1	5·8	7·9
4	All nutrients	11·2	13·2	16·7	20·0	9·8	8·9	20·0	23·4
	+N*	9·0	10·5	11·9	12·6	6·8	12·0	17·8	13·3
	−N†	2·9	2·6	2·2	7·8	0·6	0·7	3·4	8·8

* Nitrogen added, alone or with phosphate or potassium.
† No added nitrogen.

Field scale experiments

In parallel with the successful experiments at Ironbridge, some large plots were laid down by the profiling of an ash disposal lagoon near Newark (Gillham & Morley Davies 1972) to receive soil depths of 0, 8 cm, 15 cm and 30 cm of the original topsoil; half of each plot was then cultivated to a depth of 30 cm in order to mix the ash into the soil whilst the remainder was cultivated only to 8 cm depth to preclude mixing. Thirty-two such plots were fitted into an area measuring approximately 92 m^2; each plot was approximately 6 m wide and 46 m long. These plots formed part of a commercial farm holding and were of sufficient size to enable them to be farmed by conventional methods. The layout of these plots is shown in Fig. 4, from which it will be noted that each of the 16 treatments is duplicated in a different part of the field.

At the start of the experiment, the whole plot area (including the headlands) was seeded on 19 September 1966 with a conventional seeds mixture consisting of perennial ryegrass, timothy and white clover at a rate of 35 kg/hectare. In each of the three years 1967, 1968 and 1969 the grass was cut for hay in order to estimate yields and the aftermath was subsequently grazed. In 1967, the yields of dry matter ranged from 1,660 kg/hectare (on a deep-cultivated plot with no soil) to 5,625 kg/hectare on a shallow-cultivated plot

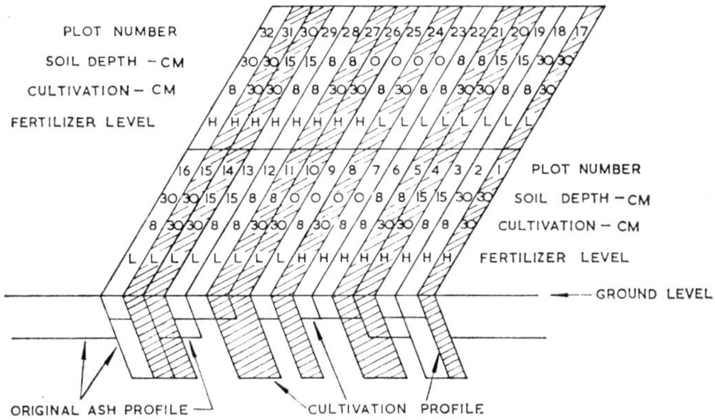

Figure 4. Diagrammatic layout of plots at Besthorpe.

with 8 cm of topsoil. In the two subsequent years, the range of yields was from 2,360 to 6,860 kg/hectare, on the same plots two previously noted. Yields on the intermediate plots have remained in roughly the same merit order but it is of interest to note the progressive increase in yield from the plots without soil. These results are shown graphically in Fig. 5. As in the case of the small plot-scale experiments, there was a marked response to the addition of nitrogen. Figure 5 illustrates strikingly that the the optimum yields of grass were obtained with only 8 cm of topsoil.

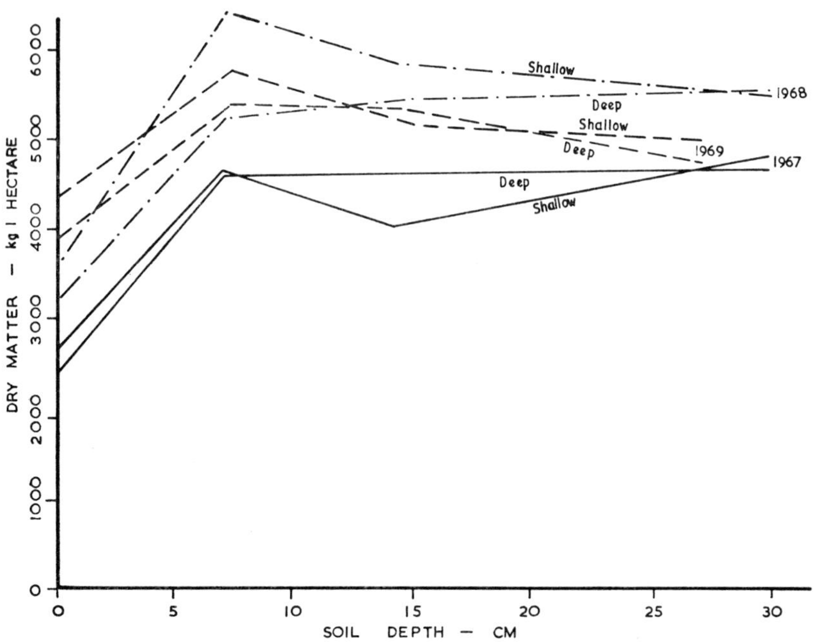

Figure 5. Yields of dry matter from grass.

After 3 years, the entire plot area was cultivated thoroughly to a depth of 3 inches (using a rotovator) in an attempt to destroy the grass and was then sown with winter wheat, which was grown in three consecutive seasons. Once again, yields were measured in each year by taking a cut along the centre of each strip with a combine harvester and weighing the grain extracted (Plate 5). There was again a marked response to nitrogen; the yields ranged from 1,750 to 6,990 kg of grain per hectare. The results are shown graphically in Fig. 6 which, as in the case of the grassland phase of this experiment, shows that the optimum yield is obtained with a soil depth of only 8 cm.

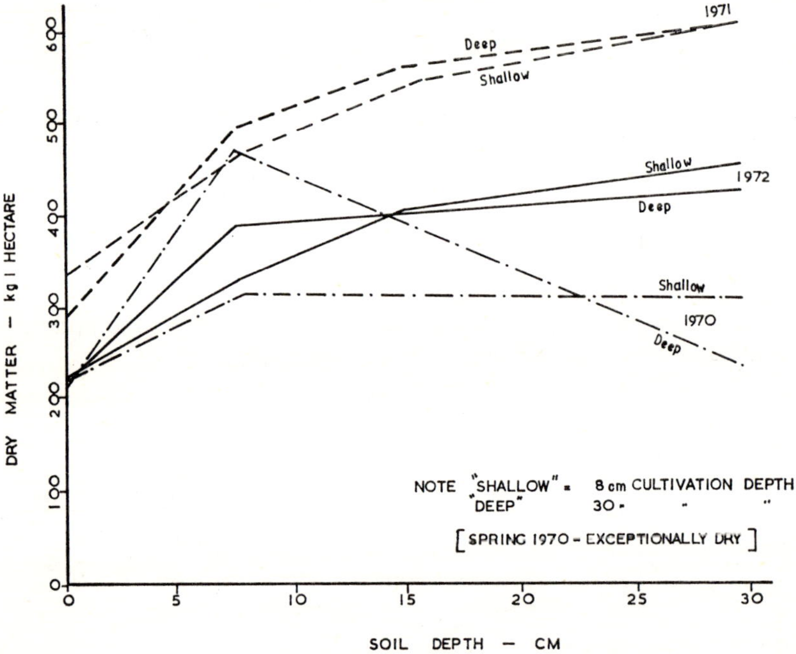

Figure 6. Yields of dry matter from wheat grain.

There are however differences between the patterns of grass and wheat yields. In the case of grass, the yields of the various plots follow a similar pattern in the three years of the experiment and show a slight decrease in all cases as the soil depth is increased beyond 8 cm. The wheat figures are however considerably influenced by a striking difference in yield pattern in 1970 (a year in which the early summer was exceptionally dry) in which the highest yield was obtained with a soil depth of 8 cm cultivated to a depth of 30 cm; with greater depths of soil, the yield decreased markedly. In the later years, yields increased progressively with greater soil depths with both deep and shallow cultivations. This marked difference in 1970 probably represents a

Plate 5. Harvesting winter wheat at Besthorpe, September 1971. (Sampling cuts along the plots—varying soil depths.)

striking illustration of the increase in available water capacity brought about by incorporating ash into the topsoil (Salter *et al.* 1971).

On the farm scale, a survey of about 30 sites restored to farm use by filling with p.f.a. was conducted in 1963 to 1968 (Morley Davies *et al.* 1971). Most of the sites examined had been restored to grassland and yields were measured either as hay or as grazing days for cattle or sheep, but a few of the sites were used intermittently for arable crops. The results indicated that with good management and the generous application of nitrogen, the crop yields were in no way inferior to those obtained on similar land before mineral extraction. No adverse effects on the health of grazing animals were observed during this survey.

Plant composition

Apart from its major constituents derived from clays, shales, sandstones plant nutrients and other infiltrating minerals, pulverized fuel ash is known to contain traces of many other elements, some of which are known toxins. Because of the presence of these elements, numerous analyses of plant material have been made and are summarized in Tables 5 & 6.

Boron, is known to act as a plant growth inhibitor and has been found to occur in plant material at concentrations up to 400 ppm; the other principal elements of interest appear to be arsenic, cobalt, chromium, copper, manganese, molybdenum, nickel, lead, selenium, mercury and zinc.

A statutory limit of 1 ppm arsenic has been fixed on food for human consumption. The 96 analyses which have been made for arsenic show results uniformly below this level in potential human foods. The highest arsenic concentration recorded was 3·5 ppm on the dry matter of a group of grass samples but even this reduces to below 1 ppm on a green matter basis. The pasture concerned was grazed by sheep and no ill-effects were recorded.

Many analyses of pasture herbage have been made for cobalt, whose absence gives rise to a deficiency of vitamin B12 in ruminants. The average of the recorded values is 0·17 ppm but the health of animals grazing the pastures concerned did not appear to have been affected adversely by upward or downward departures from the optimum cobalt level of 0·1 ppm for sheep (A.R.C. 1965).

The highest chromium concentration recorded was 8·7 ppm in one sample of lucerne grown on old ash at Ironbridge. This falls well within the safe range for cattle, corresponding to a daily intake of about 160 mg by a dairy cow. Chromium levels generally were in the range up to 4 ppm but apart from the early years at Ironbridge, figures have generally been much below this level.

Copper and molybdenum are of interest in the nutritional balance of ruminants and some low blood copper levels have been observed in young cattle grazing on restored pastures. In spite of this however, no clinical symptoms of hypocupraemia have been observed. The average copper concentration in pastures was 9·2 ppm, which compares favourably with the optimum value of 10 ppm for cattle and 5 ppm for sheep. The observed levels of molybdenum, ranging from 1 to 12 ppm and averaging 3·9 ppm, are on the high side but it is recognized that the adverse effects of molybdenum in herbage can readily be offset by dosing with copper.

Table 5. Summary of trace elements in plant materials

	Parts Per Million								
Crop	Al	As	B	Be	Cd	Co	Cr	Cu	F
Barley (leaf only)			1·6						
Barley straw on ash		0·64	27·0			0·07	0·30	5·8	7·18
Barley straw on soil		0·61	7·8			0·08	0·52	5·5	5·90
Barley whole plant on ash			85·0						
Barley whole plant on soil			16·0						
Barley grain on ash	5	0·68	20·7			0·03	0·18	3·7	1·36
Barley grain on soil		0·46	2·4			0·05	0·22	2·3	0·86
Rye grain	4		29					25·5	
Wheat grain on ash		0·13	5·1		0·2	0·023	Nil	5·4	
Wheat grain on soil		0·03	3·5		0·3	0·021	Nil	6·1	
Pastures on ash		2·3	16·2			0·15		9·0	3·2
Pastures on soiled ash		0·87	27·9			0·20		9·7	3·5
Timothy			56	0·97		1·6	3·0		
Lucerne		0·53	152	1·24		1·87	3·5		
Rapc			68·6						
Kale			122	0·24		0·52	1·5	5·6	
Savoy and cabbage			68					20	
Lettuce			91						
Dwarf bean			46						
Pea (leaf)			72						
Pea (seed)			12						
Swede (leaf)			106						
Swede (root)					0·16		2·2		
Parsnip (leaf)			55						
Parsnip (root)			30						
Potato (leaf on ash)		0·5	97·5						
Potato (leaf on soil)			30·3						
Potato (tuber)		0·07	11		0·12		2		
Carrot					0·07		3		
Sugar beet (leaf on ash)		0·52	66·8			0·23	0·45	15·9	6·1
Sugar beet (leaf on soil)		0·50	45·2			0·20	0·63	9·7	6·9
Sugar beet (root on ash)		0·30	13·3	0·06	0·33	0·08	1·3	4·6	0·5
Sugar beet (root on soil)		0·28	16·6			0·06		4·8	0·5

The availability of manganese, an essential plant nutrient, can be reduced by the high pH of the ash. Deficiency symptoms have occasionally been noted in plants and where controls are available, the indications are that the uptake of manganese from ash plots is lower than from corresponding soil plots. The recorded analyses for manganese show random variations between 10 and 160 ppm. The average manganese in pasture herbage is 77 ppm, comparing favourably with the A.R.C. recommendation (A.R.C. 1965) for a minimum level of 40 ppm in the diet of animals.

Few analyses are available for nickel but the highest recorded value was

Parts Per Million

Ga	Hg	I	Mn	Mo	Ni	Pb	Se	Sn	Ti	V	Zn
			22								
			12·5	2·79		3·5					34
			13·0	1·10		3·0					24
			8·3	1·78		0·9					56
			11·0	0·87		1·0					35
				11·5							92
				0·90	0·8	<1					32·0
				0·38	0·9	<1					43·3
		0·16	83	5·14							65
		0·16	62	1·53			0·20				37
·3				2·3	6·7	28		1·1	50	5·4	103
·94				15	6·4	33		1·6	53	4·4	99
·43				5·0	4·2	11·0		2·06	16·4	1·3	72
			28·8	1·8							
			13								
			12								
			25								
			12·8	13·7							69·1
			25								
	Nil				1·6	<1					
			22								
			4								
	Nil		9		1·0	<1					
	0·06				1·9	<1					
			54·7	2·97		5·1					52·1
			44·0	0·60		5·0					47·1
07	0·004		12·1	0·22	1·3	2·2		0·99	3·6	0·23	28·8
			17·8	0·12		0·4					13·6

12·0 ppm at Ironbridge in 1965. The highest values were found to occur in lucerne and timothy; sugar beet root showed consistent values just over 1 ppm and kale, 2 to 3 ppm.

Table 6. Summary of trace elements in plant material.

	Percent							
Crop	Ca	K	Mg	N	Na	P	S*	Cl
Barley (leaf only)		5·3		2·6		0·45		
Barley straw—on ash	0·30	0·99	0·10	0·98		0·25		
Barley straw—on soil	0·36	0·87	0·07	0·78		0·21		
Barley whole plant—on ash	0·45							
Barley whole plant—on soil	0·43							
Barley grain on ash	0·09	0·68	0·17	1·94		0·54		
Barley grain on soil	0·08	0·58	0·16	1·90		0·47		
Wheat grain on ash		0·58	0·15	1·54		0·39		
Wheat grain on soiled plots		0·56	0·13	1·61		0·37		
Pastures on ash	0·74	2·69	0·17	1·51	0·13	0·33	0·43	1·05
Pastures on soiled ash	0·56	2·13	0·16	1·46	0·25	0·31	0·51	0·92
Lucerne	1·9	2·3	0·25	2·8		0·19		
Rape		3·3	0·24	4·7		0·53		
Kale	2·16	2·75	0·23	2·46		0·34		
Savoy		5·0		4·8		0·47		
Lettuce		6·5		2·4		0·41		
Pea (leaf only)		3·3		2·9		0·38		
Swede (leaf)		4·8		3·9		0·42		
Swede (root)								
Potato (leaf)		2·2		6·0		0·39		

* Sulphate Sulphur

Numerous analyses have been made for lead and show a maximum of 53 ppm in lucerne grown on new ash at Ironbridge; timothy showed values in the range 20 to 40 ppm and kale, concentrations below 20 ppm. These values are well below the levels at which cumulative poisoning would occur in cattle (Allcroft 1951). Later samples of barley and wheat show lead concentrations of 1·0 ppm or less, with little difference between control and ash plots.

Analytical difficulties have restricted the number of analyses available for selenium but the highest value found was 0·24 ppm in pasture herbage. This is not thought to be dangerous; the deficiency level in New Zealand pastures is said to be 0·05 ppm whilst the tolerance limit for animals is thought to be 4 ppm.

A few analyses have been made for mercury, the highest recorded concentration being 0·14 ppm in the dry matter of carrots grown in 1972 at Ironbridge. Mercury was completely absent from a number of the samples.

Most of the available samples have been analysed for zinc and with one exception (probably a rogue) the results were uniformly below 200 ppm, averaging 60·9 ppm. This compares favourably with the normal range of 15 to 60 ppm for zinc in pasture herbage and the A.R.C. recommendation (A.R.C. 1965) of 50 ppm. It thus appears unlikely that there will be any harmful effects from zinc on grazing animals. Wheat samples taken at Besthorpe showed the concentrations entirely below 50 ppm with indications that more zinc was derived from the soil than from the underlying ash.

Analyses for cadmium were also made on the wheat samples; the highest recorded concentration was 0·4 ppm and again the indications were that more cadmium was derived from the soil than from the ash.

General conclusions

The experiments recorded in this paper have shown that pastures can be successfully laid down on sites used for the disposal of pulverized fuel ash. Subsequent weathering and good farm management enable a wider range of crops to be grown. In composition, these crops appear to be in no way inferior to similar crops grown on undisturbed land. The disposal of pulverized fuel ash can thus make a contribution to the area of farm land available in this country, continually eroded as it is by the needs of industrial and housing developments.

Acknowledgments

The work recorded in this paper forms part of a comprehensive programme of research sponsored or directly carried out by the Central Electricity Generating Board. The authors greatly acknowledge the encouragement and help which they have received from the Board and record also their thanks to to the farmers and other organizations who have collaborated in this work.

Summary

Pulverized fuel ash will support the growth of many plants. Most trace elements are adequate. Although nitrogen is deficient, boron may be excessive for sensitive crops. Appropriately treated land has been restored to pastoral use and has subsequently grown other crops successfully. Good yields of wheat have been recorded.

References

ALLCROFT R. (1951) Lead poisoning in cattle and sheep. *Vet. Rec.* 63, 583–90.

A.R.C. (1965) *Nutrient Requirements of Farm Livestock, No. 2: Ruminants.* Agricultural Research Council, London.

COPE F. (1961) *The Agronomic Value of Power Station Waste Ash.* Ph.D. thesis, Univ. of Leeds.

GILLHAM E.W.F. & MORLEY DAVIES W. (1972) Land restoration with pulverized fuel ash. *J. Br. Grassld Soc.* 27, 13–15.

HOLLIDAY R. & TOWNSEND W.N. (1959) Soils and farming. In *York: a Survey* (Ed. by G.F. Willmot, J.M. Biggins & P.M. Tillott), Brit. Ass. Adv. Sci., London.

MORLEY DAVIES W., GILLHAM E.W.F. & SIMPSON D.T. (1971) An economic investigation into farming on land restored with pulverized fuel ash. *J. Br. Grassld Soc.* 26, 25–30.

SALTER P.J., BERRY G. & WILLIAMS J.B. (1966) The influence of texture on the moisture characteristics of soils III. Quantitative relationships between particle size, composition and available-water capacity. *J. Soil Sci.* 17, 93–8.

SALTER P.J., WEBB D.S. & WILLIAMS J.B. (1971) Effects of pulverized fuel ash on the moisture characteristics of coarse-textured soils and on crop yields. *J. agric. Sci., Camb.* 77, 53–60.

A practical approach towards the establishment of trees and shrubs on pulverized fuel ash

D. R. HODGSON and G. P. BUCKLEY* *Department of Plant Sciences, University of Leeds, U.K.*

Introduction

Previous papers (Hodgson & Holliday 1966; Townsend & Hodgson 1973; Townsend & Gillham 1974) have described in detail the chemical and physical properties of pulverized fuel ash (PFA) in relation to their effects on plant growth. Much information is now available on techniques which can be applied successfully to deposits of PFA for the purpose of returning land to agriculture. Hitherto little was known about the growth of trees and shrubs on this waste product but from the point of view of amenity, landscape design and the overall planning of ash disposal it is essential to incorporate woody species into a scheme of revegetation. The Central Electricity Generating Board (C.E.G.B.) has been responsible for financing a programme of experimental work on the establishment of trees and shrubs on PFA and this paper is an attempt to summarize some of the salient facts which have emerged from this programme.

It was recognized at the outset that commercial timber production on deposits of PFA was unlikely but that the major role of trees and shrubs in revegetation schemes would be in landscaping, stabilization of steep slopes, provision of shelter, amenity and possibly as temporary cover until the ash was required for commercial use. In all our work on revegetation or reclamation for agriculture it has been clearly recognized that objectives will ultimately play a large part in the actual reclamation technique adopted. There are two basic approaches to the problem. Either one recognizes the ecological limitations of ash sites and with the minimum of modification to the growth medium selects species which are already adapted to such limitations in their natural habitats, or by appropriate techniques one can alter the growth

* *Present address: Department of Horticulture, Wye College, University of London, Ashford, Kent.*

medium and provide conditions which are suitable for a wider range of plant species. The second approach is often more costly, initially, than the first but the resulting land can support a wide range of species which will allow greater flexibility in the planning of land use.

The tolerance of species to PFA

The factors mainly responsible for limiting plant growth on PFA are excessive boron, high pH, high soluble salt concentrations, deficiency of nitrogen and, to some extent, phosphorus. The content of the other macro-nutrients and essential trace elements is adequate although manganese deficiency has been recorded (Gillham & Morley Davies 1971). In addition to the chemical properties affecting plant nutrition indurated layers of ash produced by compaction and pozzolanic action may inhibit normal development of root systems. All these factors must be considered in relation to general site conditions such as exposure, variable water table depths, degree of slope and atmospheric pollution in selecting suitable species for establishment on PFA. Agricultural and horticultural crops have been classified on the basis of their tolerance to PFA by Hodgson (1961) and Cope (1961) and will not be discussed further except to point out that crops belonging to the Chenopodiaceae are all highly tolerant to PFA whereas within the Leguminosae, Cruciferae and Gramineae the species of economic importance show considerable variation in their capacity to tolerate conditions in PFA. For example in the Leguminosae species of *Melilotus* are very tolerant, the forage legumes lucerne, red clover, white clover are moderately tolerant while the pulse crops are rather sensitive. It is significant that a grouping of species in relation to their ash tolerance is similar to groupings either on the basis of boron tolerance (Eaton 1944) or boron requirements (Berger 1949). Although the growth of a species on PFA is largely a reflection of its capacity to grow in the presence of excessive quantities of boron, tolerance is in addition the result of a species reaction to soluble salts, unstable crumb structure after sowing and macro-nutrient deficiencies.

No comprehensive study of the boron tolerance or boron requirement of woody species has been reported in the literature which might serve as a guide to the selection of trees and shrubs for planting in ash. As a first step, therefore, two groups of experiments were initiated. One group was designed to assess the performance of species growing in media containing different concentrations of PFA so that species could be listed in order of their tolerance to PFA. It was recognized that the establishment and growth of a tree or shrub on an ash site would depend on many factors which could not be adequately reproduced in the glasshouse but this approach enabled selection of those species which had some inherent tolerance to the specific factors in ash which inhibit plant growth. The second group of experiments was designed to examine more precisely the direct effect of boron on a selection of species using nutrient solution and sand culture techniques.

METHODS OF ASSESSMENT

The procedure for assessing degree of tolerance was to derive a toxicity curve for each species by growing established plants in pots containing a medium of fine sand to which increasing concentrations of PFA were added to give a range of growth media containing 0, 6, 12, 25, 50 and 100% by volume of PFA. Two samples of ash were used. One sample was obtained from a lagoon where the ash had been mixed with water to form an aqueous slurry and pumped to the disposal site; the second sample was collected dry from the electrostatic precipitators through which the flue gases pass before discharge into the atmosphere. The water soluble boron content, pH, soluble salt content and electrical conductivity of the two types of ash were as shown below.

Type and source of PFA	Boron* ppm	pH†	Soluble‡ salts %	Conductivity§ mmhos/cm
Precipitator (Thorpe Marsh)	57	8·0	2·159	2·35
Lagooned (Wakefield)	54	8·9	0·892	1·13

* boron soluble in water after boiling ash: water ratio of 1 : 10 for five minutes under a reflux condenser;

† Ash: water of 1:2·5;

‡, § Extract of ash: water of 1:10 after boiling for 5 minutes under a reflux condenser.

The ashes were not too dissimilar in terms of boron content or pH but the precipitator ash was much more saline than the lagooned ash. Nitrogen, phosphorus and potassium were applied to each treatment as equal quantities of a compound fertilizer containing 8% N, 12% P_2O_5 and 8% K_2O and in addition to the pure sand control a second control of a John Innes compost (7 parts loam, 3 parts sand, 1 part peat) was included. Treatments were replicated twice where possible and in four separate experiments covering growth periods of between 8 and 16 months, 34 tests were made. Assessment of performance was by means of a visual score on a precisely defined 0–5 scale which was carried out by the same person on a number of occasions during any one growth period. Scores were progressively accumulated and at the end of the period expressed as a percentage of the theoretical maximum.

John Innes compost values were usually at or near the theoretical maximum. The values were used to derive toxicity curves for each species, some examples of which are given in Fig. 1. To obtain an integrated measure of performance which could be related to a control an index of tolerance was derived which was calculated as the following ratio:

$$\text{Index of tolerance} = \frac{\text{Area enclosed by abscissa and ash toxicity curve}}{\text{Area enclosed by abscissa and J. I. control}}$$

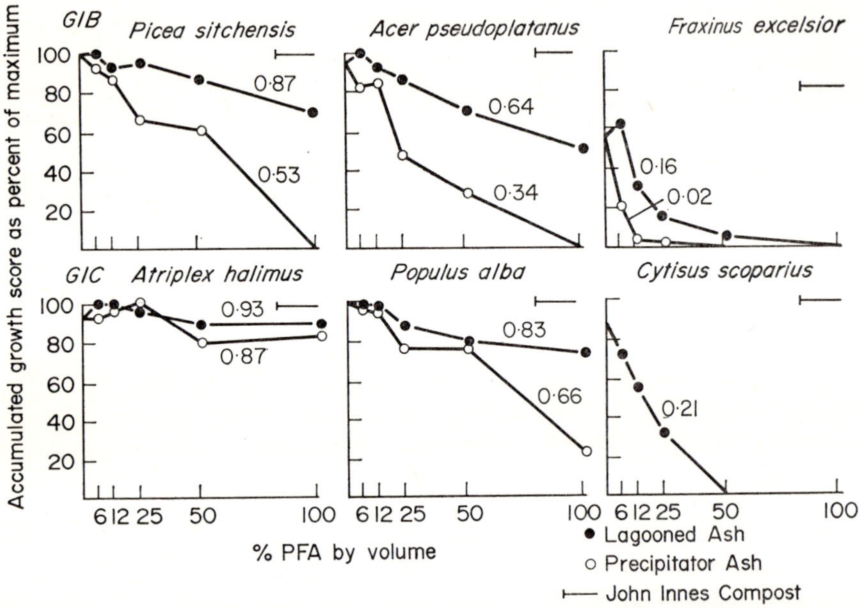

Figure 1. Toxicity curves based on accumulated growth scores for six species grown in increasing concentrations of lagooned and precipitator ash, and compared with the John Innes control. (Overall tolerance indices are shown above each curve.)

The sand control was excluded and the calculation based on the treatments containing ash. An index of tolerance was calculated for both the lagooned and precipitator ash. The results of the analysis are presented fully in Table 1 for four separate experiments, the species being listed in order according to the tolerance index for lagooned ash (column 1). The tolerance index is given for precipitator ash (column 2) and also the highest percentage of lagooned or precipitator ash on which a species survived is given in columns 3 and 4, respectively. The greater the value of the index the greater the tolerance of the species to PFA although there are some instances where a species with a high degree of tolerance as assessed by the value of the index, did not survive in the higher concentrations of ash.

Table 1. Tolerance rating for tree and shrub species growing in PFA.

	Tolerance index		Highest percentage of ash tolerated	
	Lagooned ash	Precipitator ash	Lagooned	Precipitator
Experiment G1				
Populus nigra italica	0·53	0·42	100	50
Salix britzensis	0·51	0·17	100	25
Alnus glutinosa	0·33	0·20	50	25
Picea sitchensis	0·32	0·28	50	25
Robinia pseudoacacia	0·30	0·23	100	25
Betula pendula	0·17	0·22	25	25
Fagus sylvatica	0·05	0·10	12	12
Mean	0·32	0·23		
Experiment G1A				
Picea sitchensis	0·86	0·11	100	12
Elaeagnus angustifolia	0·76	0·47	100	50
Ribes aureum	0·74	0·46	100	50
Ulex europaeus	0·72	0·07	100	6
Berberis linearifolia	0·71	0·04	100	6
Alnus glutinosa	0·69	0·13	100	50
Hypericum calycinum	0·67	0·08	100	6
Mean	0·74	0·19		
Experiment G1B				
Picea sitchensis	0·87	0·53	100	50
Hippophaë rhamnoides	0·85	0·31	50	25
Artemesia arboratum	0·77	—	100	—
Forsythia ovata	0·72	0·08	100	12
Clematis vitalba	0·68	0·26	100	25
Veronica buxifolia	0·67	—	50	—
Acer pseudoplatanus	0·64	0·34	100	50
Chamaecyparis Lawsoniana	0·56	0·13	50	12
Veronica angustifolia	0·30	0·16	25	12
Fraxinus excelsior	0·16	0·02	50	25
Mean	0·62	0·23		
Experiment G1C				
Gleditsia triacanthos	0·98	—	100	—
Erica carnea	0·93	0·73	100	100
Atriplex halimus	0·93	0·87	100	100
Colutea arborescens	0·89	0·50	100	50
Tamarix gallica indica	0·88	0·95	100	100
Populus alba	0·83	0·66	100	100
Berberis thunbergi	0·75	0·49	100	100
Picea omorika	0·73	0·24	100	50
Ceanothus azoreus	0·63	0·12	100	12
Ailanthus glandulosa	0·62	0·27	100	50
Ribes sanguinea	0·62	0·22	100	25
Cytisus scoparius	0·21	0	25	0
Mean	0·75	0·46		

— indicates no information

Because the experiments were not conducted under similar environmental conditions and the lengths of the growth period varied it is not legitimate to make comparisons between experiments. However, within each experiment apart from some exceptions in experiment G1 it is clear that the species were less tolerant to precipitator ash than lagooned ash and that this was due to the higher salinity as boron contents differed little. Some species may have high index values in lagooned ash but very low values when grown in precipitator ash, e.g. *Hypericum calycinum*, *Berberis linearifolia*, *Picea sitchensis* (G1 A only), *Ulex europaeus*, *Forsythia ovata*, *Ceanothus azoreus*. The order of tolerance is not the same for the two types of ash but some species were outstanding for a consistently good performance in each type e.g. *Populus nigra italica*, *Elaeagnus angustifolia*, *Picea sitchensis*, *Atriplex halimus*, *Erica carnea*, *Populus alba*, *Tamarix gallica indica*. On the evidence of these screening trials, species which seem particularly sensitive to PFA are *Fagus sylvatica*, *Chamaecyparis lawsoniana*, *Fraxinus excelsior*, *Veronica angustifolia* and *Cytisus scoparius*. These observations suggest that ash which has been subjected to water washing during disposal and subsequent weathering, so that much of the readily soluble material is removed and the available boron content reduced to 25–30 ppm or less, i.e. the 50% dilution, does not prevent reasonable growth of a wide selection of woody species. *Picea sitchensis* displays a surprisingly good tolerance to PFA and this has been confirmed in field trials. The halophytes *Hippophaë*, *Elaeagnus*, *Tamarix* and *Atriplex* appear to be suitable for planting directly into PFA, provided they are adapted to the prevailing climatic conditions of the locality where the ash is deposited. Included within the tolerant and semi-tolerant groups are a number of nitrogen-fixing species *Alnus*, *Elaeagnus*, *Ceanothus*, *Colutea*, *Hippophaë* and *Gleditsia* which may serve as useful pioneer species in the early stages of reclamation and also as components of mixed species plantings designed to increase the quantity of nitrogen within the nutrient cycle of the plant association. Species of *Populus* and *Salix* may also be considered for their high potential growth rate and when planted as rooted cuttings grow vigorously in PFA even though the leaves may develop symptoms of boron toxicity.

TOLERANCE OF SPECIES TO EXCESS BORON

The response of species to boron *per se* has been further investigated by growing a selection of species in sand irrigated with complete nutrient solution containing variable levels of boron. In 1971 eight species (*Alnus glutinosa*, *A. incana*, *Ulmus glabra*, *Picea sitchensis*, *P. englemannii*, *P. omorika*, *Abies sibirica*, *Colutea arborescens*) were grown in sand receiving complete nutrient solutions but with boron concentrations ranging from 0·2 to 20 ppm. The experimental technique proved satisfactory although planting material was variable, consequently standard errors were high. Toxicity symptoms, a marginal leaf chlorosis leading to necrosis, appeared on the leaves of all deciduous species receiving 5 ppm or more; growth, however, was not severely retarded at 20 ppm but abscission of lower leaves occurred in *Alnus* and *Ulmus*. These tests confirmed the high tolerance of *P. sitchensis* to boron, a characteristic not shared by the other species of *Picea* or *Abies* tested. In 1972 seeds of the woody species *Amorpha fruticosa*, *Colutea arborescens*, *Cytisus nigricans*, *Elaeagnus umbellata*, *Picea sitchensis*, *Robinia pseudoacacia*, *Spartium junceum*, *Tsuga canadensis* were germinated in sand to provide more uniform planting material. Seedlings were then transplanted into pots of Garside sand and two legumes, *Trifolium repens* (S. 100), *Melilotus alba* and two gramineous species, barley (Proctor) and *Lolium perenne* (S. 23) were included in the tests. Boron concentrations in the nutrient solutions were 0·54, 7, 14, 21, 30 ppm. The dry weights of the species growing in the different nutrient solutions were expressed as a percentage of the control (0·54 ppm B) and plotted against the boron content of the nutrient solution (Fig. 2). The agricultural crops show an order of tolerance to boron which might be expected from previous classifications (Hodgson 1961; Cope 1961). It was also confirmed that *Picea* is little affected by boron and as a seedling showed a higher tolerance than *Melilotus*. *Cytisus* which was previously shown to be ash sensitive was also very much reduced in growth by excess boron. *Elaeagnus umbellata* also proved sensitive to boron whereas *E. angustifolia* was reasonably tolerant in the original tolerance tests (Table 1). This suggests that within a genus there are differences between species in their reaction to boron but such a difference can only be confirmed by comparing each in the same type of test. The intermediate position of *Colutea* and *Robinia* confirms the semi-tolerant classification given to these species.

Three species were taken as representative of tolerant (*Picea*), semi-tolerant (barley), and sensitive (*Amorpha*) groupings according to their relative status in the highest boron treatment and their shoots analysed for boron. The content of boron in the dry matter of *Picea* increased with boron content of the solution and reached a maximum value of over 1,000 ppm (Table 2). *Amorpha* and barley on the other hand had maximum boron contents of 594 and 486 ppm, respectively, with some evidence in the case of barley that contents would not be significantly increased by higher concentrations of boron in solution. It has been observed previously (Hodgson 1961; Cope 1961) that species which are tolerant to boron may have high or low tissue boron contents suggesting that the physiological mechanisms responsible for tolerance to boron may differ according to species. The level of boron recorded in the shoots of *Picea* would lead to extensive necrosis of leaf tissue

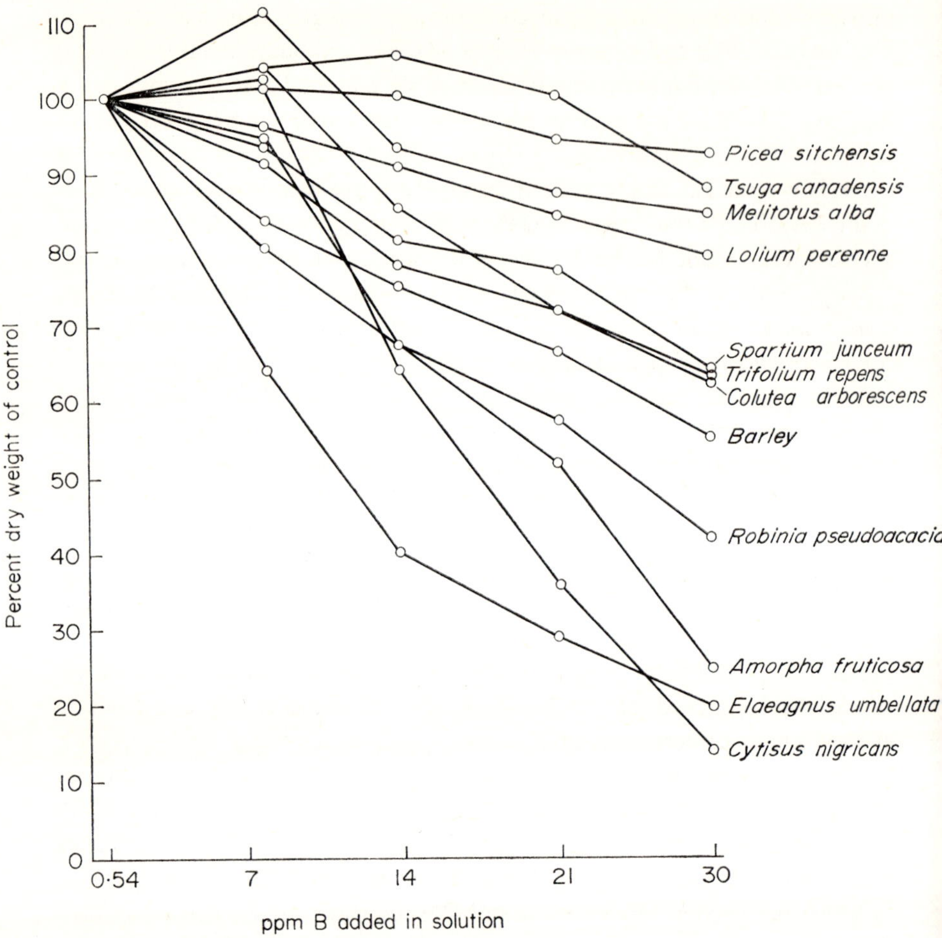

Figure 2. The effect of adding increasing quantities of boron in a standard nutrient solution on the performance of a selection of one-year woody seedlings.

Table 2. Boron content (ppm) in shoot dry matter of three species grown in nutrient solutions containing increasing quantities of added boron. (Means of six replicates and standard errors.)

ppm boron in solution	Barley	*Amorpha fruticosa*	*Picea sitchensis*
0·54	21±2	40±2	47±3
7	148±7	185±11	312±9
14	285±8	332±11	588±13
21	425±10	433±10	763±34
30	486±23	594±19	1,051±36

in many other species and there appears to be no clear predictable relationship between the performance of a species and the boron content of the dry matter. Although some species do limit boron uptake whilst others accumulate large quantities of the element, this in itself is not a clear indication of the tolerance of that species.

Much of the evidence which is being accumulated from field experiments confirms tolerance testing of species which has been attempted by the two methods described.

The establishment of trees on PFA deposits

The revegetation of much industrial waste is based on the application of a layer of soil which 'blankets' out many undesirable properties of the waste in relation to plant growth. Much work has been reported on the effectiveness of this method for restoring to agriculture land which has been covered with PFA. Because the replacement of soil is a costly procedure experiments and surveys have attempted to determine the minimum depth of soil required for the economic yield of agricultural crops. Hodgson, *et al.* (1963) from the results of a five years' experiment constructed yield isoquants which predicted the combinations of soil depth and fertilizer rate required to produce a given yield level. For all crops tested 30 cm of soil and 1·5 times normal fertilizer rates seemed a satisfactory combination although less soil and greater rates of fertilizer produced acceptable yields. More recently Morley Davies *et al.* (1971) conducted an economic survey of restored sites and recommended 30 cm of soil to ensure flexibility in farm operations particularly in cultivations. However, as little as 8 cm of soil will produce good yields of grass whilst in dry years a depth of 8 cm of soil associated with deep cultivation can produce higher yields of wheat than 30 cm of soil due to the favourable moisture-holding properties of the PFA (Townsend & Gillham 1975; Simpson 1973).

A relatively shallow application of soil to PFA aids the establishment and growth of trees and shrubs. In Table 3, the results of an experiment conducted on artificially constructed beds of different substrates are presented. Sixteen species were planted in lagooned PFA (pH 8·9); acid shale (pH 3·3), a medium loam originating from a limestone soil (pH 7·8), and on PFA covered with a 15 cm layer of either shale or soil which was partially incorporated into the surface 10–15 cm of ash. Heights of surviving trees were measured two-and-a-half years after planting and have been expressed as a percentage of height on the pure soil control. It is interesting to note the differences in growth of the species between PFA and shale. Two species failed on both, six species grew better on shale than PFA, especially *Alnus glutinosa* and *Picea sitchensis* and eight species survived and grew on PFA but not shale. Amelioration of PFA with either a layer of shale or soil improved survival and growth, and for some species (*Crataegus*, *Alnus*, *Picea*, *Elaeagnus*, *Populus*) produced a growth medium as good as soil. An extremely significant result is the mutually beneficial effects that shale and PFA have on each other when mixed together. Other experiments involving ash/shale mixtures (Hodgson & Townsend 1973) have demonstrated a similar effect which can be partly explained by a reduced boron content in the plants and the development of a neutral pH in the mixture.

Table 3. Relative heights of 16 species on different growth media two-and-a-half years after planting (soil = 100).

	PFA	Acid shale	15 cm shale on PFA	15 cm soil on PFA
Group A				
Crataegus oxycanthoides	0	0	51	127
Euonymus europaeus	0	0	48	77
Group B				
Forsythia ovata	0	47	40	64
Alnus glutinosa	53	77	104	144
Fraxinus americana	35	43	53	82
Picea excelsa	69	89	100	122
P. sitchensis	91	147	177	126
Robinia pseudoacacia	56	57	70	81
Group C				
Elaeagnus angustifolia	102	0	113	120
Fraxinus excelsior	20	0	32	44
Hippophaë rhamnoides	75	0	70	90
Ligustrum vulgare	24	0	26	83
Populus nigra	46	0	90	94
Salix britzensis	42	0	53	81
Syringa vulgaris	35	0	63	90
Ulmus glabra	24	0	50	83

Group A—no survival on PFA or shale.
Group B—survival and growth greater on shale than PFA
Group C—survival on PFA but not on shale.
Figures should be compared horizontally, not vertically.

'Blanket' applications of layers of soil or other materials with or without subsequent incorporation into the surface of the ash have proved effective means of establishing crops and grass and will certainly improve the growth of trees and shrubs although the method may be costly. Because trees are grown at relatively low densities economy in the use of soil is possible by planting into 'pockets' of soil of suitable dimensions. Small-scale experiments with holes of varying size demonstrated the value of this technique for *Acer pseudoplatanus*. Transplanted directly into PFA this species remained stunted after four years whereas a small addition of 1,350 cm^3 of soil in a 'pocket' increased mean dry weight six-fold—a response which could not be attained by fertilizer alone. The largest soil volume (8,440 cm^3) gave the best growth but relatively small volumes of soil minimized the harmful effects of the ash during the early stages of establishment and enabled the plants to develop a vigorous rooting system which extended into the surrounding ash (C.E.G.B. 1969).

DEVELOPMENT OF FIELD EXPERIMENTS

The soil 'pocket' planting technique has been tested in two experiments, one at Dunston power station (north-east England) and the second at Stourport (west Midlands). Apart from climatic differences between the regions the PFA at the two locations differed in respect of boron content and degree of consolidation. The ash at Stourport had been tipped dry to a depth of 4·5 m on low-lying marshy land and was very hard below a loose friable surface of 10 cm. Disposal of ash at Dunston was by lagooning and consolidation was less severe than at Stourport. In 1966 Stourport ash contained 100 ppm of available boron and Dunston 14 ppm. Topographically the sites were at the average level of surrounding land and did not suffer from a degree of exposure atypical of the locality.

The species under test are *Acer pseudoplatanus*, *Betula pendula*, *Fraxinus excelsior*, *Picea sitchensis*, *Populus robusta* and *Salix caprea*. All are common British species which on the basis of information from the tolerance testing might be expected to cover the range from tolerant to sensitive (Table 1). The planting technique was to drill holes with a tractor-powered 30 cm diameter auger to a depth of 40 cm, a task which was easily accomplished at Dunston but was extremely difficult at Stourport because of indurated ash. The trees, 1+1 transplants or 1 year rooted cuttings, were then planted in the auger holes back filled with either ash or soil. Planting distances were 1·1×1·1 m, each main plot consisting of five rows of trees 16·5 m in length. The experimental design was a randomized block with four replications, species forming main plots and the ash and soil comparison incorporated as a split plot treatment. A third factor, with or without a ground cover of *Trifolium repens* (S. 100) was also included in the design by undersowing two blocks with inoculated seed in the same year the trees were planted at Stourport but two years after the plots were established at Dunston. An overall application of 800 kg/ha of 12:24:0 was given at or soon after planting. Planting dates were April 1966 at Dunston and March 1968 at Stourport.

Since 1968 height measurements in the autumn have been recorded for each individual tree, but to eliminate possible edge effects results were assessed on the three centre rows only.

To illustrate the effects of site and planting method on the establishment and growth of the trees *Populus robusta* and *Picea sitchensis* have been selected for more detailed discussion. Table 4 gives the percentage survival of these species. Soil improved survival of *Picea* at both sites, had no effect on *Populus* at Dunston but at Stourport increased it four-fold. The growth curves in terms of mean tree height are presented in Figs 3 & 4 for the factorial combination of the soil and clover factors. Standard errors of the mean have been calculated for each treatment on the basis of the individual tree measurements rather than obtaining a pooled estimate of error based on the analysis of variance of the original design. Large differences between species and differential survival between plots necessitated this approach.

Table 4. Percentage survival of trees on two ash sites (September 1971).

	Dunston	Stourport
Populus—soil	88·0	80·0
PFA	90·0	21·2
Picea —soil	59·4	57·4
PFA	36·8	35·8

The two experimental sites represent the extremes likely to be encountered in reclamation programmes. Dunston ash is relatively low in boron and compaction is not severe. At Stourport, on the other hand, the ash is both indurated and contains a large excess of boron. It is of special interest, therefore, to compare tree growth on the two sites. Both *Populus* and *Picea* responded to the soil pocket planting technique and apart from *Populus* in the absence of clover this response is being maintained or increased with the passage of time (Fig. 3). The rate of increase in height of *Populus* appears to be declining in the absence of clover due to exhaustion of nitrogen. Although not indicated on the graph an addition of 35 g N/tree in March 1971 arrested this decline and tended to establish an exponential trend in growth. The introduction of clover, which only persisted for two years, injected nitrogen into the system and it is significant that recycling must be sufficient to maintain a reasonable growth rate even when most of the clover has been eliminated by competition from grasses and the tree canopy. The presence of soil also enabled the trees, especially *Picea*, to use the mineralized nitrogen more effectively, due most likely to a more extensive root system.

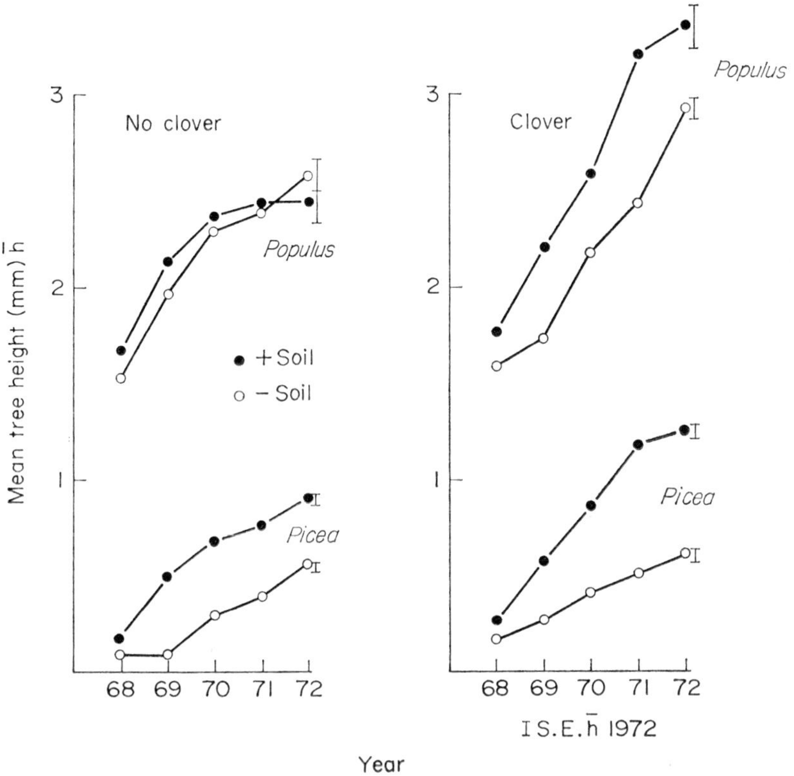

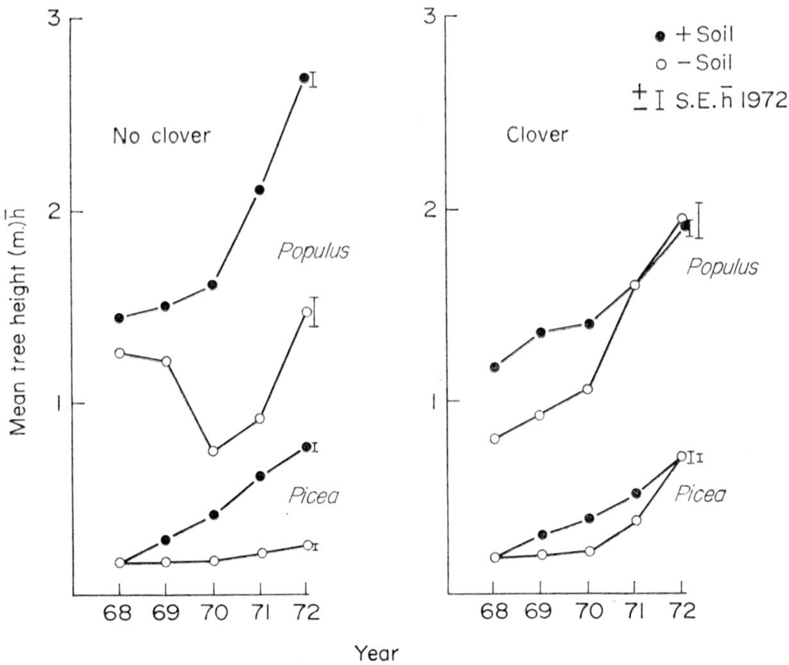

Figure 3 and Figure 4. Annual height growth of *Picea sitchensis* and *Populus robusta* at Dunston Power Station (Fig. 3) and Stourport Power Station (Fig. 4) in response to soil and clover (S.E. for treatment comparisons refer to 1972 only).

The response of *Populus* to nitrogen fertilizer applied in March 1971 was also greatest on the soil plus clover treatment, again indicating that more favourable physical conditions had been created for the root system (Table 5).

Table 5. Response of Populus in height increment (cm) to 35g N/tree applied in March 1971 (Dunston Power Station).

	TREATMENT			
	− Clover		+Clover	
	− Soil	+Soil	− Soil	+Soil
1971	23	25	73*	46*
1972	80*	102*	102*	130*

* significant response ($P \leqslant 0{\cdot}05$) to addition of nitrogen fertilizer over control.

Picea did not respond to nitrogenous fertilizer on any of the planting treatments even though the plants growing in the absence of soil would be at a very low nutritional status. *Picea* may, however, have suffered some competition from the grassy undergrowth which grew rapidly in the Spring before it was cut down in June.

A further quantitative assessment of treatment effects was made in early March 1971 when the poplar stand was thinned by removing every other tree and measuring total fresh weight. Mean tree weights are presented in Table 6 and demonstrate that a soil 'pocket' was, after five years, at least equivalent to a clover sward of only two years' effective duration and that there was a positive interaction between soil and clover, the response to clover being much greater in the presence than absence of soil.

Table 6. Mean fresh weight (kg/tree) of *Populus* thinnings in March 1971 (Dunston).

	No soil	Soil
No clover	0·492±0·061	0·680±0·053
Clover	0·704±0·08	1·619±0·170
Response to clover	+0·312*	+0·939*

* significant at $P \leqslant 0{\cdot}05$

The results from the same species grown at Stourport are, in some respects, a complete contrast to those from Dunston (Fig. 4). In the absence of clover the soil pocket planting technique has been equally beneficial for both species and *Populus* is showing a well defined exponential type growth curve to 1972. The apparent reduction in height between 1968 and 1970 for trees planted in pure ash is due to death of the 'leader' followed by regrowth from basal shoots in the spring of 1970. Soil, therefore, has proved very effective in mitigating the worst effect of the ash. Clover, however, significantly reduced the growth rate of *Populus* planted in soil pockets. This negative response to clover was probably due to excessive competition from a very vigorous stand in 1968 and 1969 and the fact that the clover seed was sown soon after the trees were planted whereas, at Dunston, the trees had been established for at least one year before sowing. The root systems of trees at Stourport may not, therefore, have been sufficiently well developed to use nitrogen mineralized from clover organic matter and intense competition for moisture could have reduced the growth rate of the trees. Evidence from current pot experiments on competition between clover and young trees suggests that clover can bring about a significant reduction in growth compared with a tree in a non-competitive situation. There was a significant residual effect of clover on the no soil treatments for *Picea* in 1971 and 1972 and *Populus* in 1972.

Root development

The influence of the auger hole planting technique on the development of the root system was studied at Stourport in 1972 on a small number of selected trees of *Populus*. It was questioned whether the roots would fail to penetrate the ash and 'ball up' like those of a pot-bound plant. Excavations of root systems of trees originally planted in soil or ash did not confirm this. Neither the tap root (t.r.) nor the laterals penetrated the consolidated ash but grew horizontally, radiating in all directions from the planting hole in the loose friable surface ash. The roots were never deeper than 7–8 cm and more usually 3–5 cm below the surface. Diagrams of the root systems are presented in Fig. 5 and a measure of degree of compaction (using a soil assessment penetrometer) is shown graphically in Fig. 6. These diagrams illustrate the way in which Poplar roots ramify in the surface ash, extend beyond the planting pocket and even beyond plot boundaries. The penetrometer readings quantify in arbitrary units the uniform nature of the compaction. The soil or ash which was backfilled into the auger hole is relatively uncompacted and only at depths of 30–45 cm, the approximate depth of the hole, does the resistance to penetration increase markedly. On the undisturbed ash surface, however, the mean values of forty readings show the ash to be impenetrable beyond 8–9 cm. The form of the root system considered in relation to the patterns of physical compaction of the ash and boron contents at increasing depths below the surface is significant in explaining how a tree can survive and grow reasonably well on what appears to be a most unfavourable site

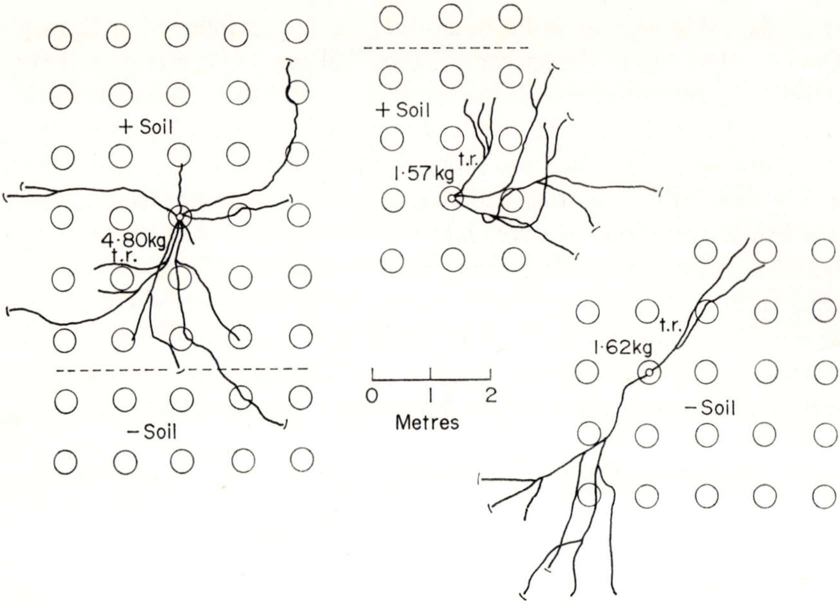

Figure 5. Excavated root systems of *Populus robusta* ramifying in the surface layers of PFA at Stourport. (Circles represent augered planting holes; fresh weights of individuals shown in kg; t.r. = tap root.)

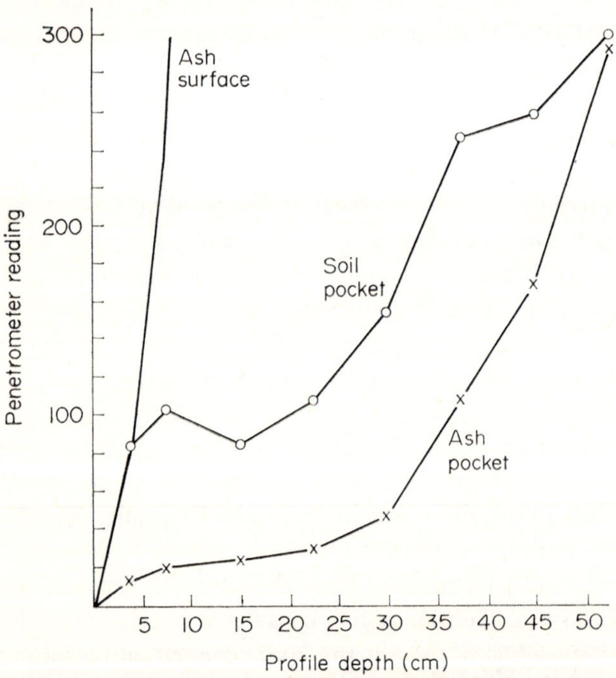

Figure 6. The relative compaction in arbitrary units with increasing profile depth in soil and ash-filled pockets and in the undisturbed surface ash at the Stourport Power Station site.

for trees. Gillham (priv. comm.) sampled the Stourport ash to a depth of 112 cm and analysed for available boron in successively deeper horizons of the profile. These results, shown in Fig. 7, indicate low values at the surface, a maximum value at approximately 15 cm and then a steady decline to 30–40 cm. Much of the root system of Poplar is confined to a zone of low toxicity which probably accounts for an absence of severe toxicity symptoms on the leaves and a growth vigour much greater than might be expected from a boron analysis of the bulk ash. The extensive root system will also ensure maximum exploitation of the organic rich surface layers although there will be some liability to drought and perhaps instability when the trees obtain greater heights.

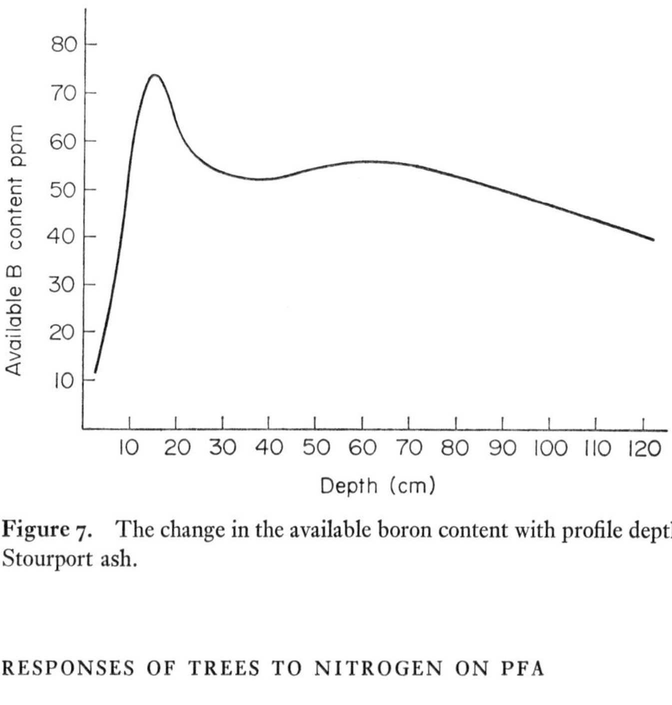

Figure 7. The change in the available boron content with profile depth in Stourport ash.

RESPONSES OF TREES TO NITROGEN ON PFA

In common with many other mining and industrial wastes PFA is extremely deficient in nitrogen and this is exemplified by the evidence obtained from the Dunston and Stourport experiments. The need for nitrogen is further substantiated from similar experiments at two other power stations, Drakelow, Burton-on-Trent (Midlands) and Tir John, near Swansea (South Wales). The PFA at Drakelow had been lagooned initially, then excavated and tipped on low-lying marshy land. Much of the surface was contaminated with the gravel which had formed the floor of the lagoon resulting in much variation in the growth medium. In 1966 the available boron content averaged

50 ppm and was thus intermediate in toxicity between Dunston and Stourport ashes. At Tir John the ash had been lagooned and left undisturbed for 12–18 months prior to laying down the experiment. The boron content was relatively low at 7 ppm, a property common to PFA derived from coals of the South Wales coalfield. The experiments were similar in design to those at Dunston and Stourport with the exception that no soil was used at either site and due to cultivation difficulties at Drakelow the clover treatment was replaced by a nitrogen fertilizer treatment of 35 g N per tree in March 1971. In Fig. 8 *Populus* and *Picea* are compared at each site. It seems that both

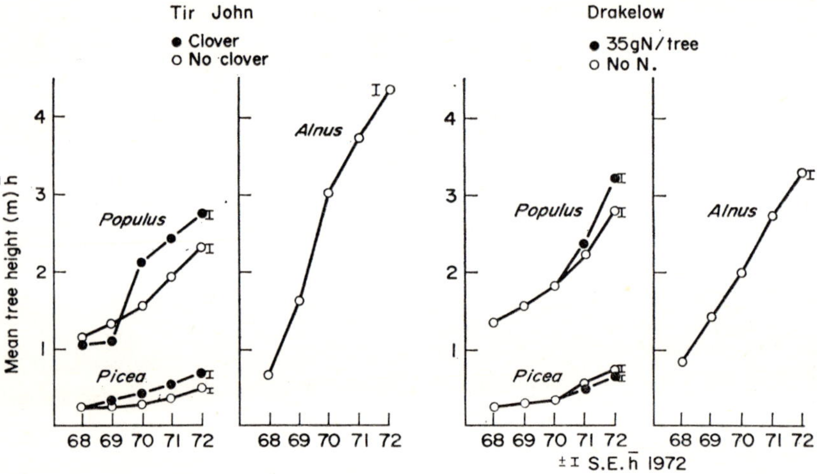

Figure 8. Responses of *Populus robusta* and *Picea sitchensis* to clover at Tir John and nitrogen fertilizer at Drakelow Power Station, and the performance of a nitrogen-fixing species, *Alnus glutinosa* at each site.

species at Tir John were still responding to the nitrogen introduced in 1968 by the clover sward, which only persisted for 2 years. It is interesting to note that *Populus* did not appear to have benefited from clover sown in 1968 until 1970 suggesting competitive effects in the early years but considerable nitrogen recycling later. Mean tree weights of *Populus* determined from thinnings in March 1971 were 0·708 kg without clover and 0·981 kg with clover; with another species *Salix caprea* this effect was much greater and highly significant statistically ($P < 0{\cdot}001$); with weights of 0·604 kg and 3·600 kg, respectively.

At Drakelow *Populus* gave significant responses by 1972 to the application of nitrogen fertilizer made in March 1971. As recorded at Dunston responses to a single application of nitrogen increased with the years thus ensuring a well defined exponential type growth curve. How long this nitrogen effect is likely to persist will depend on the efficiency of recycling and the fraction retained each year in woody tissue. As at Dunston, *Picea* did not respond to nitrogen; the quantity may have been too high for this species as one or two deaths were recorded.

The growth curves of *Alnus glutinosa* from a separate experiment at these two sites are given to illustrate the growth of a nitrogen-fixing species which theoretically is not likely to be deficient in this nutrient. *Alnus* is much superior to all species at Tir John but at Drakelow *Populus* which received nitrogen is comparable to the *Alnus* in performance so far (Fig. 8).

Finally, to emphasize the importance of nitrogen, Fig. 9 shows the response of a fertility demanding species, *Acer pseudoplatanus*, to nitrogen fertilizer at two sites and to clover, at Tir John. In the latter situation despite the absence of toxicities and a dressing of fertilizer at planting, without clover, growth was nil; the trees remained stunted with leaves small and chlorotic. In contrast the clover effect was large each year apart from the first but the decreasing magnitude of the increment with time indicates that there is insufficient nitrogen being recycled to maintain the initial rate of growth.

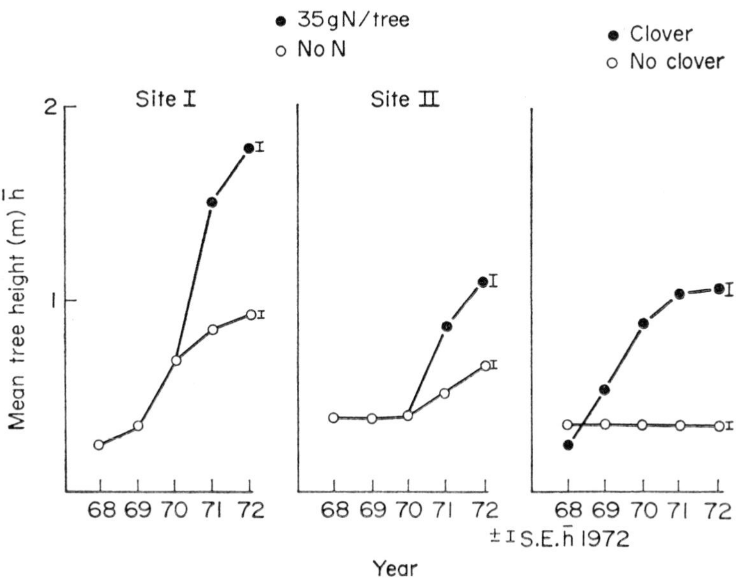

Figure 9. The growth of *Acer pseudoplatanus* in the presence and absence of clover or nitrogen fertilizer.

Woody species as a long-term source of nitrogen

The possibilities of providing the trees with an initial supply of nitrogen from clover have already been discussed, but shading caused by the upper canopy means that this form of biological fixation must be considered as essentially short-term, and tree growth responses therefore limited in their duration. A potentially much longer-term source of biological nitrogen is to grow woody nitrogen fixing species either in pure sand, or in mixture with other woody non-nitrogen fixing plants. This latter type of situation poses silvicultural problems relating to the mixing pattern, proportion of the species, and spacing of individuals which may often vary in growth potential, form and habit. Very often the nitrogen-fixing species are of the pioneer, light-demanding, leguminous type and may be in sharp contrast to non-fixing species which are more affected by the fertility of the site. The former tend to grow rapidly, whereas the latter are initially slow but have a much greater growth potential and possibly commercial value.

An example of a natural plant succession described by Crocker & Major (1955) serves to illustrate the point. They studied plant succession on glacial moraines in Alaska which had gradually been exposed by large-scale glacial retreat over the past 200 years. Pioneer vegetation, which included *Dryas drummondii*, a nitrogen-fixing species, developed on freshly exposed mineral surfaces and rapidly gave way to a complex of Arctic *Salix* species. *Alnus crispa* dominated the next stage of succession and eventually formed almost pure thickets, which are finally invaded by *Picea sitchensis* as the subclimax vegetation. The complete succession from bare moraine to spruce forest takes about 250 years, during which time the most significant effect of the vegetation is to build up soil nitrogen, which itself is closely related to a reduction in pH from 8·0 to 5·0 in 30 to 50 years. Both effects are largely due to *Alnus crispa*, which fixes nitrogen at the rate of about 68 kg ha^{-1} a^{-1}, so that when spruce is established there may be as much as 3,000 kg ha^{-1} nitrogen in the top 45 cm of the profile. Under spruce the nitrogen pool falls to some 1,750 kg ha^{-1} in 100 years, but an equilibrium point is eventually reached when the annual growth rate equals the rate of litter decomposition.

There are a number of parallels in this example that might be applied to the PFA situation. Firstly woody nitrogen-fixing species are capable of accumulating a large bank of nitrogen, whilst at the same time, on alkaline parent material they will reduce the pH considerably. Provided the pool of nitrogen is allowed to develop for long enough for it to be incorporated into the growth medium, the system should eventually become self-supporting. The time of natural succession can, however, be considerably speeded up by carefully managing the nitrogen fixing stage and including with this vegetation, species of the desired 'climax' vegetation. Most forestry silvicultural systems seek to do this.

Tarrant (1961) compared the performance of Douglas Fir growing in pure stands and also mixed with Red Alder, and showed that after 20 years growth at 240 cm spacing, the firs growing in association with alder were significantly greater in height and girth. The greatest contribution was to the dominant firs which had developed a more vigorous upper crown in the association with alder. Soil nitrogen was increased by 65% under the mixed stands and this was reflected by the increased levels of leaf nitrogen in the case of fir mixed with alder.

In order to examine the possibilities of developing a self-supporting canopy of woody species on PFA, associations of *Picea sitchensis* and three other nitrogen-fixing species *Alnus glutinosa*, *Robinia pseudoacacia* and *Elaeagnus angustifolia*, have been established at Drakelow and Tir John power stations. Pure stands of *Picea* were planted adjacent to mixtures of *Picea* alternating with the three nitrogen-fixing species at a spacing of 1·2 m. In the past five years since establishment, the spruce has tended to suffer competition in the mixed plantings from the more rapidly growing nitrogen-fixing trees, which have had to be regularly pruned to prevent severe shading at this close spacing. Consequently no significant differences have been observed between the pure plots of *Picea* and *Picea* in the mixtures at Drakelow, but at Tir John a significant increase in height has been recorded in *Picea* growing with *Elaeagnus* (Fig. 10). *Elaeagnus* probably offers the least light

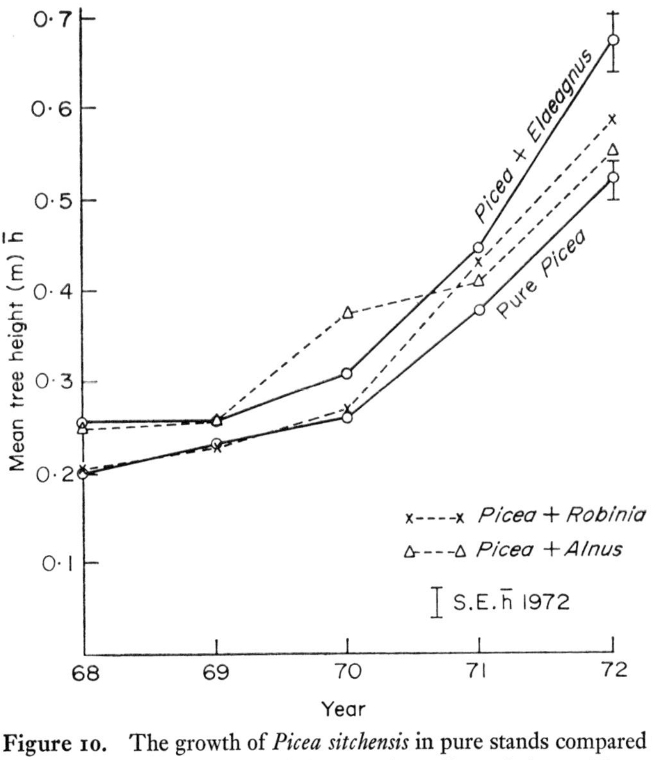

Figure 10. The growth of *Picea sitchensis* in pure stands compared with *Picea* grown in mixture with three nitrogen-fixing species, *Alnus glutinosa*, *Elaeagnus angustifolia*, and *Robinia pseudoacacia*.

competition with *Picea* by virtue of its growth habit, and it may be that this significant trend is a direct result of the improved nitrogen nutrition of *Picea* in this association. Tir John is probably a more nitrogen-deficient site than that at Drakelow (which is based on a gravel-ash mixture) judging by the performance of nitrogen-demanding trees such as *Fraxinus* and *Acer*. The effect of improved nitrogen nutrition is therefore more likely to be seen first at the former site, whereas it may be some time before *Picea*, which is not a strongly nitrogen-demanding species, shows improvement fr m mixed plantings at Drakelow.

POTENTIAL YIELD OF TREES ON PFA SITES

Height increments of the three species *Populus*, *Acer* and *Picea* over the 4-year growth period were taken as a basis for comparison with standard yield tables produced by the Forestry Commission for commercially grown trees on more conventional sites (Forest Management Tables, H.M.S.O. 1966). In these yield tables, standard height/age graphs are given for each separate yield class, the latter being expressed as the maximum mean annual volume increment, irrespective of the age at which this culminates. Standards vary from 3·6–14·2 m^3 ha^{-1} in *Populus* and 3·6–10·7 m^3 ha^{-1} in *Acer* to 5·3–24·9 m^3 ha^{-1} in the case of *Picea*, the highest yielding species.

Corresponding to the yield classes given in the tables, height measurements are entered against each age of tree so that height increments or absolute growth rate can be calculated at a given age. The absolute growth rates for each ash-grown species on the best treatments were then calculated for the four years' period 1968–72 and compared with each yield class standard. In the case of *Populus*, after 7 years the maximum growth rate occurred in the soil plus clover treatment at Dunston and corresponded approximately to the lowest standard yield class of the Forestry Commission tables. Maximum values at Drakelow, Tir John and Stourport were all well below this (Table 7). According to these same tables, the trees at Dunston have the potential to reach 24 m at sixty years when each tree would have achieved a volume of approximately one cubic metre, given the appropriate thinning regime. Estimates of *Acer* and *Picea* have to be more tentative because the standard tables do not record heights for these species until the trees are about 15 years old, when the first opportunity to calculate the absolute growth rate arises. Extrapolation to an age of 5 years, however, suggests that *Acer* in the presence of clover, soil or added fertilizer nitrogen would fall in the middle range of the yield classes at about 7 m^3 ha^{-1} maximum mean annual increment, whilst at Dunston something very near the optimum of 10 m^3 ha^{-1} would be achieved. *Picea* on the other hand, potentially the highest yielding species of the three, shows at Tir John, Stourport and Drakelow a development corresponding to a projected yield class of 14–16 m^3 ha^{-1} on the best plots, with the maximum performance approaching 18 m^3 ha^{-1} at Dunston on soil plus clover plots.

Table 7. The productivity of *Acer pseudoplatanus*, *Picea sitchensis* and *Populus robusta* at four ash sites expressed in terms of the ranges given by conventional forest management tables

	Acer	*Picea*	*Populus*
Productivity range, maximum mean annual increment $m^3.ha^{-1}$	3·6–10·7	5·3–24·9	3·6–14·2
Ash site:			
Dunston	9·8	17·8	3·6
Drakelow	5·3	15·1	—
Stourport	4·4	16·0	—
Tir John	7·1	14·2	—

It should be stressed that whenever future predictions of growth are made, the limitations of the site should first be considered. On PFA there is already mounting evidence that the long-term nitrogen supply is insufficient to maintain present growth characteristics and that other factors, such as the purely physical problem of limited rooting depth could prevent in later maturity the full expression of yield potential calculated on the present basis. Nevertheless it is encouraging that the best of the trees at present growing on pulverized fuel ash could, with careful management, be harvested possibly as a commercial proposition in future years.

Conclusions

Results based on experimental work have so far demonstrated the importance of the soil pocket planting technique in the early nutrition and establishment of woody plants in PFA deposits. Young trees planted in soil pockets can, in the first instance, rely on the intrinsic resources of the ameliorant present in the planting pocket and have been shown to respond more efficiently to nitrogen mineralized from legume activity. By the use of young transplants, which have a lower initial nutrient requirement in absolute terms and whose rooting systems are capable of adapting much more radically to site toxicities, economies in the quantities of ameliorant added can be achieved.

The key to the early improvement in soil pockets is undoubtedly the addition of organic matter which locally improves the nitrogen status, reduces the boron content and modifies the physical structure of the ash. Organic ameliorants other than soil are however, frequently more available and it has been shown that nitrogen-rich organic ameliorants rotovated into the surface of PFA have produced much greater yields from agricultural crops than inorganic fertilizers (Hodgson & Townsend 1973; Rippon & Wood 1975). As with soil, these ameliorants can be mixed with the ash and used to backfill tree planting-beds or pockets; experiments at present in progress seek to investigate the growth and establishment of trees growing in 'pockets' containing mixtures of PFA with peat, poultry manure, chopped straw and soil waste from sugarbeet factories.

The planting pocket technique, together with biological nitrogen supplied initially from associated leguminous crops and later on by nitrogen-fixing woody trees or shrubs, offers a practical solution to revegetation at a relatively low cost. Achieving a long-term supply of nitrogen and maintaining a sufficiently high level in the nutrient cycle of the plant association is a critical factor. Nitrogen-rich organic materials mixed with PFA is an alternative approach for providing an early source of nitrogen although the excess boron in the ash may impair establishment of a species with only moderate tolerance to this element. Whether species which show extreme ash tolerance such as *Tamarix* and *Atriplex* will be suitable subjects for planting in conditions where boron contents are abnormally high, will depend on the results of experiments currently being conducted in the exposed situations of ash deposits. Compaction and pozzolanic activity may be the cause of problems if superficial root systems fail to give adequate stability to trees, which may then in the long term not live up to their currently determined potential.

Summary

Tolerance of woody species to PFA is assessed and also the effect of the limiting factor, excessive boron, on the growth of a selection of species in sand culture. The value of soil and nitrogen-fixing species for improving the growth medium is discussed using the results of short and longer term experiments on trees.

References

Berger K.G. (1949) Boron in soils and crops. *Adv. Agron.* 1, 321–48.

Bradley R.T. *et al.* (1966) *Forest Management Tables.* HMSO, London.

C.E.G.B. (1969) *Tree Growth on PFA.* Progress report to the C.E.G.B., Leeds University.

Cope F. (1961) *The Agronomic Value of Power Station Waste Ash.* Ph.D. thesis, University of Leeds.

CROCKER T.L. & MAJOR J. (1955) Soil development in relation to vegetation and surface age at Glacier Bay, Alaska. *J. Ecol.* **43**, 427–48.

EATON F.M. (1944) Deficiency, toxicity and accumulation of boron in plants. *J. agric. Res.* **69**, 237–77.

GILLHAM E.W.F. & MORLEY DAVIES W. (1971) *The Behaviour of Plants Growing on Pulverised Fuel Ash.* Scientific Services Department, C.E.G.B.

HODGSON D.R. (1961) *Investigations into the Reclamation of Land Covered with Pulverised Fuel Ash.* Ph.D. thesis, Leeds University.

HODGSON D.R., HOLLIDAY R. & COPE F. (1963) The reclamation of land covered with pulverised fuel ash. The influence of soil depth on crop performance. *J. agric. Sci.* **61**, 299–308.

HODGSON D.R. & HOLLIDAY R. (1966) The agronomic properties of pulverised fuel ash. *Chemistry and Industry*, 785–90.

HODGSON D.R. & TOWNSEND W.N. (1973) The amelioration and revegetation of pulverised fuel ash. In *The Ecology and Reclamation of Devastated Land* II. (Ed. by R.J. Hutnik & G. Davis), Gordon & Breach, London.

MORLEY DAVIES W., GILLHAM E.W.F. & SIMPSON D.T. (1971) An economic investigation into farming on land restored with pulverised fuel ash. *J. Br. Grassland Soc.* **26**, 25–30.

RIPPON J.E. & WOOD M.J. (1975) Microbiological aspects of pulverised fuel ash. In *The Ecology of Resource Degradation and Renewal* (Ed. by M.J. Chadwick & G.T. Goodman), Blackwell, Oxford.

SIMPSON D.T. (1973) *A General and Economic Survey of the Use of Pulverised Fuel Ash as a Soil Ameliorant.* Scientific Services Department, C.E.G.B.

TARRANT R.F. (1961) Stand development and soil fertility in a Douglas Fir/Red Alder plantation. *For. Sci.* **7**, 238.

TOWNSEND W.N. & HODGSON D.R. (1973) Edaphological problems associated with deposits of pulverised fuel ash. In *The Ecology and Reclamation of Devasted Land* I. (Ed. by R.J. Hutnik & G. Davis) Gordon & Breach, London.

TOWNSEND W.N. & GILLHAM E.W.F. (1975) Pulverised fuel ash as a medium for plant growth. In *The Ecology of Resource Degradation and Renewal* (Ed. by M.J. Chadwick & G.T. Goodman), Blackwell, Oxford.

Microbiological aspects of pulverized fuel ash

JOAN E. RIPPON and M. JOCELYN WOOD
Central Electricity Research Laboratories, Leatherhead, Surrey, U.K.

Introduction

The properties of PFA as they affect plant growth have been described elsewhere (Townsend & Gillham 1975; Hodgson & Buckley 1975). It has also been shown that good growth of grass and other crops can be obtained on PFA if suitably treated. For example, a thin covering of soil greatly increases plant growth compared with that on plain ash.

Microorganisms play an important part in soil fertility. Among their activities are the break-down of plant residues, the fixation of atmospheric nitrogen and by binding soil particles, the improvement of soil structure. As it leaves the boiler PFA will, of course, be sterile, but it will acquire a microbial population from the water used for lagooning or conditioning and also from the atmosphere.

The work to be described in this paper commenced as an investigation of the natural microbial populations of PFA. When it was found that microorganisms could live and multiply in fresh ash consideration was given to the value of adding bacteria to ash as 'fertilizers', and because they would require a nutrient to stimulate growth the addition of waste organic materials was also evaluated.

The microbial colonization of precipitator ash

The rate of colonization of PFA by microorganisms was studied by filling bunkers with ash taken from the electrostatic precipitators and exposing them to the atmosphere. The changes in the numbers of bacteria and fungi in the top 8 cm of the ash were followed over a period of a year. Ashes from six power stations were used, each station burning coal from a different coal field. Four of the ashes were received dry—these were tipped into the bunkers and damped down with town mains water. The remaining ashes were received already damped with mains water. The ashes were then left to weather naturally and the microorganisms counted at intervals.

The ashes from the Kent and South Wales coals remained soft throughout the period of test, but the other ashes, particularly those from the West and East Midlands coals, formed hard layers which made sampling difficult. The South Wales coals and therefore their ashes, are always low in boron, whereas those from the Midlands are high in boron. The South Wales ashes will therefore become colonized by plants quite readily, whereas those ashes from the Midlands are generally phytotoxic. Measurement showed that as weathering occurred the pH of the ashes fell, and also the amounts of available metals and boron. The rates of decrease were irregular, with the layers of concretion tending to delay the weathering processes.

The numbers of microorganisms increased rapidly in the first few months. The results shown in Fig. 1 are for one of the Midlands ashes. There were only minor differences in the rate of colonization of the ashes by the various microorganisms; the less phytotoxic South Wales ashes did not support a larger or faster growing microbial population than the Midlands ashes.

The bacterial numbers found after one year were about 1/10 to 1/100 of the number found in an average soil. The bunkers were sited on open ground and nutrients probably accumulated quite rapidly. The microorganisms with greater energy requirements, e.g. the fungi and cellulolytic organisms, the phosphate dissolving and the denitrifying bacteria, only appeared in appreciable numbers as the organic matter increased. Although *Nitrosomonas* were found in the last few months, *Nitrobacter* were rarely isolated.

The results obtained from these tests were confirmed by a survey of ash fields and lagoons in different parts of the country. The samples, taken from five sites, ranged in age from a few months to thirty years. They included one soil sample from a South Wales site and four ash/soil mixtures from the Besthorpe experimental plots (Townsend & Gillham 1975). The plant cover varied, some fields had no vegetation and others had a dense cover.

The samples were divided into four groups by age and the average number of different types of microorganism in each group calculated.

The age groups were:

A	Less than 1 year	Average pH 9·9 (9 samples)
B	1–4 years	Average pH 8·6 (8 samples)
C	5–30 years	Average pH 7·4 (6 samples)
D	Soil (1) and ash/soil mixtures	Average pH 6·9 (5 samples)

The microorganisms counted included the total bacteria, Actinomycetes and fungi and some more specific groups, e.g. the cellulolytic organisms, the phosphate-dissolving bacteria, and the denitrifying and nitrifying bacteria.

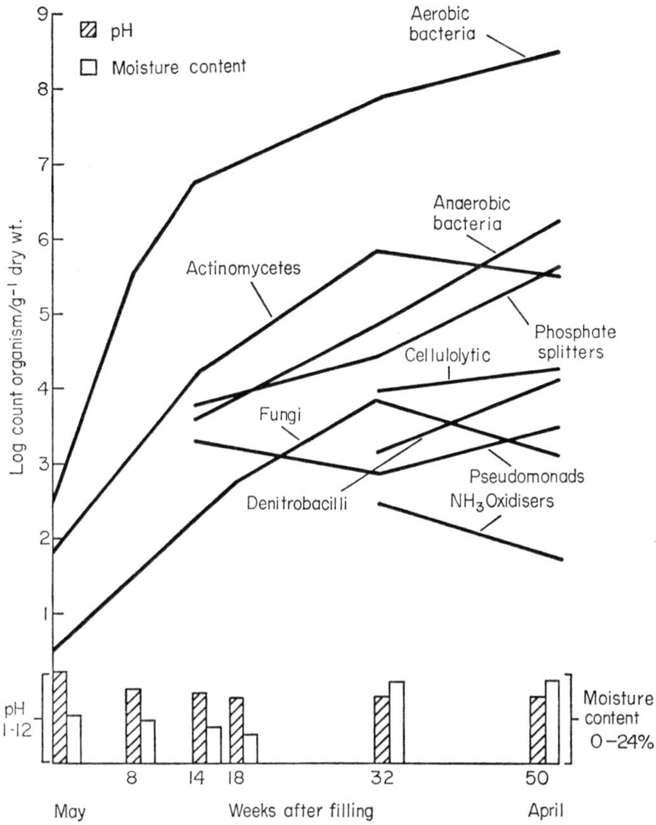

Figure 1. The rate of colonization of ash by various microorganisms.

The total bacterial numbers increased in the order A < C < B < D, but most of the organisms requiring an energy source increased in the order A < B < C < D. The denitrifying bacteria were rarest in the 1–4 year age group whereas the nitrifying bacteria increased through the groups as D < B < A < C. As the plant cover increased so usually did the bacterial numbers. The free-living nitrogen-fixing *Azotobacter* were only found in one ash sample and three of the samples with soil. They were never found in numbers greater than 100 g^{-1} dry weight.

Some of the results for one site are given in Table 1. These ashes came from north-east coals and they had been well leached by flooding so that they were low in available boron (average 10 ppm). One lagoon had been hydro-seeded with a mixture of grass and clover seed, fertilizer and cellulose fibre, which had given a good cover of grass on ash only one year old. The bacterial numbers below the grass were higher than with the two-year-old unseeded ash, but only at the surface. Profile samples showed a rapid fall in microbial numbers with depth.

Table 1. Microbial numbers in ash samples from Ferrybridge Power Station.

Sample Age (mths)	Microbial Numbers Thousands g^{-1} dry wt				
	Bacteria	Actinomycetes	Fungi	pH	Remarks
1	17,260	?0·1	1·5	10·1	Same lagoon
3	24,550	1	2	12·8	still filling
12	589,000	?	39	9·7	Hydroseeded 3 months earlier
19	1,941,000	27	99	8·3	
Top*	224,400	3	65	7·9	Same plot
Middle	23,500	<1	0·6	8·4	sampled
Bottom	2,028	<1	<1	10·1	7 months later
24	97,700	43	68	10·1	Patchy moss a few Atriplex
31	1,538,000	101	60	9·4	Same plot, sampled 7 months later

* Top = 0–1·5 cm   Normal sample top 15 cm
Middle = 7·5 cm
Bottom = 23 cm

There was therefore little field evidence that the constituents of PFA were inhibitory to microorganisms. Laboratory tests also have shown that many species can grow on culture media containing 100 ppm boron, as much as is likely to be found in most ashes.

Pot experiments

The survey described above showed that microorganisms could survive in PFA and that, as the ash weathered and accumulated nutrients, so the population increased in variety and numbers.

Considerable success has been reported from Russia in the use of bacterial fertilizers (Cooper 1959), especially those prepared from the free-living nitrogen-fixing bacteria, *Azotobacter*, and from a species of *Bacillus* that releases organically bound phosphorus: attempts to repeat these experiments in this country have met with little success (Brown *et al.* 1964). It was possible however that if a nutrient supply and bacterial fertilizers were added to relatively fresh PFA, plant growth would be possible at an earlier stage. Some experiments were therefore carried out to test this hypothesis.

If the experiment was ever to prove an economic proposition the nutrients provided would have to be cheap and in ready supply. Several waste organic materials, such as sewage sludge, domestic refuse and poultry manure were therefore considered as a source of nutrients.

In each experiment three bins of PFA were set up; one was left as a control (A) and the waste organic material was added to the other two (S and F). A cellulolytic fungus was added to one of the bins containing the waste material (F). The fungus should then degrade the cellulose in the waste organic matter, thus releasing nutrients for those microorganisms without this capability. About one month was allowed for the waste material to be broken down before adding the first bacterial fertilizer (P), a phosphate-dissolving bacterium of the type described by Louw & Webley (1959). This requires an energy source in order to produce 2-keto-gluconic acid, which brings insoluble inorganic phosphates into solution, thereby increasing the plant-available phosphate.

After a further month the mixtures were seeded and *Azotobacter* added (N). This bacterium requires an adequate supply of phosphate in order to fix atmospheric nitrogen and it was intended that the phosphate-dissolving bacterium should provide this. Each bacterial fertilizer was added to half of each mix, thus providing controls at each stage.

In the first experiment S1 (Table 2), sewage sludge from the Leatherhead works was used as the organic additive. There was little risk of excess contamination of this sewage sludge by heavy metals, which could be deleterious to plant growth. The material had been oven-dried before use and therefore contained few microorganisms. Barley was used as the test plant, as it is relatively sensitive to boron.

Germination was good in all pots. After 43 days the tops were harvested as inflorescences were forming; all were showing some toxicity symptoms which were less obvious in the presence of the cellulolytic fungus.

The dry yields are shown in Fig. 2. A statistical analysis of the results (Table 3) showed that none of the bacterial treatments had any significant effect, but that the sewage sludge gave a significantly ($P = \cdot 05$) increased yield compared with the plant ash, and that the addition of the cellulolytic fungus caused a further highly significant ($P = \cdot 001$) increase in yield.

The addition of the sewage sludge gave significantly increased levels of total nitrogen and plant available phosphate in the substrate ($P = \cdot 001$) and corresponding increases in the plant content of nitrogen and phosphorus ($P = \cdot 001$).

Table 2. Details of pot experiments.

	Experiment	
	S1 (March 1967)	G1 (February 1968)
Ash	Ferrybridge lagooned ash pH 9·3	Ferrybridge precipitator ash weathered at CERL pH 8·9
Organic additive	Leatherhead sewage sludge, oven-dried ~3% w/w	Chopped grass from a Ferrybridge PFA lagoon cut from area equal to area of bin
Cellulolytic fungus	Fungus from an ash site (*Chaetomium* spp) 1×10^3 viable units g^{-1} d.w. PFA Added day 0	As S1 1×10^5 viable units g^{-1} d.w. PFA Added day 0
Phosphate-dissolving bacteria	Isolated from an ash site—a mixed culture 2×10^7 cells of mixed culture g^{-1} d.w. PFA Added day 33	Isolated from an ash site 1×10^7 cells g^{-1} d.w. PFA Added day 49
Azotobacter	Strain from PFA site Added day 61 2×10^3 bacteria/seed ~20 g^{-1} d.w. PFA	As S1 Added day 98 1×10^3 cells g^{-1} d.w. PFA
Inorganic fertilizer	None	Sodium nitrate and superphosphate to give ~200 ppm N and ~670 ppm P in PFA Added day 98
Replication	9	4
Plant	Barley 20 seeds per pot Planted day 61 Harvested day 104	Barley 20 seeds per pot Planted day 105 Harvested day 147

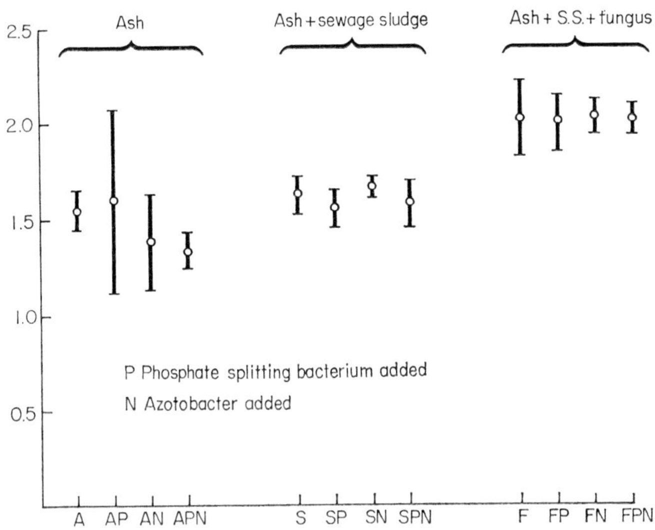

Figure 2. Experiment S1: average dry-weight yields (g per pot) of Barley.

Table 3. Experiment S1 Results at harvest.

Variable	A	S	F	No P	P	No *Azotobacter*	*Azotobacter*
Dry-weight yield g per pot	1·47	1·61	2·24	n.s.d.		n.s.d	
	A ← S: *	S ← F: ***	A ← F: ***				
Nitrogen in ash ppm	103	434	365	319 ← ***	336	322 ← **	333
	A ← S: ***	S → F: n.s.d.	A ← F: ***				
in barley ppm	12797	17435	14837	n.s.d.		n.s.d.	
	A ← S: ***	S → F: ***	A ← F: ***				
Phosphorus plant available ppm	155	195	192	n.s.d.		n.s.d.	
	A ← S: ***	S → F: n.s.d.	A ← F: ***				
in barley ppm	1656	2931	2161	n.s.d.		2203 ← **	2295
	A ← S: ***	S → F: ***	A ← F: ***				

n.s.d., no significant difference
* p = 0·05–0·01
** p = 0·01–0·001
*** p = > 0·001

The phosphate-dissolving bacteria apparently caused a significant (P = ·001) increase in the substrate nitrogen by the end of the experiment. This was possibly due to the increased growth of the *Azotobacter* in the presence of the phosphate-dissolving bacteria, as shown in Table 4.

Table 4. Numbers of *Azotobacter* at 104 days (log no. g^{-1} dry wt).

Treatment Mixture	*Azotobacter*	*Azotobacter* and Phosphate dissolving bacterium
A	3·9488	4·2379
S	3·1032	4·7515
F	3·1056	5·0833
Average with 95% C.I.	3·386±·470	4·691±·403

In another experiment, G1, grass was used as the organic additive to simulate the ploughing in of a grass ley. The grass was roughly chopped before adding and a further variation was introduced when extra nitrogen and phosphorus were added in the form of sodium nitrate and superphosphate (Table 2). The fertilizers were added at the same time as the *Azotobacter*. Barley was again the test plant.

The addition of the inorganic fertilizer to the pots as the seeds were sown, significantly (P = 0·05) decreased germination; no other treatment had any effect.

The statistical analysis of the dry yield results and of some of the chemical analyses are given in Table 5. Although adding chopped grass (G) significantly increased the substrate nitrogen level, the overall levels in the ash were lower than in the previous experiment. The lagooned ash used for S1 had initially a higher nitrogen content (100 ppm) than the ash used for G1 (< 10 ppm) which had been weathered at C.E.R.L. by exposure to rain. The sewage sludge also added more nitrogen than the grass. The nitrogen levels increased with all treatments throughout the experiment to the levels shown in Table 5. The increase was least in the plain ash with no treatment.

The addition of the chopped grass to the PFA caused an initial reduction in the level of plant available phosphate. During the experiment all the levels increased, although the available phosphate in the grass mixes never reached the level in the plain ash (Table 5).

All the treatments and additives significantly improved the yield (Fig. 3), although the effect of the added *Azotobacter* was less marked than that of the phosphate-dissolving bacteria or the inorganic fertilizers. Even with added inorganic fertilizer the yields on the plain ash were lower than in experiment S1, but this was probably due to the higher level of available boron in G1,

46 ppm on average compared with 10 ppm in S1. Despite this the nitrogen content of the barley from all the bins, even without inorganic fertilizer, was higher than in S1. The phosphorus content of the barley did not reflect the measured available phosphate in the substrate; the barley grown on the plain ash contained the least phosphorus, although there was apparently more phosphate available.

In other experiments it has been found that grass added to ash may even lower the yield of barley and perennial rye grass, although the further addition of a cellulolytic fungus will normally give an increased yield. This

Table 5. Experiment G1: results at harvest.

Variable	A	G	F	No P	P	No *Azotobacter*	*Azotobacter*	No Fertilizer	Fertilizer
Dry-weight yield g per pot	0·89 ←*** (from G); 0·89 ←*** (from F)	1·70 ←*** (from F)	2·54	1·49 ←***	1·92	1·59 ←**	1·83	1·40 ←***	2·01
Nitrogen in ash ppm	92 ←*** (from G); 92 ←*** (from F)	155 ←*** (from F)	229	148 ←***	169	152 ←**	165	119 ←***	198
in barley ppm		n.s.d.		26038 ←***	27459	26363 ←*	27134	19487 ←***	34009
Phosphorus plant available ppm	440 →*** (to G); 440 →*** (to F)	395 → n.s.d. (to F)	387	417 →*	387	394 ←*	410	253 ←***	551
in barley ppm	3318 ←* (from G); 3318 ←* (from F)	4108 ← n.s.d. (from F)	4200		n.s.d.	3786 ←**	3970	2695 ←***	5061

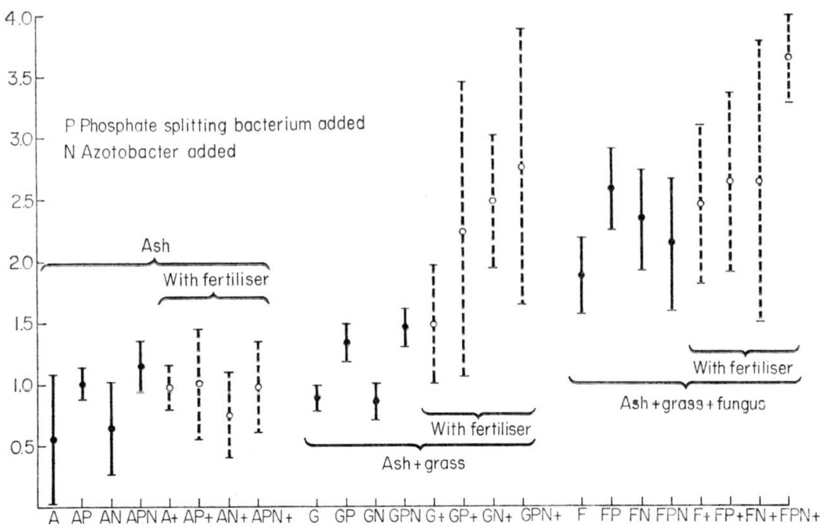

Figure 3. Experiment G1: average dry-weight yields (g per pot).

may be due to the binding of available phosphate by the grass before it is decomposed.

Other organic wastes have been tested in ash mixes, using perennial ryegrass as the test plant. The waste materials tested in pot experiments have included raw and composted domestic refuse, beet washings, mushroom compost and peat, with and without poultry manure or clay. The results of one trial are shown in Fig. 4. A second harvest gave improved yields, probably as a result of further leaching of boron. Inorganic fertilizer (Fison's 52) giving 14 g N m^{-2} significantly improved the yield of all the mixes except that containing poultry manure. In this trial the peat and the domestic refuse did little to improve the yield.

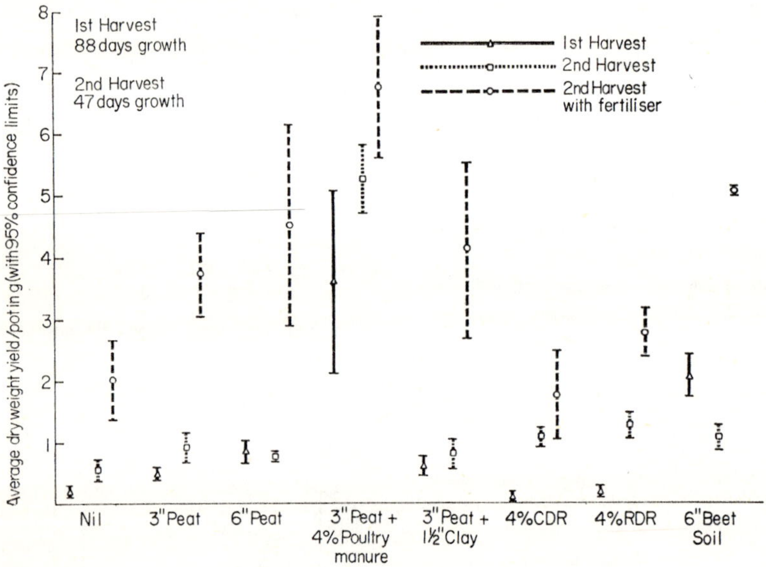

Figure 4. Pot trials with Thorpe Marsh ash.

Field trials

The experiments described above showed that in pot experiments the fertility of PFA could be improved by the addition of organic matter and 'microbial fertilizers'. It was of interest therefore to test the value of these additives in the field.

The first trial was set up on an ash field at Drakelow Power Station. The waste organic materials selected were some of those that could be obtained in large quantities near the site of a proposed ash disposal scheme. They were pulverized domestic refuse, sewage sludge and poultry manure.

Many composted domestic wastes would be unacceptable as soil conditioners because of the sizeable pieces of glass that they contain. These could be a hazard, particularly if ingested by cattle. However, in 1969, a new plant at Wetherby, Yorkshire was producing a compost in which the maximum size of solid particles was said to be 2×10^{-3}m. The compost was prepared by passing domestic refuse through two hammer mills and a vibrating screen, after the removal of ferrous objects. Sewage sludge was then mixed with the pulverized material at the rate of 900 l per 1,000 kg (200 gallons per ton). The final mixture was then tipped into windrows for composting.

A considerable part of the cost of this material would be the labour charges for composting and therefore in this field trial the composted domestic refuse (C.D.R.) was compared with the raw mixture as it left the plant (R.D.R.). Each material was used at two levels: approximately 1,004 Mg, and 168 Mg ha^{-1}, roughly 25% and 4% by weight, assuming the additive is mixed to a depth of 0·25 m. Two other additives were also used, the Wetherby sewage sludge used in the compost, and a poultry manure specially prepared as a fertilizer; both were used at 168 Mg ha^{-1}.

The plots were set out on an area 53×23 m at the corner of an ash field that was still being extended by tipping lagooned ash. The site was very compacted by the lorries and difficulty was experienced in ploughing to a uniform depth of 0·25 m. The ploughed area was divided into seven rows, the organic materials were rotovated into six, leaving one as a control. Each row was divided into 16 plots on which two treatments were tested alone and in combination. The two treatments tested were a cellulolytic fungus as used in the pot tests and calcium lignosulphonate. Owing to unexpected production difficulties the amount of fungus watered on gave less than 200 viable particles per cm^2 of ash surface. This could not be expected to produce an effect and none was found. Laboratory studies at Leeds University had shown that boron was leached more rapidly from fresh ash in the presence of calcium lignosulphonate. As a second treatment therefore, calcium lignosulphonate was used at a rate of 1,121 kg ha^{-1}, but no effect was found on plant growth, or on boron levels in the substrate or grass. It is probable that only fresh ash will benefit from a lignosulphonate treatment.

The plots were seeded with S23 perennial rye grass in August 1968. At the time of the first harvest in June 1969 there was very good growth on the 4% poultry manure and the 25% R.D.R., with the yield from the former only just significantly higher. Growth was good on the 4% R.D.R., and poor on the 25% C.D.R. Growth was so poor on the plain ash, the 4% C.D.R. and the 4% sewage sludge that it was not possible to cut it.

Immediately after this harvest Fison's Nitro '26' fertilizer was applied to one half of each row. The amount applied was equivalent to a rate of 678 kg ha^{-1} or 17·6 g nitrogen m^{-2}. The effect of this fertilizer treatment was soon apparent in the improved colour of the grass on all the rows except those with the poultry manure and the 25% raw domestic refuse.

In October 1969 the fertilized and unfertilized strips were harvested separately from all the rows using a quadrat of 0·06 m^2. The yield from the unfertilized poultry manure row was larger than before but was smaller than that from the 25% R.D.R., although not significantly so. The organic fertilizer increased the yields on every row except the 25% R.D.R. The increases were significant on the 4% R.D.R., the 4% C.D.R. and the sewage sludge. The yields obtained at each harvest are illustrated in Fig. 5, and the chemical analyses of the grass in Table 6.

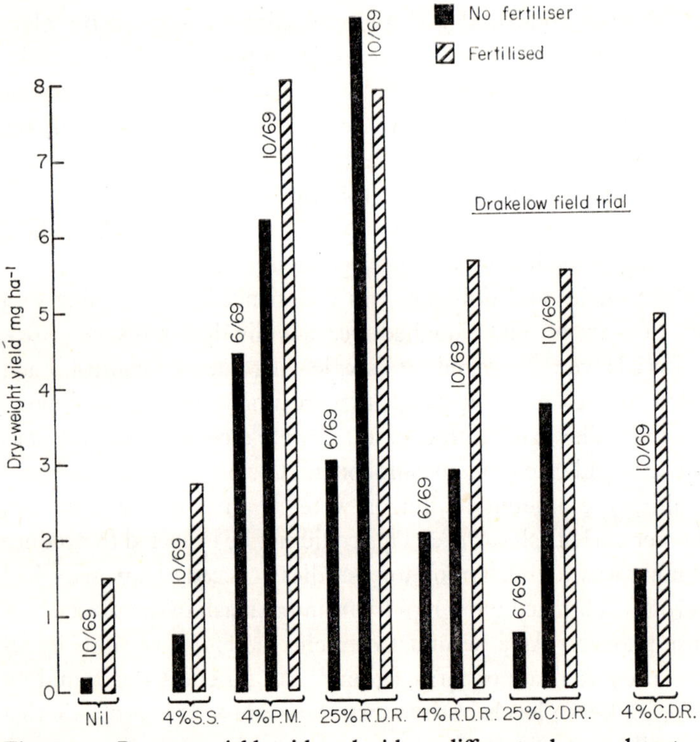

Figure 5. Rye grass yields with and without different ash amendments.

The lowest yields at Drakelow came from the four adjacent rows nearest the middle of the field, the plain ash, the two C.D.R.'s and the sewage sludge. The raw and composted domestic refuse had been tested in pots prior to use in the field, and in these and later confirmatory tests, the composted material had always given better growth than the raw waste. It seems possible that there was some factor at Drakelow which increased across the rows and inhibited plant growth. This could explain why the sewage sludge and the composted wastes gave such poor yields.

Table 6. Drakelow—Plant analyses

Substrate	Inorganic Fertilizer	Harvest June 1969				Harvest October 1969			
		Nitrogen		Phosphorus		Nitrogen		Phosphorus	
		In plant ppm	Removed kg ha^{-1}	In plant ppm	Removed kg ha^{-1}	In plant ppm	Removed kg ha^{-1}	In plant ppm	Removed kg ha^{-1}
Plain ash	−	No harvest				10,150	2	1,565	0·3
	+					10,900	17	1,561	2·0
4% Poultry manure	−	23,959	107	5,253	24	17,620	110	3,981	25
	+					18,980	153	3,829	31
4% Sewage sludge	−	No harvest				9,660	7	1,840	1·0
	+					9,140	25	1,756	5
4% Raw domestic refuse	−	14,541	30	2,771	6	10,340	30	2,479	7
	+					10,650	60	2,468	14
25% Raw domestic refuse	−	17,769	54	3,323	10	10,860	97	2,375	21
	+					12,370	98	2,574	20
4% Composted domestic refuse	−	No harvest				9,070	14	1,845	3
	+					10,720	53	1,851	9
25% Composted domestic refuse	−	17,709	13	3,445	26	9,750	37	2,479	9
	+					10,620	59	2,336	13

A second field trial was set up at Littlebrook Power Station in July 1969. The area used, 200 × 80 m, had been filled with PFA from several power stations and it is unlikely that it had all been lagooned before transporting to Littlebrook.

In this trial only two organic additives were used; poultry manure and sewage sludge. The poultry manure was obtained from a local farmer and the sewage sludge from the neighbouring Dartford works. Analysis showed the poultry manure to contain 4% nitrogen and the sewage sludge 2·3%, with plant-available phosphorus at 1·4 and 0·3% respectively. All the microbial treatments used in the pot experiments were used in this field trial (Table 7). To allow adequate replication each additive was spread on four rows, and there were four 'nil' rows. Each row was divided into sixteen plots, each 5 × 4 m. The treatments were then applied following a statistical design.

Table 7.

LITTLEBROOK SITE Treatments
1. Cellulolytic fungus—*Chaetomium* sp. $6{\cdot}7 \times 10^{3}$ viable units cm^{-2} ash Added 7.7.1969
2. Phosphate dissolving bacterium $1{\cdot}2 \times 10^{7}$ bacteria cm^{-2} ash Added 7.8.1969
3. *Azotobacter vinelandii* Strain O 5×10^{8} bacteria on seeds for one plot Added 24.9.1969
4. Inorganic fertilizer 7 g nitrogen m^{-2} 3·5 g phosphorus m^{-2} 3·5 g potassium m^{-2} Added 23.9.1969

The plots were seeded with perennial rye grass in September 1969 but unfortunately a long dry spell followed, germination was poor and growth was delayed until the spring. It was therefore unlikely that any of the treatments would have any marked effect. The sewage sludge had produced a large crop of tomato plants and these were removed before seeding. These, and the nil rows then remained relatively weed free, but the poultry manure, particularly at one end of the area, supported a heavy growth of *Chenopodium urbicum*. Harvests were taken in July and October 1970. The yields with and without fertilizer are shown in Fig. 6 and the chemical analyses in Table 8.

Table 8. Littlebrook—Plant analyses

Substrate	Inorganic Fertilizer	Harvest July 1970 Nitrogen In plant ppm	Harvest July 1970 Nitrogen Removed kg ha⁻¹	Harvest July 1970 Phosphorus In plant ppm	Harvest July 1970 Phosphorus Removed kg ha⁻¹	Harvest October 1970 Nitrogen In plant ppm	Harvest October 1970 Nitrogen Removed kg ha⁻¹	Harvest October 1970 Phosphorus In plant ppm	Harvest October 1970 Phosphorus Removed kg ha⁻¹
1st Trial									
Plain ash	–	No harvest				22,240*	2	1,718	0·2
	+					24,583	11	1,606	0·7
4% Poultry manure	–	20,825	73	2,018	7	22,710	40	2,538	4
	+	21,786	61	2,047	6	23,722	57	2,543	6
4% Sewage sludge	–	13,843	29	1,640	3	17,756	20	2,524	3
	+	14,816	38	1,679	4	22,624	54	2,587	6
		Harvest August 1972							
2nd Trial									
Plain ash	–	14,195	7	2,219	1				
	+	14,238	14	2,025	2				
4% Poultry manure	–	12,688	54	2,263	10				
	+	9,850	59	2,188	13				
4% Sewage sludge	–	10,426	52	2,293	11				
	+	10,563	68	2,025	13				

* Ammonium sulphate added to all plain ash plots

The sewage sludge in this trial gave yields comparable with those from the poultry manure, whilst the growth on the plain ash was too poor to harvest in July 1970, and was still very slight at the second harvest.

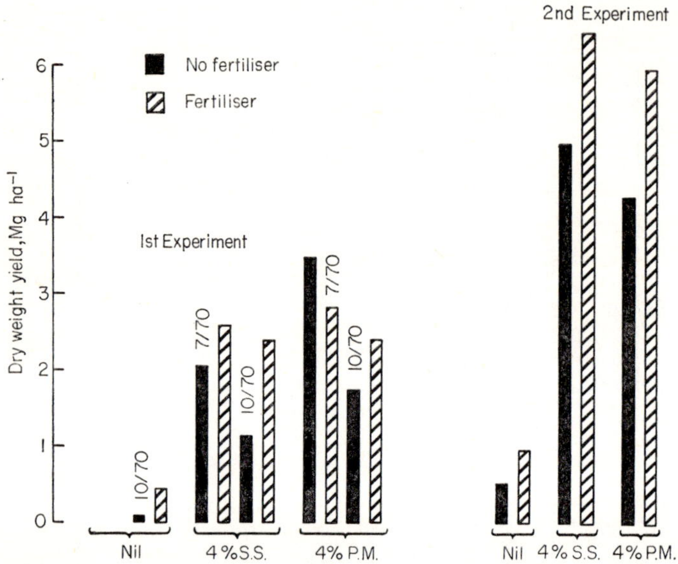

Figure 6. Littlebrook field trials.

Access to the site was restricted in the spring of 1971, but in July it was possible to set up a second smaller experiment. The rows were rotovated along their length after removing all top growth, and then redivided into four plots each 4×15 m. The treatments used were inorganic fertilizers and white clover (S100), (Table 9). Perennial rye grass S23 was again the test plant.

The results of this experiment are also shown in Fig. 6 and Table 8. Growth was best on the sewage sludge and least on the plain ash throughout

Table 9. Littlebrook second experiment—treatments per plot.

Treatment	Inorganic Fertilizers				
	Potassium sulphate 378 g	Super phosphate 1,058 g	Sodium nitrate 2,634 g	Sodium nitrate 1,317 g	Clover seed 20 g
1	o	o	o	o	o
2	+	+	+	o	o
3	o	o	o	o	+
4	+	+	o	+	+

the growing period. The added clover grew very patchily and many weeds including legumes appeared with most on the poultry manure.

The clover had no effect on the yield, and the growth on the plots with clover and fertilizer was less than that on those with the fertilizer alone, probably because less nitrogen was applied. In this experiment the yields without fertilizer were much better on every substrate than they had been in the first Littlebrook experiment, with inorganic fertilizer further increasing the yields.

Although the addition of waste organic materials to PFA has generally improved yields at both sites there have been differences. Some of these have almost certainly been caused by natural differences in the ashes, in particular the unknown factor that appeared to have affected yields on the C.D.R. and sewage sludge at Drakelow.

The poultry manure used at Drakelow had been prepared as a fertilizer, and gave very high levels of nitrogen and phosphorus in the substrate which were reflected in the levels in the grass, especially at the first harvest. This may have accounted for the better yields from poultry manure at Drakelow than at Littlebrook. The sewage sludge at Drakelow however, gave very poor yields, possibly because of the unknown 'toxic' factor. The nitrogen levels were similar in the sewage sludge treatments from the two sites at the start but improved at Littlebrook by the final harvest (Table 10).

Table 10. Substrate analyses at final harvest

Substrate	Total Nitrogen ppm	Plant Available Phosphorus ppm	Plant Available Boron ppm	Boron in Grass ppm
DRAKELOW				
Plain ash*	300	57	50	150
4% Poultry manure	4,600	904	56	85
4% Sewage sludge	800	43	47	126
4% Raw domestic refuse	1,600	55	36	70
25% Raw domestic refuse	1,600	51	34	85
4% Composted domestic refuse	1,000	51	32	72
25% Composted domestic refuse	2,900	39	34	103
LITTLEBROOK				
Plain ash	787	29	48†	>300†
4% Poultry manure	1,151	104	21†	170†
4% Sewage sludge	1,676	81	19†	148†

* All results for substrate without inorganic fertilizer except

† Pooled samples, with and without inorganic fertilizer.

Analysis of the grass showed a wide range of nitrogen and phosphorus levels (Tables 6 & 8), most of which were within the normal range for grass. The mineral content of grass is known to change during growth and, even at one harvest, the plants were not always at the same stage of maturity, as shown by differences in flowering. The substrate nitrogen and phosphorus levels were also important in influencing the levels in the plants although the measured levels of plant available phosphorus were not a particularly good guide to plant uptake.

A further factor influencing yield may have been the boron uptake (Table 10). The level of available boron was similar in all the substrates, but there appeared to be a greater uptake at the Littlebrook site, particularly at the earlier harvest.

The yields from Drakelow and Littlebrook may be compared with those from the Besthorpe experiment already described by Townsend & Gillham (1975) and especially as the first harvests from each site were all taken after approximately the same growth periods. The crop was pure perennial rye grass at Littlebrook and Drakelow, but a mixture of grasses, including perennial rye grass and a little clover, at Besthorpe. The best and worst crops taken from each site at the first harvest are compared in Table 11. No cut was possible from the plain ash at Drakelow or Littlebrook. The table shows that the best yields at the three sites are comparable, both in quantity and quality, except that the grass from the Littlebrook site contained more boron. The higher yield that was obtained from the plain ash at Besthorpe than from the plain ash at either Drakelow or Littlebrook could have been caused by the greater degree of weathering of the ash at Besthorpe after flooding.

Table 11. Yields from Besthorpe, Drakelow and Littlebrook

Site	Removed at Harvest (kg ha^{-1})			
	Dry matter	Nitrogen	Phosphorus	Boron
BESTHORPE (June 1967)				
Plain ash	2,058	25	4	0·15
12 in. soil	3,702	56	10	0·09
DRAKELOW (June 1969)				
25% CDR	742	13	3	0·02
4% poultry manure	4,470	107	24	0·08
LITTLEBROOK (July 1970)				
4% sewage sludge	2,072	29	3	~0·38
4% poultry manure	3,483	73	7	~1·13

Besthorpe—low fertilizer level = ~68 kg N ha^{-1}

Drakelow and Littlebrook—no fertilizer.

Conclusions

The results have shown that the addition of some types of organic waste to PFA can improve the fertility, so that the addition of top soil is unnecessary. The value of bacterial fertilizers has not been proven.

Of the organic wastes poultry manure was particularly good, but domestic refuse and sewage sludge, if taken from the right source where contamination of heavy metals did not occur, were also valuable.

Acknowledgment

This paper is published by permission of the Central Electricity Generating Board.

Summary

As pulverized fuel ash ages it is colonized by microorganisms until the population approaches that of soil. Microorganisms used as 'fertilizers' in pot experiments gave improved growth of plants. The results have been applied in field trials using waste organic materials as microbial nutrients. The waste materials alone improve ash fertility.

References

Brown M.E., Burlingham S.K. & Jackson R.M. (1964) Studies on *Azotobacter* species in soil III. Effects of artificial inoculations on crop yields. *Pl. Soil* **20**, 194–214.

Cooper R. (1959) Bacterial fertilizers in the Soviet Union. *Soils & Fertiliz.* **22**, 327–33.

Hodgson D.R. & Buckley G.P. (1975) A practical approach towards the establishment of trees and shrubs on pulverised fuel ash. In *The Ecology of Resource Degradation and Renewal* (Ed. by M.J. Chadwick & G.T. Goodman), Blackwell, Oxford.

Louw H.A. & Webley D.M. (1959) A study of soil bacteria dissolving certain mineral phosphate fertilizers and related compounds. *J. appl. Bact.* **22**, 227–33.

Townsend W.N. & Gillham E.W.F. (1975) Pulverised fuel ash as a medium for plant growth. In *The Ecology of Resource Degradation and Renewal* (Ed. by M.J. Chadwick & G.T. Goodman), Blackwell, Oxford.

Renewal of china clay strip mining spoil in southeastern United States

JACK T. MAY *University of Georgia, Athens, Georgia, U.S.A.*

Introduction

Site conditions on strip mine spoil are so unique that a distinctive classification of factors influencing ecological stability is needed. The big difference between site conditions on undisturbed and on strip-mined land lies in the rooting medium. On undisturbed sites, the soil is an organized natural body resulting from weathering of parent material near the earth's surface. On strip-mined land the rooting medium is an unorganized mass of material derived from the overburden.

Some characteristics of spoil material from kaolin clay strip mining were described by May *et al.* (1973a).

Classification of spoils

Strip mine spoils may be classified in many different ways. However, for reclamation it is important that they be classified according to factors or features that affect plant growth and equipment operation. Since spoils are a mixture of all strata in the overburden, the characteristics of the spoil can be partially controlled by the mining technique. In general, the range in concentration of elements and in particle size distribution within the spoil is similar to that found within the combined strata of the overburden. All spoil material is deficient in or devoid of nitrogen, organic matter, and micro- and macroorganisms. Major plant nutrients such as P, K, Ca and Mg are deficient in about 80% of the spoil. The soil reaction of spoil material ranges from about pH 1·8 to 8·0. Major toxicity is associated with a low pH and high iron concentrations.

There is a wide diversity in particle size distribution and elements within short distances and all depths. The percentage of sand ranges from 4 to about 90% and clay from about 10 to 85%. Generally the surface of the spoil contains a much higher percentage of clay than the undisturbed sites.

The soil moisture holding capacity of the spoil is extremely diverse in any single mining area. The infiltration capacity and erodibility of spoil material is related to length and steepness of the slope, the sand-clay ratio, and the extent of crusting and compaction in the surface layer.

In the south-eastern United States, a broad classification based on slope, texture, colour and acidity is feasible.

SLOPE GROUPS

1. 0 to 3%.
Slopes are level or nearly level. Water erosion is slight except on very long slopes of highly erodible spoils. Some broad terraces may be desirable. Sites can be seeded to grasses or planted to trees.
2. 3 to 8%.
Slopes are gently sloping or undulating. Runoff is slow to medium. Rill erosion can develop into gully erosion. Slopes should be short or be broken by catchment basins. Sites can be seeded to grasses. Trees can be planted on non-erodible spoils.
3. 8 to 16%.
Slopes are strongly sloping or rolling. Runoff is medium to rapid. Sheet, rill and gully erosion will develop unless there is a grass cover. Blow-outs occur where water, through internal drainage, emerges down slope.
4. 16 to 35%.
Slopes are moderately steep or hilly. Runoff is rapid or very rapid. Sheet, rill and gully erosion develop under normal rainfall. Blow-outs are common.
5. 35%+.
The landscape is steeply sloping or very hilly. Runoff is very rapid. Slopes are highly erodible. These spoils are developed by dragline mining.
6. Highwalls.
These are vertical walls of mines or extremely steep backslopes of some spoils.

Slope classes 1, 2, 3 and 4 can be cultivated, seeded and fertilized with farm machinery. Slope class 5 presents serious problems in reclamation. Highwalls are physical hazards.

COLOUR AND TEXTURE GROUPS

Colour and texture are closely associated and can be merged in a classification system.

1. Surface light coloured.
The surface material consists of rug material or low grade kaolinitic clay.
2. Surface light coloured.
The surface material consists of light textured sands and sandy loams.
3. Surface reddish to brown.
The surface material is sandy to clayey.
4. Surface greenish to olive brown.
The surface material is a mixture of spoil containing more than 60% fuller's earth.
5. Surface whitish to greyish or brown.
Surface material consists of sandy to clayey marl or limestone containing marine fossils.
6. Surface yellowish, grey or brownish.
Surface material consists of sands to clays containing a high percentage of pyrite.
7. Surface grey to reddish or brown.
Surface material consists of a sandy to clayey mixture of the A and B horizons that has been spread over spoil material; and is about 15 to 30 cm thick.

ACIDITY GROUPS

1. pH is above 6·0.
This material consists of fuller's earth, marl or other calcareous material. A few native pioneer plant species seed-in naturally. Establishment of natural vegetation is accelerated by application of nitrogen and sometimes phosphorus.
2. pH is between 4·5 and 6·0.
This group may contain a wide range of particle size distribution. Complete fertilization is required for establishment of all agronomic crops. All tree species require nitrogen and sometimes phosphorus.
3. pH below 4·5.
Concentrations of pyritic materials results in hot spots with pH as low as 1·8.

Reclamation

Selection and use of vegetation in reclamation depends primarily on three considerations, namely: characteristics of spoil material, characteristics of species, and ultimate objectives of the owner. All strip-mined lands should be restored to productive uses compatible with adjacent land use.

Vegetation may be established for a variety of uses, such as erosion control, water control, food and cover value, wildlife, grazing, agronomic crops, aesthetic values and to produce wood products such as pulpwood, Christmas trees, posts, poles and sawtimber.

The two major elements that create problems in reclamation are (1) the rooting media (spoil material) and (2) the climate. The nature of the spoil material can be modified by grading, redistribution of top soil, use of fertilizers and the use of surface mulches. But major problems can still be caused by the one uncontrollable factor—the weather. The mean annual rainfall along the Fall Line of Georgia is 109/45 cm per annum. Devastating rains in the spring are often followed by excessive periods of drought in summer and autumn. Land that has been graded, fertilized and seeded can be completely destroyed by sheet erosion resulting from torrential rainfall prior to establishment of a vegetative cover.

SPECIAL PROBLEMS OF THE SPOIL MATERIAL

1. Spoils consisting of high irregular conical steep-sided piles.
These can be made plantable by pushing the top of the pile down hill and creating a concave or saucer-shaped catchment basin at the highest elevation. The catchment must be deep enough to hold all precipitation from the heaviest rains. Water then moves through the spoil by gravity. Long slopes can be broken by wide benches or catchment basins that will hold all water from the slope above. Generally the slopes should be on less than a 35% grade and less than 15 m long. The landform must provide for water control and the use of agricultural equipment for fertilizing, seeding and mulching.

2. Spoils that are rich in kaolin.
Vegetation is more difficult to establish in heavy clay spoils than the more sandy material or the red clay-sand mixtures of the Barnwell formation. Where mining conditions permit, Fuller's earth or top soil from adjacent cuts or from stockpiles should be spread over the kaolin clay spoil.

3. Acidic sands and clays.
They are toxic to vegetation. Heavy applications of lime have not reduced the acidity. Fuller's earth or top soil should be spread over these acidic spoils.

SPECIES SELECTION

If an adequate cover of vegetation would develop naturally on strip mine spoils, there would be no reclamation problem. But so many adverse conditions prevail that very few spoils have been successfully vegetated in this way. Site conditions, climate and inherent requirements of a species limit the successful invasion of the species that can be grown in any particular area. May *et al.* (1973b) described some of the problems and techniques involved in establishing grasses and tree vegetation on spoil from kaolin clay strip mining.

MIXED VEGETATIVE—TREE COVERS

A permanent vegetative cover, except for pasture, will normally include a mixture of grasses, herbaceous plants and woody plants (shrubs and trees). Four possible establishment techniques are described.

1. Herbaceous vegetation or grasses on graded rolling sites.
2. Trees on relatively level sites.
3. Trees on sites with established vegetation.
4. Herbaceous vegetation and trees during the same year on graded sites.

Herbaceous vegetation or grasses on graded rolling sites

This technique has been successful on a wide range of sites. Mixtures that have produced good vegetative covers are:

Secale cereale L.
Lespedeza cuneata (Dumont) G. Don
Cynodon dactylon (L.) Pers.

Secale cereale
Paspalum notatum Var. Saurae Parodi
Cynodon dactylon
Panicum ramosum L.

Lespedeza cuneata
Lespedeza striata Thunb.
Eragrostis curvula (Schrad.) Nees

Secale cereale
Festuca arundinaceae Schreb
Lespedeza cuneata

Paspalum notatum
Cynodon dactylon
Trifolium repens L.

The most successful treatments are obtained with autumn sowing of

Secale cereale. *Lespedeza* spp., *Festuca* sp., *Paspalum* sp., and *Cynodon* sp. can be sowed with the *Secale* sp. or seeded in the spring.

Spring sowing is preferable for *Lespedeza* spp., *Panicum* sp., *Eragrostis* sp., and *Paspalum* sp. Seeds should be drilled into the spoil and packed with a cultipacker. Mulching is essential for protection of the site and germinating vegetation.

Fertilizer requirements vary with soil reaction and nutrient levels of the spoil. Where pH is between 4·0 and 5·5, 4,500 to 6,000 kg of lime per ha are required. Nutrient requirements range from 56 to 112 kg of N per ha, 112 to 195 kg of P_2O_5 and 56 to 200 kg of K_2O per ha.

A fertilizer supplement, especially nitrogen, is needed the second year, except for Lespedezas. *Cynodon* sp. and *Paspalum* sp. begin to disappear after four years and only one maintenance fertilization. A third fertilization after the fifth or sixth growing seasons will result in vigorous growth of the existing residual cover.

Trees on relatively level sites

This technique is relatively simple on all level sites except where the soil pH is above 6·5 or below 4·5.

Tree species adapted to spoil sites are *Pinus elliottii* Engelm., *P. taeda* L., *P. virginiana* Mill., *Platanus occidentalis* L., *Quercus acutissima* Carruthers, and *Alnus glutinosa* (L.) Gaertn. Machine and hand planting have given comparable survival and growth rates.

Fertilization is essential for the successful establishment of trees on bare sites. A complete fertilizer with about 28 g of N per seedling is optimum for both survival and growth. Mortality increases with an increase of nitrogen. Applications of Ca and K have not significantly affected survival or growth of most of the species tested on clay spoil. Maintenance fertilization may be required about the third year for most sites and for most species. It is possible that a third or fourth treatment may be required before nutrient recycling becomes effective.

Fertilizers have been applied in a circle or a spot on the ground surface near the seedlings, in slits adjacent to the seedlings, and as pellets in the planting slot and in adjacent slits. In general, surface fertilization and slit fertilization provide comparable growth response. The application of pellets in the planting slits has an adverse effect on survival of conifers but not hardwoods. Growth response is slower from pellets in adjacent slits than from surface fertilization.

Where spoils are predominantly mixtures of fuller's earth or of a highly calcareous nature, *Alnus* sp., *Platanus* sp., and *Lespedeza bicolor* Turcz. have given better survival and growth than other species. Fertilization of these sites stimulates the germination and growth of invading native grasses and Lespedezas which crowd out planted seedlings.

Trees on sites with established vegetation

This technique presents special problems, namely:
a. The established vegetation provides too much competition for space, moisture and nutrients.
b. Fertilization of tree seedlings within a complete cover or a disintegrating cover stimulates growth of herbaceous vegetation, which subsequently crowds out the tree seedlings.

Herbaceous vegetation and trees during the same year on graded sites

This procedure has been tested on a few sites. It shows promise when a forest cover is the ultimate objective.

Soil genesis

The ultimate objective of reclamation is to develop a soil that has physical, chemical and biological properties equal to or better than the original overburden. Soil is defined as the collection of natural bodies on the earth's surface, containing living matter, and supporting or capable of supporting plants. Each of the natural bodies have a unique morphology resulting from a unique combination of climate, living matter, parent material, relief and time. Mining spoil is the parent material that will at some future time become soil. The results of soil formation will be stabilization of sites through aggregation of mineral and organic matter, increased nutrient status, initiation of nutrient cycling and reduced erosion.

Current studies deal with such phases of soil formation as effects of vegetation on spoil material, invasion of micro- and macroorganisms, nutrient cycling, and physical and chemical weathering of spoil material.

EFFECTS OF VEGETATION

The initial impact of vegetation on spoil is associated with development of a root system, providing a surface cover, the accumulation of litter, and the translocation of nutrients.

The heterogeneity of spoil material, to some extent, favours the development of an extensive rooting system. Two-year-old *Pinus elliotti* seedlings have developed both extensive lateral roots parallel to the surface (some extending more than six metres from the tree), and tap roots (some more than one metre long). *Lespedeza cuneata* and *Eragrostis curvula* develop a mass of deep fibrous roots within two to three years.

Secale cereale and mixtures of other grasses provide a dense vegetative cover which protects the surface from the impact of rain, wind, and solar radiation. Both herbaceous and woody vegetation provide a surface litter.

SOIL ORGANISMS

New spoil is almost completely devoid of any form of soil organisms. A copious and diverse population of microorganisms invades the spoil with the establishment of plants. The objective of a present study is to determine rates of invasion of symbiotic, saprophytic and parasitic organisms in natural and artificially established plant communities on kaolin clay spoil.

Many species of Pinus will not thrive on a medium lacking in mycorrhiza-forming fungi. *Thelephora terrestris* Ehr. that develops on seedling roots in the nursery is not adapted to adverse sites. However, during the first year after pine seedlings are planted, there is a proliferation of a drought resistant fungus, *Pisolithus tinctorius* (Pers.) Cok. & Couch, on the seedling roots. This symbiotic relationship contributes to the vigour and growth of planted seedlings.

Cenococuum graniforme (Sow.) Ferd. and Winge, is a drought resistant mycorrhizal-forming fungus that occurs throughout the southeastern United States. Seedlings treated with these fungus are being planted on different types of spoil materials. Plantings made in 1973, indicate that seedlings will not survive on spoil with a low pH.

CHANGES IN SPOIL MATERIAL

Troth (1972) reported on an examination of six sites where spoil material had been exposed to weathering and invasion of vegetation for periods of 15 to 65 years. Clay content was found to increase with depth while sand content decreased. The organic matter content ranged from 1·1% in the upper 0–2 cm of the 15-year-old spoil to 6·2%+ in the upper 0–2 cm of spoil deposited over 40 years ago.

A profile description of two sites shows that the effects of vegetation and weathering extend to a depth of about 26 cm.

Site No. 2. Vegetation consists of a dense cover of pines, mixed hardwoods and honeysuckle. The site had been undisturbed for approximately 50 years.

Layer	*Depth* (cm)	*Description*
1	0–3	Very dark greyish brown (10YR 3/2) loam containing a few, faint yellowish brown (10YR 5/4) mottles; weak to moderate, fine and medium granular structure; very friable; numerous small roots; abrupt, wavy boundary; range in thickness: 1–4 cm.
2	3–14	Dark greyish brown (10YR 4/2) sandy clay loam with few diffuse brown (10YR 5/6) mottles; weak, fine and medium granular structure; friable; numerous small and large roots; clear, wavy boundary; range in thickness: 2–20 cm.
3	14–26	Very dark grey (10YR 3/1) sandy clay loam with brown–dark brown (7·5YR 4/4), yellowish red (5YR 4/8), red (2·5YR 5/8) and white (10YR 8/2) mottles; weak fine granular and weak medium subangular blocky structure; firm, numerous small and large roots; many pore spaces; appears to be a mixture of layers 1 and 2; clear, wavy boundary; range in thickness: 2–12 cm.
4	26–34	Pale brown (10YR 6/3), very dark greyish brown (10YR 3/2), strong brown (7·5YR 5/6) and yellowish brown (10YR 5/4) clay loam containing red (2·5YR 5/8) fragments; coarse subangular blocky; firm; clear, wavy boundary; range in thickness: 11–14 cm.
5	34–60+	Yellowish brown (10YR 5/6) and very pale brown (10YR 7/3) sandy clay loam with yellowish red (5YR 4/8) and strong brown (7·5YR 5/6) mottles; massive; firm.

Site No. 3. Vegetation consists of scattered pines, occasional hardwoods and grasses and sedges as ground cover. This area appears to be very stratified as a result of mining. Mining ceased at least 15 years ago.

Layer	*Depth* (cm)	*Description*
1	0–2	Brown (10YR 5/3), pale brown (10YR 6/3), greyish brown (10YR 5/2) and very dark grey (10YR 3/1) loamy sand with few faint yellowish brown (10YR 5/4) mottles; weak fine granular to single grain structure; very friable;

numerous small roots; appears to be overwash; abrupt, smooth boundary; range in thickness: 1–3 cm.

2 2–7 Very dark grey (10YR 3/1), dark greyish brown (2·5YR 4/2) and greyish brown (10YR 5/2) sandy loam containing some bleached white sand pockets; weak, fine and medium granular structure with some platy units; friable; many fine roots; abrupt, smooth boundary; range in thickness: 4–6 cm.

3 7–26 Brown-dark brown (10YR 4/3) and dark yellowish brown (10YR 4/4) sandy loam with lenses of very dark greyish brown (10YR 3/2) material; weak, fine and medium granular structure with some weak fine subangular blocky; friable; many fine roots; illuvial material from layer 2 present in root channels and worm holes; clear, wavy boundary; range in thickness: 10–20 cm.

4 26–75^{+} White (10YR 8/1) and pale red (10YR 6/3) clay loam; massive; firm; few roots.

Other observations from a study of the six sites reveal that:

1. Heterogeneity of spoil colours increased with depth.
2. Structural development decreased with depth.
3. Amount of clay in the subsurface layer increased with time.
4. Organic matter was incorporated to the greatest depths at those sites undisturbed for the longest periods of time.

In a study of effects of fertilization on yield of *Paspalum notatum* and *Cynodon dactylon*, Troth (1971) found that yields could be sustained with maintenance fertilizations (Table 1). In general, spoil, vegetation and fertilization interactions indicated that:

Table 1. Bahia grass (*Paspalum notatum*) Yields (kg/ha).

Fertilization 1967			Yield	Cumulative Fertilization 1967–1972			Yield
N	P	K		N	P	K	
56	0	0	205	224	0	0	442
56	0	37	46	224	0	157	699
56	20	0	493	224	84	0	1,869
56	20	37	1,178	224	84	157	2,305
112	20	37	1,922	448	84	157	4,677
112	45	93	3,327	448	183	372	4,274
224	45	93	4,859	896	183	325	5,817
224	90	186	5,196	896	315	604	6,766
224	90	0	2,515	896	366	186	6,207
448	90	186	8,855	1,568	366	744	7,879

1. Spoil pH reached a maximum two years after application of limestone.
2. Most extractable P was found in the upper 5 cm of spoil.
3. Extractable P was increased down to 15 cm by applications of 45 and 90 kg of P/ha.
4. The amount of extractable K in the 0–5 cm layer generally increased with an increase in the rate of K application.
5. After five years, Ca and Mg were well distributed in the upper 15 cm of spoil, decreasing sharply at depths of 15–30 cm.
6. Organic matter accumulation was closely correlated with rate of N application.

These current studies reveal that the combination of natural physical and chemical weathering, fertilization, establishment of plant communities, and subsequent decomposition of organic residues creates a nutrient cycle and accelerates the process of soil genesis.

Summary

Renewal involves: shaping the landscape for control of water and erosion, fertilization, establishment of vegetation and soil organisms, and soil genesis. The best species are *Cynodon dactylon*, *Lespedeza cuneata*, *Paspalum notatum*, *Alnus glutinosa*, *Pinus taeda*, and *Platanus occidentalis*. *Pisolithus tinctorius* colonizes on *Pinus* roots. Soil genesis is measurable within five years.

References

May J.T., Johnson H.H., Perkins H.F. & McCreery R.A. (1973a) Some characteristics of spoil material from kaolin clay strip mining. In *Ecology and Reclamation of Devastated Land* (Ed. by R.J. Hutnik & G. Davis), Gordon and Breach, London.

May J.T., Porks C.L. & Perkins H.F. (1973b) Establishment of grasses and tree vegetation on spoil from kaolin clay strip-mining. In *Ecology and Reclamation of Devastated Land* (Ed. by R.J. Hutnik & G. Davis), Gordon and Breach, London.

Troth J.L. (1971) *Evaluation of Movement and Effectiveness of Lime and Fertilizer Materials on Kaolin Spoil Sites.* M.S. thesis, University of Georgia.

Troth P.M. (1972) *Soil Formation on Kaolin Spoil.* M.S. thesis, University of Georgia.

The biology of land revegetation and the reclamation of the china clay wastes in Cornwall

A. D. BRADSHAW, W. S. DANCER, J. F. HANDLEY and J. C. SHELDON *Department of Botany, University of Liverpool, Liverpool, U.K.*

Introduction

The successful and economic restoration of vegetation on derelict or degraded land depends on overcoming the environmental factors limiting or restricting plant growth on the material. There are a large number of edaphic factors, any of which could by itself be preventing vegetation growth:

Physical:	Water	too dry—too wet
	Surface temperature	too high
	Texture	too hard—too coarse—too fine
	Stability	too unstable
Nutrient:	Macronutrients N, P, K, Ca, Mg, S; Micronutrients Fe, Cu, Zn, Mo, B Mn, Co	too little present—too little available—too rapidly lost
Toxicity:	pH	too high—too low
	Metals	too high in heavy metals
	Salinity	too high due to excess Na, Mg, SO_4, Cl etc.

If these can be corrected then successful establishment of vegetation should be possible. But it is necessary to identify which particular factors are critical and pay special attention to these, otherwise effort and money can be wasted attending to factors which are not critical or attending insufficiently to factors which are critical.

The identification and assessment of critical factors involves systematic investigation. The object of this paper is to show how this is being done for one particular type of material about which little was known. It is an example of how such investigations proceed. It also throws light on general problems of land restoration.

China clay wastes

China clay in Britain is extracted from deep deposits in the middle of the granite masses of south-west England. The deposits were formed by the decay of the granite under the influence of water and volcanic gases forced up from below (Exley 1959), so that they are of almost unlimited depth although somewhat funnel shaped. The feldspars in the granite decayed to give kaolin, so that the final deposit is a mixture of quartz particles, mica, kaolin and undecayed blocks of granite in the ratio of approximately 6:1:1:1. The deposits are worked by high pressure water. The quartz separates out readily and is dumped by conveyor or bucket in large flat or conical heaps; the mica and kaolin are separated in settling tanks and the mica is pumped out into large lagoons. The overburden which varies in thickness from one to three metres is usually dumped separately or beneath the sand.

Since the clay deposits are bottomless and the demand for kaolin is increasing it is impractical to put the wastes back into the pits; so they must accumulate on the surface, where the sand heaps in particular form a spectacular addition to the landscape. In restricted amounts the workings are quite attractive, but with the passage of time in areas which have been heavily worked, such as in the 30 sq. miles round St Austell, they dominate and destroy the original landscape (Plate 1). Production is increasing. Current sand production is at the rate of approximately 15 million tons per year; if it was all dumped in one place in the next 50 years it would form a heap 1,000 ft (300 m) high and 5 miles (8 km) across at the base.

A landscape solution is clearly necessary, and is now being required by the planning authorities. As the industry is in a competitive market, expensive solutions, such as the use of top soil, are not possible since it would price the industry, which is a major employer in the region, out of the market. Top soil is also very difficult to obtain in the region. At the same time it is known that the application of a layer of material structurally dissimilar to that beneath can lead to severe instability at the boundary between the materials (Fournier 1972): there has been experience of landslip involving top soil in the china clay workings already. The surrounding countryside is mainly pastoral, permanent pasture together with scrub and woodland: at higher elevations above 1,000 ft there is open moorland. Any landscape solution must therefore meet two requirements, firstly to be relatively cheap, and secondly to blend the workings into these surroundings.

The work described here was started in 1970. Since almost nothing was

Plate 1. Aerial view of China clay workings in Cornwall, east of St. Enoder (Cambridge University Collection; copyright reserved).

Plate 2. A forty-year-old heap with natural colonization.

Plate 3. An experimental grass plot on sand which has ceased growing with different aftercare treatments.

Plate 4. A good two-year-old grass/clover sward on sand

known about the problems of growing plants on the materials, the initial objective was very simple—to discover how to establish a grass sward on the wastes, especially the sand heaps. This was chosen since it was potentially one of the most important landscape solutions, and would certainly indicate the factors controlling plant growth in general.

Initial considerations

To find a solution to any landscape problem it is firstly necessary to find out:
(a) the constitution of the material;
(b) what sort of natural colonization can occur;
(c) what has been achieved in the past.

Table 1. Particle size analysis of material from a sand heap at Maggie Pie.

Fraction	Size limits (particle diameters in mm)	Composition (% air dry soil)
Gravel	>1·0	57
Coarse Sand	1·0–0·5	15
Medium Sand	0·5–0·25	14
Fine Sand	0·25–0·05	11
Silt	0·05–0·002	2
Clay	<0·002	1

The main constituent of the waste is a gravel fraction (Table 1). It is therefore an open free draining material and could become very dry. However, the rainfall is high and well distributed and potential evaporation (M.A.F.F. 1967) is low (Fig. 1) so that severe drying out is not to be expected except during dry periods in the summer when water may be deficient in the surface layers (Fig. 2). The mica is much finer and the particles are flat: it is deposited in lagoons. It therefore retains water well, but lagoons once they are disused are well drained and show no signs of waterlogging. The sand is almost pure silica: its mineral nutrient content is therefore very low indeed (Table 2): the mica which is a low potassium, muscovite type, is little different: all major plant nutrients are in short supply in both materials. If nutrients were added in a soluble or semi-soluble form the sand would be unlikely to have any satisfactory retention capcity, but the mica might have because of its physico-chemical and structural configuration.

The pH of the materials is quite low (Table 2), but the lime requirement, the amount of $CaCO_3$ required to raise the pH to 6·5, is small (about one-tenth) in comparison with an equivalent well developed sandy loam soil of the same pH. This is advantageous but unfortunately reflects a low cation

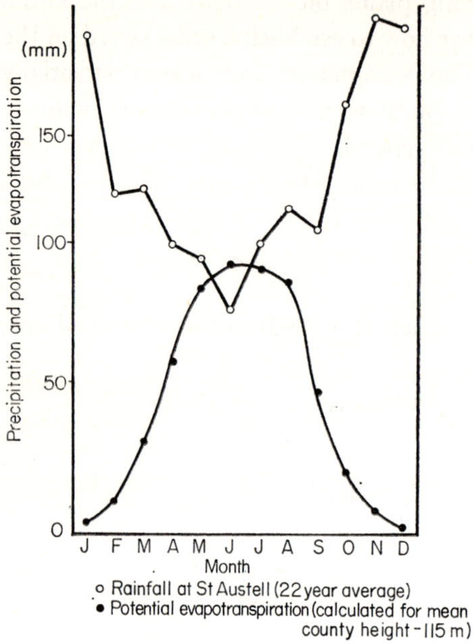

Figure 1. Mean rainfall and potential evapotranspiration in Cornwall.

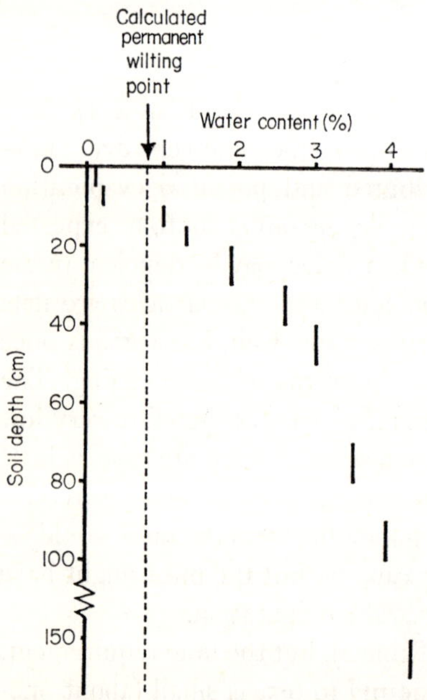

Figure 2. Water content of the surface layers of a large sand heap at Maggie Pie after 4 weeks late summer drought.

exchange capacity (C.E.C.). The C.E.C. of productive soils frequently exceeds 10 m.equ./100 g; it is less than 2 m.equ/100 g in sand wastes. The C.E.C. of the mica is somewhat higher, about 5 mequ./100 g.

Table 2. Nutrient analyses of sand and mica wastes and neighbouring pastures.

Site	Material	Exchangeable[1] cations K	Mg	Ca	Exchangeable[2] P	Total N	pH[3]
Maggie Pie	sand	10	16	85	2·0	9	4·5
	mica	13	20	115	2·8	18	4·0
Lee Moor	sand	8	14	90	4·6	11	4·5
	mica	12	28	110	5·6	24	3·9
Whitmore	ryegrass ley	176	130	990	46	1,560	4·7
	old pasture	370	170	1,700	70	1,283	4·8
Bodmin Moor	*Agrostis*/ clover pasture	110	71	880	42	1,275	4·2

[1] extracted with IN ammonium acetate.
[2] extracted with sodium bicarbonate (Olsen *et al.* 1945).
[3] measured in 0·01M $CaCl_2$.
} all nutrients measured in ppm.

Because of the coarseness of the sand most rain passes through the heaps. But in times of storm there can be surface run off leading to erosion and gully formation on steep slopes. Gullies may become about 1 m in exceptional circumstances but are usually only about 30 cm deep even after several years.

The industry has been in existence since 1770 (Barton 1966). There are therefore an enormous series of heaps of different ages, the oldest of which are now well colonized. They do not however form a perfect time series since the methods of extraction have changed. Initially the overburden was often mixed with the sand; and the technique of separation was not perfect so that some of the kaolin and the mica remained mixed with the sand. However, there are sufficient examples to show that the colonization of both sand and mica is what would be expected on an acid nutrient-deficient site (Plate 2). There is a slow invasion of calcifuge grasses and shrubs (Table 3). Leguminous species such as *Ulex gallii* and *U. europaeus* are very common and in some areas the tree lupin *Lupinus arboreus* has invaded very successfully. In later stages of the succession *Salix caprea* and *Rhododendron ponticum* grow extremely well. The climax appears to be an acid oakwood.

So there does not appear to be any fundamental factor limiting colonization in the long term. But the process is slow despite an abundance of seed from potential colonists in the surrounding areas. A few experiments had been carried out in the past to see if the process of vegetation establishment could be assisted and could be directed towards a more productive grass sward. In one area a general agricultural grass seed mixture together with a low level of fertilizer (probably 20 kg/ha of N, P and K) were sprayed on the side of a sand heap. There was a good establishment and initial growth but subsequently growth ceased completely and four years later there were only a few scattered plants of *Festuca rubra* and *Lolium perenne* to be seen, about 2 cm in diameter and almost dead.

Table 3. Common vascular plants on naturally colonized sand waste.

Blechnum spicant	*Calluna vulgaris*
Polypodium vulgare	*Erica cinerea*
Pteridium aquilinum	*Erica tetralix*
	Vaccinium myrtillus
Digitalis purpurea	
Jasione montana	*Lupinus arboreus*
Potentilla erecta	*Sarothamnus scoparius*
Rumex acetosella	*Ulex europaeus*
Rubus fruticosus	*Ulex gallii*
Agrostis setacea	*Salix aurita*
Agrostis tenuis	*Salix cinerea* ssp. *atrocinerea*.
Festuca ovina	
Holcus lanatus	*Rhododendron ponticum*

More systematic trials by the Ministry of Agriculture (M.A.F.F. 1971) were established in two sites on 30 cm of sand spread on top of mica. *Lolium perenne*, *Festuca rubra* and *Trifolium repens* were sown with lime and fertilizer at the following rates:

N	60 kg/ha
P	100 kg/ha
K	100 kg/ha
limestone	5,000 kg/ha

Growth was satisfactory at both sites, and clover became well established. Grazing by sheep and cattle on one site seemed to have helped keep it in good condition.

All this suggests that of the many factors which can limit plant growth on derelict land materials, those most likely to be operating on china clay wastes, are:

	sand	*mica*	*comments*
macronutrient lack	*yes*	*yes*	very definite—but which is most important?
micronutrient lack	?	?	possible but no clear evidence.
pH	*yes*	*yes*	will need correcting—but easy to do.

	sand	*mica*	*comments*
drought	?	*no*	no real evidence—but what happens on steep sided heaps?
instability	*yes*	*no*	serious on steep slopes—but might only be important at early stages of growth.

Laboratory experiments

Since most of the problems of derelict land are connected with soil, simple pot experiments can go a long way to demonstrate what factors are restricting growth and how they can be overcome (Fitter, Handley, Bradshaw & Gemmell 1974). An experiment was therefore set up to examine the effect of addition of lime, balanced complete fertilizer containing N, P and K, extra additions of each of these nutrients, and micronutrients and magnesium. A simple seeds mixture of *Festuca rubra* and *Trifolium repens* was used. Five different waste materials were included.

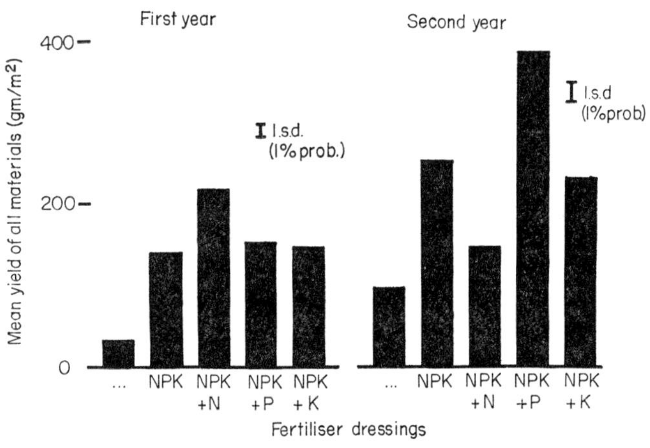

Fertiliser dressings	N	P	K	in kg/ha
NPK	50	22	41	
NPK+N	100	22	41	
NPK+P	50	44	41	
NPK+K	50	22	82	

Figure 3. Mean yields of different fertilizer treatments applied to grass/clover swards established on various sand and mica wastes in preliminary pot experiment.

The results, summarized in Fig. 3, showed at the first harvest that in all cases the effect of fertilizer was considerable; without it growth was very poor. Extra complete fertilizer, or only nitrogen, further improved growth, whereas extra phosphorus or potassium had no effect. Limestone had little effect. There was some difference between materials, growth with fertilizer being better on mica. There was no effect of micronutrients, but there was some effect of magnesium.

By the time of second harvest maximum growth was achieved by the treatment where extra phosphorus had been added. This could be correlated with a strong development of clover, and suggested that maintenance of nitrogen input could be important.

Field trials

The one weakness of laboratory experiments is that they do not duplicate the conditions of climate, exposure and drainage as they occur in the field. A series of simple field trials were therefore established in such a pattern that they could be set up by unskilled labour on a wide variety of sites and materials. They were repeated at three times during the year, July, October and November, to include the effect of season on establishment. Complete fertilizer was applied at three levels, giving respectively 0, 33 and 66 kg/ha of each of N, P_2O_5 and K_2O. These were combined with three lime levels 0, 660 and 1,320 kg/ha $CaCO_3$ in a factorial design. A slightly more complex seeds mixture was sown including *Lolium perenne*, *Festuca rubra*, *Agrostis tenuis* and *Trifolium repens*.

Quantitative assessment of these trials was not possible but from careful observations several conclusions could be drawn. In all cases growth was negligible without fertilizer, but excellent with it. The higher level of fertilizer was the most effective. Lime, at least at the lower level, was essential for clover but not for grass; however as the plots became older it was apparent that *Lolium* and *Festuca* would not flourish in the absence of lime: these plots became dominated by *Agrostis*.

Growth with the addition of fertilizer was distinctly better on the mica sites than on sand. Some plots were on steep sand slopes: on these growth was just as good as on the flatter sand plots. There was little sign of erosion: establishment of a grass cover was rapid and prevented any further erosion of the surface. Although the first series of field trials were sown in July conditions were such that there was no serious inhibition due to surface dayness.

Taken as a whole the field plots fully confirmed the conclusions that the main factor controlling growth was lack of macronutrients which could readily be added.

However, as the plots each reached the age of about 4 months, growth on them ceased; the grass became brown. The possibility of serious water deficits during dry spells has been demonstrated (Fig. 2) and water shortage might be suggested to be the cause of the browning. However, in one or two cases where animals had accidentally got in and excreted there were bright green patches. This suggested that there was nitrogen deficiency. The cessation of growth was more marked on the sand than on the mica. The plots all then remained in a more or less moribund state thereafter, although the plants did not die. The situation resembled that of the earlier attempts at reclamation.

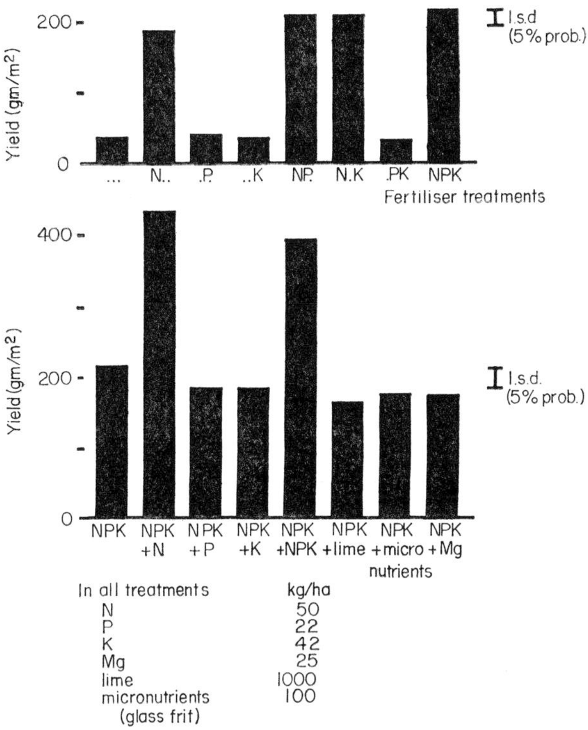

Figure 4. Regrowth of moribund grass swards on mica following application of fertilizers.

It was conspicuous that where clover had become established in the plots treated with lime and fertilizer the grass recovered, firstly in the immediate vicinity of the clover plants. All this pointed to an acute nitrogen deficiency. This was borne out by two complex aftercare experiments (Plate 3) where nitrogen, phosphorus and potassium were added in factorial combinations to selected single plots of the previous experiment, and also to plots of another experiment, which had stopped growing and turned brown for at least six months.

In all cases growth resumed in plots which received nitrogen, either by itself or in combination with other nutrients (Fig. 4). There was only a nitrogen effect, and little or no interaction with other nutrients. These effects were not permanent and again six months later growth had more or less ceased, although the swards which had received nitrogen were now in a substantially better state than they had been previously.

The fate of applied nitrogen

The rainfall of the area is high and evapotranspiration low (Fig. 1). This leads to considerable excess rainfall and leaching even during summer months (Table 4).

Table 4. Rainfall and evapotranspiration data for two sites in Cornwall (in mm).

	Site	Rainfall	Evapotranspiration (calculated)	Excess rainfall
April–Sept.	St. Austell	584	424	160
	Lee Moor	685	424	261
Oct.–March	St. Austell	899	119	780
	Lee Moor	931	119	812

The colloidal minerals have been removed from the wastes by the extraction processes. There is no organic matter. Because of these deficiencies the wastes absorb only small amounts of rain and water percolates rapidly through the sand, so leaching of nutrients could be severe.

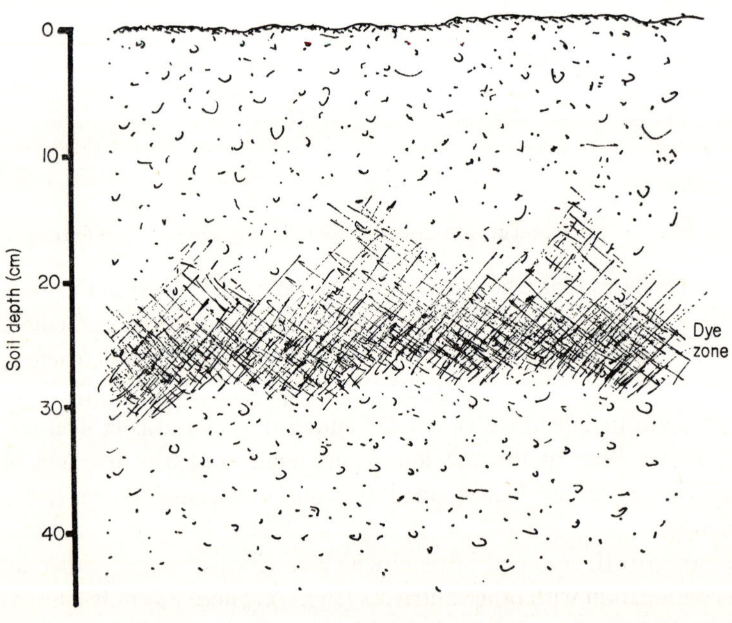

Figure 5. The effect of 2·5 cm of precipitation on the movement of a dye (acid fuchsin), with properties similar to nitrate, sprinkled dry on the surface of sand waste.

Nitrate fertilizers are readily soluble, and nitrate is an anion and not absorbed by the predominant cation exchange sites of the soil particles. As a result nitrate nitrogen is especially susceptible to leaching (Gardner 1965). Ammonium nitrogen is better since ammonium will be retained by cation exchange sites. But the cation exchange capacity of these wastes is low anyway. So all forms of nitrogen will be lost readily by leaching.

The rapidity of leaching was demonstrated in an experiment in which a dye with similar properties to nitrate, acid fuchsin, was sprinkled on the surface of a sand heap. The area was then given 25 mm of artificial rain over a period of about one hour. Twenty-four hours later a pit was excavated. The dye was found to have moved down in a narrow band to a depth of 200 mm (Fig. 5).

The movement of nitrate and ammonium nitrogen was subsequently compared in a similar manner but in this case the materials were applied to the surface and left for 23 days during which time there was 58 mm of rain. Assays for nitrogen showed that the downward movement of nitrogen was considerable for both ammonium and nitrate nitrogen (Fig. 6). Very little is left in the surface layers except a little ammonium.

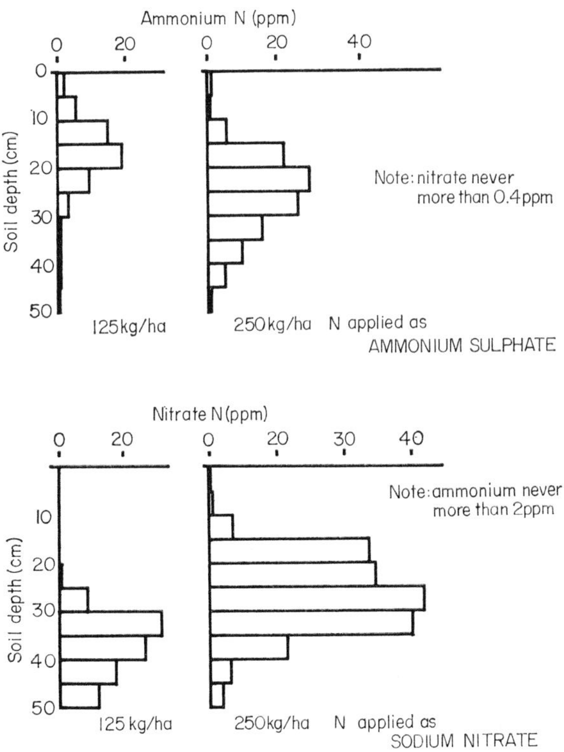

Figure 6. The distribution of ammonium and nitrate nitrogen, originally applied to the surface of sand waste, after a period of 23 days during which there was 58 mm of rain.

Loss of nitrogen in this way might be prevented by slow release fertilizers (Lunt 1971). A comparison of a range of fertilizers including both traditional and synthetic slow release materials gave interesting results. For several weeks after establishment, rainfall was higher than average, and the wet spring was then followed by a very dry summer. On sand the performance of sodium nitrate was very poor in contrast to ammonium sulphate. This must be attributed to leaching since the contrast disappears on mica where leaching would not be expected to be a problem. Urea formaldehyde gave poor results on both materials. Sulphur-coated urea was disappointing on the sand; in a separate experiment urea was shown to be most unsatisfactory on sand perhaps due to volatilization (Gasser 1964). Organic slow release materials such as fish manure and dried blood maintained reasonable growth throughout the first year.

In the second season regrowth was very poor in all treatments. Experiments in progress indicate that as much as 300 kg N/ha of even a good slow release material may be necessary for a significant carry-over effect on the sand material. Only 100 kg/ha was used in the experiments described.

In general terms the experiments suggest that ordinary fertilizers are little use although ammonium is better than nitrate, and that slow release materials, besides being expensive, are of little value in such situations. Other more effective long-term means of supplying nitrogen to the swards must be found.

Use of legumes

Since what is needed is a slow continuous release it is obvious that legumes are the only sensible way of providing nitrogen to the developing sward. This is well understood by agriculturalists and by some reclaiming authorities (e.g. Dilz & Mulder 1962; Rees & Warwick 1967), but not appreciated by others.

A field trial to assess the yield and nitrogen accumulation of a range of legumes species was set up on a sand area. It included large legumes such as lupins as well as more normal pasture species such as clover. The legumes were sown both in the presence and absence of grass, with 44 kg/ha of P as superphosphate, 41 kg/ha of K, and 5,000 kg/ha of lime dressing. All were inoculated with their appropriate strains of *Rhizobium* supplied on agar slopes from Rothamsted Experimental Station.

Yields and N accumulation are given in Table 5. Satisfactory rates of accumulation of nitrogen up to 100 kg/ha/year were achieved, equivalent to a heavy fertilizer dressing from which little was lost by leaching. The most conspicuous result was the performance of *Trifolium repens* despite a poor yield above ground: much of its yield and nitrogen is in stems and roots (Troughton 1963) which are not harvested by normal procedures and also

Table 5. Yield and total nitrogen accumulation of legume stands on sand and mica wastes after one year.

Species	Total dry weight (kg/ha)		Nitrogen accumulated (kg/ha)	
	sand	mica	sand	mica
Trifolium repens (S184)	9,670	5,070	95	78
Trifolium pratense (early)	7,350	7,280	71	103
Trifolium hybridum	5,580	5,450	87	84
Lupinus angustifolius	2,620	9,210	40	115
Vicia sativa	1,110	2,210	21	53

Dry weight includes roots in first 10 cm of waste: nitrogen accumulated is increased over control sampled to depth of 10 cm (control sand contained 14 kg/ha, control mica 36 kg/ha).

prevent it from appearing productive to the eye. When combined with grass it gives a very satisfactory sward (Plate 4).

In a separate experiment crown vetch, *Coronilla varia*, was investigated. It behaved poorly to begin with, but it improved later. It is widely used in reclamation in the United States (Breeding 1961). How valuable it and other species really are depends on the outcome of longer term observations. But at the moment white clover appears the most valuable.

Management

The maintenance of clovers in any sward in balance with grasses depends on careful management. The most important two factors are an adequate supply of phosphorus and calcium to encourage clover growth (Snaydon 1962), and grazing or cutting to reduce the competition of the grass (Blackman & Templeman 1938).

The importance of calcium and phosphorus is clear from the subsequent performance of the first field experiments. Clover growth is best in the high lime, high phosphorus plots.

The effectiveness of different phosphorus sources (Bixby *et al.* 1964) are therefore being compared when applied to a single grass clover mixture at four lime levels. Basic slag has given good results in the absence of lime or at a low lime rate of 1,000 kg/ha. Superphosphate is better at high lime levels. Magnesium ammonium phosphate gives good results at all lime levels. Rock phosphate is not effective at the levels used.

Grazing or cutting serves two functions, to reduce the competition of the grass, and to return organic matter and nutrients to the soil. In the context of the problem of china clay waste reclamation both are very important.

In a quarrying area grazing animals, however plentiful they may be in the surrounding region, are a nuisance. It has not therefore been possible so far to examine carefully the effect of grazing. However, in one or two areas grazing of trial seedings has occurred. It is quite clear that the result is a vastly improved sward. In no case has the sward been damaged, even with heavy grazing. The growth of clover has been considerably encouraged, and the grasses although reduced in size have tillered more. The sward is generally greener, which suggests that nitrogen recycling is more rapid.

A series of aftercare trials to test these conclusions further have been established.

Alternative solutions

In some areas grazing may never be possible. Since natural colonization of the heaps includes shrubs such as *Calluna vulgaris*, *Ulex europaeus* and *U. gallii*, *Salix cinerea* and *Rhododendron ponticum*, all these could be used. However natural growth of these species is slow, presumably because of the absence of nutrients, especially phosphorus, and nitrogen in the case of the non-legumes.

It would therefore be necessary to go through a clover rich sward phase first to provide immediate cover, and a build up of organic matter containing nitrogen. Subsequently a wide variety of shrubs could be used. On some heaps the tree lupin *Lupinus arboreus*, has become naturalized. It grows rapidly even on heaps with little existing vegetation, and could be a valuable colonist if systematically sown. It is not very long lived, perhaps 10 years, and so would have to be interplanted with other material. The growth of grass round isolated bushes indicates that it is almost certainly an excellent nitrogen accumulator.

Methods of establishment and hydroseeding

All the field plots were established by hand by normal methods using gentle raking to cover the seeds. It is quite clear that normal agricultural techniques broadcasting seed and fertilizer, followed by a light harrowing, can be used for establishment of large areas. They have in fact been used very successfully even on rather steep slopes: seed germination and establishment are as good as on normal materials.

However, on the steep sides of heaps at a natural angle of rest normal techniques are not possible. Hydroseeding is an obvious alternative. In this method seed, fertilizer, lime and peat or an equivalent material are sprayed on to the surface in a wet mulch. Usually a latex material is included. The peat is supposed to act as a mulch to protect the seeds and the latex to act as a binder to prevent erosion. The technique was developed in the United States and imported into other countries (Jaaback & Muzzell 1971).

The technique is effective in obtaining establishment, and has been used on an experimental basis in a number of sand slopes. But it is very expensive when including all the components. It is also not possible to include a great deal of fertilizers in the seeding mixture—not more than 300 kg/ha of a complete fertilizer giving about 50 kg/ha N, 25 kg/ha P and 40 kg/ha K—without damaging the seed. As a result nitrogen deficiency soon appeared on those sites where hydroseeding had been attempted. To aggravate matters, very little white clover appeared to have established even when it had been included in the seeds mixture.

To discover exactly the value of the various components and whether any of them could be omitted, an experimental slope was set up outside at Liverpool in a situation as exposed as that of St Austell. To this various combinations of hydroseeding materials were given. Erosion, loss of seeds,

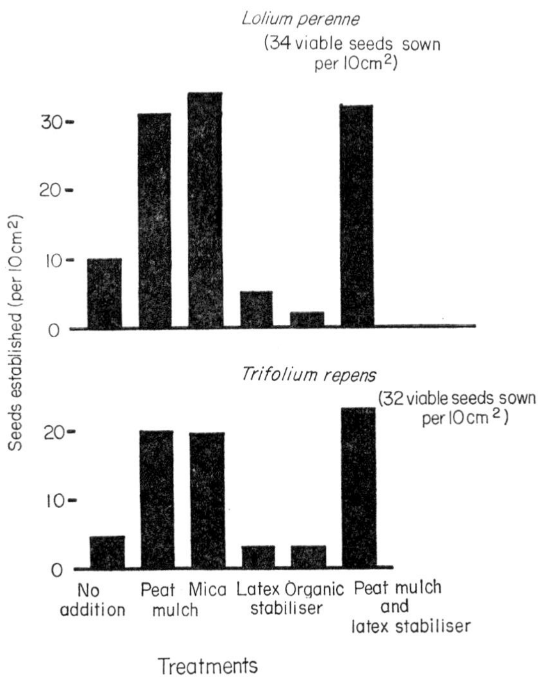

Figure 7. The establishment of ryegrass (*Lolium perenne*) and wild white clover (*Trifolium repens*) on an artificial sand slope outside, when seed was scattered on the surface and covered with various mulches and stabilizers.

and establishment of seedlings were monitored at weekly intervals. The results (Fig. 7) showed that establishment was assisted by a mulch, which could be of mica or peat, but was considerably reduced by a latex or an organic stabilizer. The stabilizers produced a hard surface but were not sufficiently effective in binding the coarse sand, and so allowed both sand and seed erosion. Further, the latex coated the surface sand particles breaking capillary rise, and so further reduced seed germination by imposing a water deficient microclimate at the surface. On the untreated surface a hard crust was quickly formed in dry conditions through which the seedlings could not force their roots. This is prevented by the presence of the mulch, which also maintains moisture round the seed.

It is interesting that the full normal mixture using both mulch and stabilizer was no more successful than one omitting the stabilizer with consequent enormous savings in costs.

Subsequent full-scale field trials have demonstrated that these conclusions are correct. Stabilizers have adverse effects, and effective establishment can be achieved without any mulch or stabilizer, although this is weather dependent. However, it is clear that the mulch will allow a higher degree of germination and more reliable establishment especially on south facing slopes. The technique without a mulch leaves the seed exposed on the surface, while a mulch will cover the seed and reduce the effect of adverse climatic changes.

The failure of clover establishment with hydroseeding could be explained either by its slow germination and establishment leaving it vulnerable to surface drying, or because of a possible sensitivity to the hydroseeding treatment. A series of greenhouse experiments have shown that it is sensitive to exposure to fertilizer, especially nitrogen (Fig. 8). It is clear that nitrogen must be kept to a very low level if establishment is to succeed. This has been confirmed in full-scale field trials. It does mean however returning after two months to provide nitrogen for the grass sward before clover nitrogen fixation contributes to the growth of the grass. It appears that there is little appreciable contribution from the clover to the grass in the first year, when the clover is sown at normal low seed rates in the mixture, and its development is restricted by the accompanying grass. Failure to become inoculated is a possible contributory factor.

The slow establishment of clover is also a problem. It can be overcome to a considerable extent by soaking the seed for one day prior to hydroseeding. This reduces the time taken to germinate from four days to one day and improves its competitive position in the sward. It is somewhat surprising that the hydroseeding technique requires so much modification from its original plan. The technique was developed for rough surfaces in arid climates where heavy rains may occur. This is different from the sand surfaces of the

Cornish china clay waste heaps, and many of the modifications stem from difference of material and of climate.

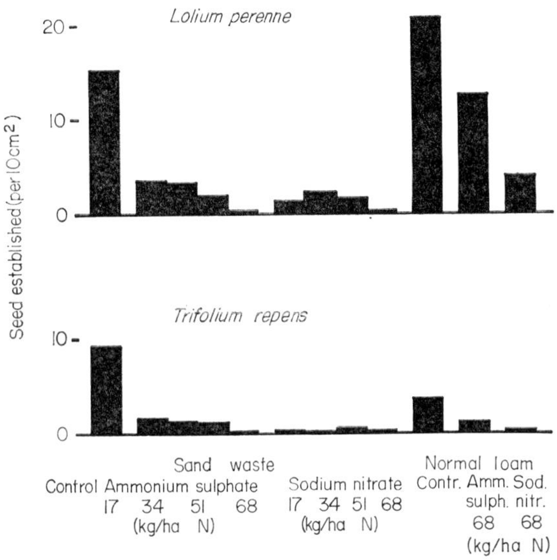

Figure 8. The establishment of ryegrass (*Lolium perenne*) and wild white clover (*Trifolium repens*) on sand and loam, when seed was scattered on surface with various nitrogenous fertilizers.

The normal cost of the full treatment is £500/ha. Omission of stabilizer reduces the cost to £250/ha. At the same time this produces a more reliable technique. This illustrates the need for proper scientific investigation of reclamation techniques to ensure that optimum solutions are being employed.

Attaining a properly functioning ecosystem

In the end the object of any reclamation programme in which vegetation is being established must be to achieve a self-reliant ecosystem, in which the soil contains sufficient nutrients in a recyclable form to provide adequate available nutrients for satisfactory plant growth. The objective is therefore really to create a fully developed soil ecosystem as soon as possible. The development of a mature soil profile, in the soil scientist's sense, is not necessary.

To do this nutrients must be accumulated, especially nitrogen, and in these china clay waste materials, calcium and phosphorus: in other materials other nutrients such as potassium may be important. For calcium, phosphorus and potassium the only natural supply is a small fallout from the atmosphere. But these nutrients can readily be added in large amounts by lime and fertilizers. They will then complex with soil minerals, or remain in a rather insoluble form, and so not be lost rapidly. Yet they will usually remain in a form available to plants.

Nitrogen is rather different. Not only is a large continuous supply required by plants but it can only be held in the soil in organic matter. For this reason artificial fertilizers are not a good method of supplying nitrogen to derelict land materials.

In natural conditions some fixed nitrogen can come from the air and some by free living N fixing organisms. But the most rapid and effective accumulation of nitrogen in both agricultural (Stevenson 1965) and natural ecosystems (Crocker & Major 1955), and derelict land (Leisman 1957), is by symbiotic nitrogen fixers. In some ecosystems rhizophere organisms may be important (Harris & Dart 1973). All these nitrogen fixers accumulate nitrogen in the soil progressively and continuously, bound in organic matter; this is very different to the way in which nitrogen must be added artificially. Once accumulated in organic matter the nitrogen should be readily available by mineralization. But only a fraction of the nitrogen can be mineralized at any one time, so that an effective supply of nitrogen will depend on the existence of a substantial organic store.

The degree to which a satisfactory pool of nutrients has been achieved in various reclaimed areas is given in Table 6. Progress is obvious, but the levels for nitrogen are not yet anywhere near those to be found in fully developed soils.

Table 6. Macronutrient levels achieved after successful reclamation with grass and clover.

Site	Material	Site age[1]	Exchangeable cations K	Mg	Ca	Exchangeable P	Total N	pH
Maggie Pie	sand	12	41	24	500	12	60	6·4
	mica	21	61	36	750	36	243	7·1
Lee Moor	sand	21	22	14	150	14	177	5·4
	mica	21	47	36	160	36	290	4·1

These areas have had various fertilizer treatments but all now have reasonable growth. They should be compared with the values in Table 2.
[1] months.

The situation for nitrogen is complicated too by the presence of unfavourable carbon/nitrogen ratios which will tend to lock up nitrogen which might otherwise be available. This can be demonstrated experimentally by examining the amount of mineral nitrogen achieved in soils when mineral nitrogen is added after the dead organic material lying on the surface has been incorporated and the whole soil incubated to obtain a measure of mineral nitrogen release or uptake (Table 7).

Table 7. Release of nitrogen from recently reclaimed sand wastes.

Site	Sward	pH	Total N	Inorganic N	Change in N after incubation* Original material	Material + plant debris	Material + plant debris +45 ppm ammonium N
Maggie Pie	hydroseeded 2 yrs (no clover)	4·0	262	12	+6	−7	−59
Park	hydroseeded 2 yrs (no clover)	4·8	244	12	+4	−7	−27
Lee Moor	sown 2 yrs (some clover)	5·4	177	16	+39	−9	−43
Stannon	sown 4 yrs (good clover and grazing)	6·9	291	53	+63	+66	+44
Maggie Pie	original sand (control)	4·5	7	2	0	—	+44

* Incubation for 14 days at 25°C (Keeny & Bremner 1967).

However, the accumulation of nitrogen in a form in which some of it is readily mineralized can occur quite rapidly especially when clover is present. The levels achieved by reclamation in rather poor hydroseeded and refertilized grass, and in good grass and clover mixtures, can be compared with those found in naturally colonized older heaps, and in adjacent natural grassland (Fig. 9). Rapid nitrogen accumulation in an available form is the key to the whole reclamation process. Although it depends on carefully overcoming a whole series of problems there seems no doubt that it can be achieved and satisfactory plant cover achieved. It is likely that nitrogen is the major limiting environmental factor in most reclamation programmes.

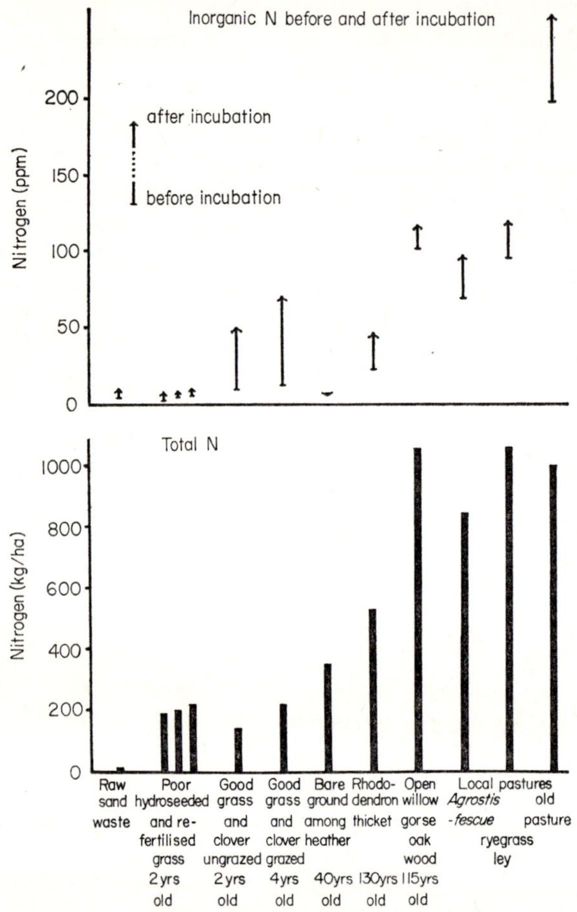

Figure 9. Inorganic and total nitrogen in sand wastes after reclamation, compared with areas naturally colonized for different lengths of time, and with typical local pastures.

Acknowledgments

We would like to acknowledge the considerable help of Mr L. D. C. Owen and his staff at English China Clays Ltd throughout this work, Dr D. White for the experiment shown in Fig. 5, Mr E. K. Cundy for the data on soil moisture, and in particular Mr W. Newnes for his unremitting technical assistance. Mr A. Pattison of Rothamsted Experimental Station kindly provided all the *Rhizobium* cultures. The natural Environment Research Council have supported some of the work described in this paper.

Summary

China clay workings produce mainly vast quantities of coarse quartz sand. The major problem in establishing a cover of vegetation is the paucity of major rather than minor nutrients. These can be added by fertilizers to permit excellent plant growth. However, there is excessive mobility of plant nutrients, especially nitrogen, so that deficiency occurs within a few months. The crucial problem is therefore to ensure adequate long-term nitrogen supply. This is true of most derelict land.

References

BARTON R.M. (1966) *A History of the Cornish China Clay Industry.* Bradford-Barton, Truro.

BIXBY D.W., RUCKER D.L. & TISDALE S.L. (1964) *Phosphatic Fertilisers, Properties and Processes.* Tech. Bull. No. 8, The Sulphur Institute, Washington.

BLACKMAN G.E. & TEMPLEMAN W.G. (1938) The interaction of light intensity and nitrogen supply on the growth and metabolism of grasses and clover. I. The effect of light intensity and nitrogen supply on the clover content of a sward. *Ann. Bot. N.S.* **2**, 257–80.

BREEDING C.H.J. (1961) Crown vetch as an aid to strip mine reclamation. *Mining Congress Journ.* **47**, 70–1.

CROCKER R.L. & MAJOR J. (1955) Soil development in relation to vegetation and surface age at Glacier Bay, Alaska, *J. Ecol.* **43**, 427–48.

DILZ K. & MULDER E.G. (1962) Effect of associatied growth on yield and nitrogen content of legume and grass plants. *Pl. Soil* **16**, 229–37.

EXLEY C.S. (1959) Magmatic differentiation and alteration in the St. Austell Granite. *Quart. J. Geol. Soc.* **114**, 197–230.

FITTER A.H., HANDLEY J.F., BRADSHAW A.D. & GEMMELL R.P. (1974) Site variability in reclamation work. *J. Inst. Landsc. Arch.* **106**, 29–31.

FOURNIER F. (1972) Aspects of soil conservation. *Nature and Environment* **5**.

GARDNER W.R. (1965) Movement of nitrogen in soil. In *Soil Nitrogen* (Ed. by W.V. Bartholomew & F.E. Clark), Amer. Soc. Agron., Madison.

GASSER J.K.R. (1964) Some factors affecting losses of ammonia from urea and ammonium sulphate applied to soils. *J. Soil Sci.* **15**, 258–72.

HARRIS D. & DART P.J. (1973) Nitrogenase activity in the rhizosphere of *Stachys sylvatica* and some other dicotyledenous plants. *Soil Biol. Biochem.* **5**, 277–9.

JAABACK G. & MUZZELL P. (1971) The establishment of vegetation in civil enigneering work. *Proc. Grassld. Soc. Sth. Africa* **6**, 181–4.

KEENY D.R. & BREMNER J.M. (1967) Determination and isotope-ratio analysis of different forms of nitrogen in soils. VI Mineralisable nitrogen. *Soil Sci. Soc. Amer. Proc.* **31**, 34–9.

LEISMAN G.A. (1957) A vegetation and soil chronosequence on the Mesabi iron range spoil banks, Minnesota. *Ecol. Monogr.* **27**, 221–45.

LUNT O.R. (1971) Controlled release fertilisers: achievements and potential. *J. agric. Fd Chem.* **19**, 797–800.

M.A.F.F. (1967) Potential transpiration. *Min. Agric. Fish and Food Techn. Bull.* **16**.

M.A.F.F. (1971) Report on an investigation into the possible use of china clay residues for agricultural purposes. Private communication.

OLSEN S.R., COLE C.V., WATANABE F.S. & DEAN L.A. (1954) Estimation of available phosphorus in soils by extraction with sodium bicarbonate *U.S. Dept. Agric. Circ.* **939**.

REES W.J. & WARWICK J. (1957) Clover establishment on power-station waste ash. *Emp. J. exp. Agric.* **25**, 256–62.

SNAYDON R.W. (1962) Micro-distribution of *Trifolium repens* L. and its relation to soil factors. *J. Ecol.* **50**, 133–43.

STEVENSON F.J. (1965) Origin and distribution of nitrogen in soil. In *Soil Nitrogen* (Ed. by W.V. Bartholomew & F.E. Clark), Amer. Soc. Agron., Madison.

TROUGHTON A. (1963) The root weight under swards of equal age in successive years. *Emp. J. exp. Agric.* **31**, 274–81.

The importance of chemical toxin detection and substrate factor interactions in the reclamation of unusual toxic sites

RAYMOND P. GEMMELL *Planning Department, Lancashire County Council, Preston, England*

Introduction

THE NEED FOR RESEARCH

The revegetation of colliery spoil is the major concern of most authorities now engaged in derelict land reclamation and has been the subject of a considerable amount of scientific research. The establishment of vegetation on certain other types of industrial waste such as pulverized fuel ash, blast furnace slag, zinc and copper smelter wastes and various types of mine spoil has been similarly but less well investigated. All this research has provided reclamationists with information on the factors limiting vegetation establishment and amelioration techniques for revegetation. However, wastes are frequently encountered whose growth potential has never been properly investigated. Good examples of such materials are certain types of blast furnace slag, some non-ferrous metal smelter wastes, and most types of chemical waste.

Recently, Lancashire County Council planned the reclamation of the site of a derelict chromate smelter but was faced with the problem that the heaps of waste were extremely inhibitory to vegetation growth. Because the site was planned for amenity and recreational use, it was imperative for the success of the scheme to establish a permanent vegetation cover. However, techniques of revegetation were not available and it was not even known what factors were causing the tip toxicity. Because no previous research had been conducted on this substrate, it was necessary to postpone reclamation until the causes of toxicity had been established and a feasible reclamation procedure devised and tested by chemical means. It is suggested that the approach adopted and described here should be employed in future investigations on unusual and toxic sites where information on revegetation of the substrates concerned is minimal or non-existent.

DESCRIPTION OF THE SITE AND INDUSTRIAL HISTORY

The site, located 2·5 km S.E. of Bolton, Lancashire, was one of the three principal centres of chromate smelting in the United Kingdom and consisted of approximately 800,000 metric tons of waste heaps covering an area of 8·1 ha. Smelting of imported chromium ores had been conducted over a period of about 80 years but other chemical processes associated with the manufacture of chromates had been carried out on and near the site. Thus, the waste consisted mainly of residues from the chromate smelter but some alkali waste from the manufacture of sodium carbonate by the Leblanc process was also present. A further contaminant was sodium sulphate which originated from the chemical separation of chromates after smelting.

Vegetation was absent from the chromate waste heaps but did occur on the flat tops of the alkali waste tips. This consisted of almost a pure but sparse stand of *Festuca rubra*. The vegetation of the natural soils surrounding the tip site was generally unaffected by the waste except where surface runoff and lateral seepage of effluent from the tips had occurred. This had caused extensive dieback in some areas, indicating that soluble toxic contaminants may be released from the waste in the presence of moisture.

CHEMISTRY OF THE SUBSTRATES

There are two basic processes of chromate smelting which involve roasting of chromite ($FeO.Cr_2O_3$) with either sodium carbonate or sodium carbonate and lime. Smelting at the present site was by the latter method, details of which are described by Gafafer (1953). Briefly, chromite ore was mixed with magnesian limestone and sodium carbonate, this being roasted in an oxidizing atmosphere to produce sodium chromate. The soluble chromate was extracted from the roast by leaching and subsequently converted, by acidification, to sodium dichromate.

Consideration of the chemistry of the process gave some clues to the toxicity of the waste. According to Gafafer (1953) the residue contains chromium in two forms, viz. trivalent cationic chromium and hexavalent anionic chromate. The latter is mainly water soluble and is generally recognized as being severely toxic to plant growth. It is present mainly as calcium chromate which is less soluble than sodium chromate and consequently less readily removed by the leaching stages of extraction. However, the residue contains other contaminants such as calcium hydroxide and various sodium, magnesium and aluminium salts. All these were considered to be possible sources of toxicity in the waste.

The trivalent cationic chromium referred to above is mainly water insoluble but acid soluble. It was regarded, therefore, as a potential source of plant toxicity in the long term.

Sampling and analysis of the waste

METHOD OF SAMPLING

From visual observations of the tip variability and knowledge that changes in smelting and extraction processes had occurred during the industrial history of the site, it was considered essential to obtain samples covering the entire area of tipping. It was also apparent that natural leaching and weathering had caused great colour and textural changes, and probably chemical changes, in the surface strata. Therefore, samples of both superficial material and underlying unchanged waste were extracted for analysis in order to obtain information on the nature and causes of residual and initial toxicities respectively.

TECHNIQUES OF ANALYSIS

The samples of waste were sieved through 1·5 mm mesh and extracted in water as described by Pitcairn (1969) and Gemmell (1971) for zinc and copper smelter wastes. Conventional types of soil extractant give serious overestimations of the levels of available metals in inorganic waste media (Street & Goodman 1967).

Chromium, aluminium, magnesium and calcium were measured by atomic absorption spectrophotometry using a nitrous oxide/acetylene flame in order to minimize the effects of interfering ions. Sodium and potassium were determined by flame photometric techniques. Hydroxide and 'bicarbonate' alkalinities and chlorides were estimated by titration, sulphate levels being obtained by a turbimetric method. Specific conductivity and pH were measured electrometrically.

RESULTS

The results of chemical analyses of surface and unweathered samples of waste from 88 site locations are presented in Table 1.

Table 1. Chemical analysis of chromate smelter waste.

Contaminant	Concentration in mE/kg waste ± S.D.	
	Surface material	Unweathered material
Chromate (CrO_4^{--})	2·3 ± 1·9	45·7 ± 44·9
Hydroxide (OH^-)	1·1 ± 4·0	83·9 ± 167·1
Sulphate (SO_4^{--})	42·2 ± 63·5	265·9 ± 199·8
Calcium (Ca^{++})	29·7 ± 68·1	228·7 ± 174·3
Magnesium (Mg^{++})	56·4 ± 31·5	4·7 ± 14·3
Sodium (Na^+)	7·1 ± 2·8	15·4 ± 14·4
Potassium (K^+)	0·8 ± 0·4	0·7 ± 0·7
Aluminate (AlO_2^-)	0	0·002 ± 0·006
Chloride (Cl^-)	0·6 ± 1·2	4·2 ± 10·9
'Bicarbonate' ('HCO_3^-')	44·7 ± 38·5	39·5 ± 26·7
Conductivity of 1:10 extract (mmhos/cm)	0·35 ± 0·32	1·42 ± 1·43
pH of 1:1 extract	8·6 ± 0·3	9·6 ± 1·2

Toxic contaminants

The levels obtained of the various waste contaminants suggested that factors other than chromate could be contributory causes of toxicity although there is little doubt from the figures that chromate is involved. In the case of the unweathered waste, the levels of hydroxide, sulphate, 'bicarbonate', conductivity and pH are all in the toxicity range. Magnesium may also be important in the weathered waste. The data also indicates that the chromate and hydroxide in the unweathered waste are mainly present as calcium chromate and calcium hydroxide respectively.

Variability

The high values of the standard deviations in relation to the means for all the contaminants indicate extreme sample variability. Thus, it was considered likely that the degree and even the nature of toxicity could be subject to the same variation. It was also appreciated that the toxic factors operative might be different in samples of different chemical status. In materials where variability of this nature exists, it is important to sample widely and take account of substrate variability. Otherwise it is likely that certain toxicity factors may be overlooked, leading to large-scale vegetation failure if the waste is mixed during earthworks or localized toxicity if planting is carried out in the presence of unsampled material.

Effect of exposure

The data shown in Table 1 confirmed the suspicion that exposure of the waste had drastically altered its chemical composition. It can be seen that the concentration of chromate was reduced to about 5% of its original value and hydroxide was almost entirely removed. With the exception of magnesium and 'titrateable bicarbonate', all other contaminants including pH and the general level of soluble salts were very significantly lowered. This was largely attributed to leaching. However, magnesium and 'bicarbonate' both showed increases although the latter was only minimal. This is probably explicable in terms of increased magnesium solubility as the hydroxide as a consequence of the reduction in pH (Breeze 1973). The presence of magnesium hydroxide would account for the levels of 'bicarbonate' detected in the weathered waste.

Because reclamation involves earthworks and consequent exposure of unweathered material, it can be seen that sampling of the superficial exposed waste only would have been largely meaningless and could have resulted in a reclamation technique unsuitable for the conditions prevailing at the time of planting. However, it is important to include analyses of surface materials because residual toxicity following weathering of the waste may be quite different in nature and causes from the initial toxicity, a factor which could affect the persistence of established vegetation.

Toxicity of the waste and relationship to chemical composition

ASSESSMENT OF TOXICITY

The degree of phytotoxicity exhibited by the waste was initially determined by sowing *Sinapis alba* and *Lolium perenne* as test plants in a range of dilutions of waste with sand. Thereafter, a standard concentration of 1% was adopted in order to quantify the level of toxicity exhibited by each of the samples whose chemical analysis has been described in the preceding section. For this investigation the test plant was *Sinapis alba* since this species rapidly produces sufficient aerial growth for quantitative dry matter assessment and is generally accepted as a good indicator plant for nutrient deficiencies and toxicities. It was considered to be of crucial importance to carry out the toxicity assessment over as short a period as possible because leaching, weathering and other factors could have altered the nature of the growth medium during the course of the investigation.

RESULTS

Nature and degree of toxicity

As expected, germination did not occur in the untreated waste and it was noticed that the addition of ammonium nitrate fertilizer resulted in the evolution of gaseous ammonia.

Initial experiments demonstrated that, in general, the concentration of the unweathered waste in sand had to be reduced to 1% for germination to occur. Highly toxic samples required still further dilution for significant growth to be obtained and even at 0·02% these samples inhibited growth by about 50%. Weathered superficial material was approximately one-tenth as toxic as freshly exposed waste.

The high degree of toxicity exhibited by the unweathered waste strongly supported the suspicion that chromate was implicated as a causal agent.

Relationship of toxicity to chemical composition

In both the unweathered and weathered samples of waste, the degree of toxicity was well correlated with chromate status at the 0·1% level of significance. Because the relationship was of an exponential nature, as is normally the case with respect to the effects of toxic soil contaminants on growth, the concentrations of the chemical factors investigated were compared with the level of toxicity by semi-logarithmic correlation.

It can be seen from Table 2 that chromate was of major importance irrespective of weathering of the waste. Certain other factors were either negatively or positively correlated with the level of growth obtained but these results must be interpreted with caution. For example, aluminate and 'bicarbonate' in the unweathered waste were significantly and negatively correlated with growth potential but may not be causal agents because both were positively correlated with chromate. On the other hand, hydroxide is likely to be a causal factor because there was no evidence of association with chromate. The positive correlations of calcium and magnesium with dry matter obtained suggest that these factors may antagonize toxicity. Robinson *et al.* (1935) have indeed reported that calcium alleviates chromate toxicity in soil.

Although chromate appears to be a source of toxicity following weathering, there is some evidence that magnesium is implicated. Magnesium and 'bicarbonate' were both significantly and negatively correlated with growth potential and neither showed any significant association with chromate. As reported by Breeze (1973), magnesium hydroxide is the likely source of these

Table 2. Semi-log. correlations of chemical status with growth potential of diluted wastes.

Contaminant	Weathered waste	Unweathered waste
Log chromate	−0·49‡	−0·68‡
Log hydroxide	−0·21	−0·25*
Log sulphate	−0·05	−0·14
Log calcium	0·38‡	0·29†
Log magnesium	−0·27*	0·37‡
Log sodium	−0·23*	−0·19
Log potassium	0·09	0·14
Log aluminate	—	−0·28†
Log chloride	−0·10	0·04
Log 'bicarbonate'	−0·41‡	−0·64‡
Log conductivity	0·06	0·20
pH	−0·37†	−0·21

* 5 per cent probability.
† 1 per cent probability.
‡ 0·1 per cent probability.

ions and also the cause of the residual high pH, both magnesium and 'bicarbonate' being significantly and positively correlated with pH. Magnesium is now recognized as a source of soil toxicity in certain situations (Proctor 1969). Calcium was again significantly and positively correlated with growth potential; in this case it may antagonize both chromate and magnesium toxicity.

Relative importance of chromate and pH as toxicity factors in unweathered waste

Although chromate was significantly correlated with toxicity in the freshly exposed waste, there was very considerable variation in toxicity in different samples of similar chromate status. This suggested that another factor was operative in determining the actual degree of toxicity.

Examination of the chemical status of the unweathered samples suggested that they could be characterized and separated into three groupings of low, medium, and high chromate content. At low chromate, 'bicarbonate' was less than 25·0 mE/kg; at high chromate 'bicarbonate' was greater than 45·0 mE/kg and sulphate was in excess of 200 mE/kg. Samples of intermediate chromate status were variable. The basis of this classification was tested and confirmed by multiple discriminant analysis (Gemmell 1973). When the three groups were examined separately for relationships of chemical composition with toxicity (Table 3), it was apparent that chromate was only related to toxicity in the Group 1 samples of low chromate. In Groups 2 and 3, hydroxide was significantly correlated with toxicity but chromate was not. In view of the high levels of chromate in these samples, it was concluded that toxicity was caused by both hydroxide and chromate, hydroxide exerting a synergistic effect on chromate toxicity. It is also possible that the peculiar solubility effects of chromate in the waste as influenced by pH (Breeze 1973) may be of importance.

Table 3. Correlations of chemical status with growth potential following group classification of substrates.

Contaminant	Group 1 N = 25	Group 2 N = 41	Group 3 N = 22
Log chromate	−0·54*	−0·09	−0·40
Log hydroxide	−0·27	−0·42*	−0·55*

Groups 1, 2, and 3 are characterized by low, medium and high chromate respectively.

Conclusions of chemical substrate evaluations and importance for revegetation

CAUSES OF TOXICITY

The substrate analytical investigations described above clearly implicated chromate and hydroxide as principal causes of toxicity. It was also clear that dilution of the waste would not permit vegetation establishment since these factors are toxic in trace amounts. Therefore, some sort of chemical treatment was necessary. However, the high levels of sulphate and other contaminants suggested that secondary toxicity factors could become operative once the principal contaminants were removed. This indicated that chemical treatment plus dilution of the waste or treatment with some other ameliorant was essential in order to eliminate all sources of toxicity.

HYPOTHETICAL RECLAMATION TREATMENT

Because chromic salts are less toxic than chromates in soil (Robinson *et al.* 1935), mainly due to precipitation of chromic hydroxide at neutral pH (Breeze 1973), it was assumed that chemical reduction of the chromate to the trivalent chromic state would alleviate toxicity. This would also allow additions of organic matter to complex any chromic ions still in solution as described for other metal ions (Lucas 1948, Himes & Barber 1957). At the same time, reduction of the high pH by hydroxide neutralization was obviously necessary.

Chemical reduction of chromate is normally achieved by reaction with ferrous sulphate which is well known as a soil acidifying agent. Therefore, the application of ferrous iron to the waste was believed to be a possible method of alleviating the two principal substrate toxicities at the same time. The method would be economically viable because ferrous sulphate was found to be available at very low cost as a chemical waste following the extraction of titanium from ilmenite ($FeTiO_3$). This could be followed up by organic matter addition in the form of sewage sludge from the 'activated sludge' process, itself available as waste in the vicinity of the chromate tips. Sewage sludge has been used on other types of smelter waste in order to complex toxic metal cations and provide nutrients (Weston *et al.* 1965; Street & Goodman 1967).

Investigations into revegetation techniques

SOURCES OF MATERIALS

Experiments on the growth potential of waste following amelioration were conducted on unweathered waste, unless stated otherwise. Waste of high toxicity containing 50–70 mE/kg water soluble chromate and high pH (11·8–12·7) was used throughout the investigations.

Sewage sludge was obtained from a sewage treatment works using the 'activated sludge' process, this material being free of toxic concentrations of metals. Subsoil was excavated from beneath uncontaminated grassland surrounding the smelter site and was sieved through 2·5 cm mesh to remove stones.

EXPERIMENTAL TECHNIQUES

All growth investigations were conducted in a glasshouse using plant pots. Experiments on top-coverings were conducted in large pots of 23 cm diameter and entailed a double-decker system with the lower pots containing waste and the upper bottomless pots filled with subsoil and/or organic amendment materials.

Lolium perenne was used as a test plant since this species was to be planted after reclamation. The seed was sown at the rate of 125 kg/ha. The three nutrients N, P_2O_5 and K_2O were each applied at the rate of 62·5 kg/ha.

Growth was evaluated in terms of dry matter production. In experiments of long-term, growth recovery following cutting was also measured on this basis. Root penetration was examined by exposure of vertical sections through the layers of amendment and waste.

RESULTS

Effect of ferrous sulphate and organic matter incorporation

As a first experiment, ferrous sulphate ($FeSO_4.4H_2O$) at a rate equivalent to 2,500 kg/ha was added to unweathered waste diluted with various levels of sand, soil, and peat (Gemmell 1972). The results shown in Fig. 1 demonstrate that toxicity was alleviated by ferrous sulphate in all the dilution media.

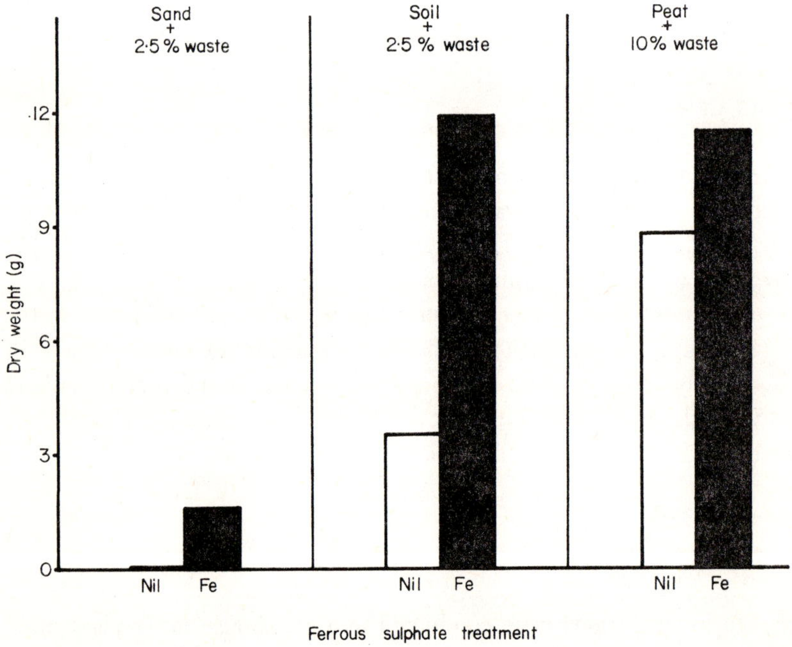

Figure 1. Effect of ferrous sulphate ($FeSO_4.4H_2O$ at 2,500 kg/ha) on toxicity of media contaminated with chromate smelter waste.

It is also apparent that the nature of the dilution medium is of importance, irrespective of the presence of ferrous sulphate. Soil was better than sand but peat was much superior to soil, presumably on account of its high organic matter content. Further experiments demonstrated that sewage sludge was comparable to peat, depending on its organic matter status.

The effect of organic matter in reducing chromate toxicity was unexpected. It was previously believed that the chromate in the waste would have to be changed to the cationic state for organic matter to form stable complexes, thereby removing chromium as a source of toxicity. However, Table 4 clearly shows that organic matter is operative against chromates in growth media although the precise mechanism is not understood. This suggests that organic materials may be useful in counteracting other forms of anionic toxicity and may even be effective against anionic forms of secondary toxicity in the waste.

Table 4. Effect of organic matter on reduction of chromate toxicity.

Chromate concentration (ppm $Na_2CrO_4.10H_2O$)	Growth inhibition (per cent)	
	Sand	Sand/20% peat
16	75%	Nil
80	89%	Nil
400	100%	Nil
2,000	100%	21%
10,000	100%	100%

Importance of dilution rate and level of ferrous sulphate addition

Due to the high levels of chromate in most types of undiluted waste and because toxicity was multifactorial, it was normally impossible to alleviate toxicity by ferrous sulphate application alone, even when very high rates up to 25,000 kg/ha were applied. Figure 2 illustrates the importance of dilution in the case of samples with high chromate status and high pH. The presence of 5% waste prevented an acceptable growth response to ferrous sulphate and it was necessary to reduce the waste concentration to 1% for successful treatment. Only when chromate concentration is low and the pH less than about 9·0 does ferrous sulphate give an acceptable growth response on undiluted waste. Because most samples were of high chromate and/or pH dilution was regarded as a strict necessity for successful reclamation.

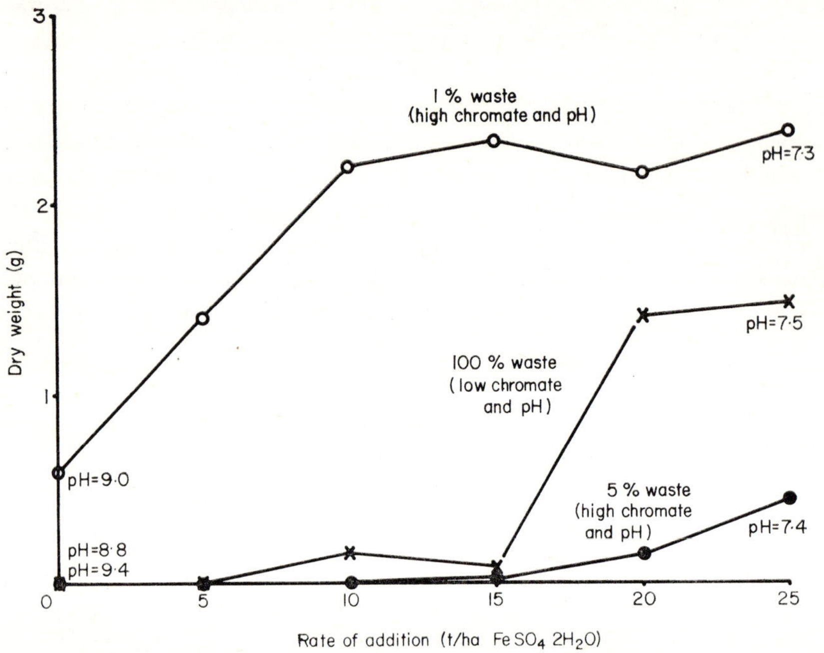

Figure 2. Effect of ferrous sulphate at different rates of application on toxicity of chromate smelter waste contaminated media. Sand was used as diluent.

Use of top-coverings and importance of chromate mobility

When very high rates of dilution are required for amelioration of toxicity as in the present situation, it is better and more economical to use top-coverings of amendment materials rather than dilution methods. However, the levels of amendment applied must be of sufficient magnitude to achieve a waste-free rooting medium. Although sewage sludge would have been ideal in some respects, the amounts required to provide a 15 cm surface covering would have been prohibitively expensive on account of transport costs. Also, the availability of suitable material at any one time could have been limited. Further, the degradation of organic matter over a period of time could have resulted in growth regression due to the reappearance of toxicity as demonstrated for zinc and copper smelter wastes (Gemmell 1971). Therefore, the effect of initially covering the waste with a layer of subsoil was investigated, suitable material being available beneath the grassland surrounding the chromate waste heaps.

Preliminary amelioration with subsoil was also considered to be necessary in view of the possible mobility of chromate and other toxins in soil. Breeze

(1973) has produced evidence that chromate is mobile in soil and suggests that upward diffusion from the underlying waste into soil coverings may eventually result in toxicity in the rooting substratum. Accumulation of zinc has been demonstrated in surface coverings of pulverized fuel ash on zinc smelter waste (Gemmell 1971) which was later followed by vegetation die-back. The subsoil used in the present experiment was of a sandy, gravelly, extremely porous nature, and therefore ideal as a barrier to upward diffusion of toxic ions. Further, the absence of an exchange complex in the layer directly overlying the waste would eliminate the possibility of chromate retention on soil exchange sites.

Figure 3 shows the initial and recovery growth of grass planted on waste amended with soil, sewage sludge and peat in the presence and absence of a

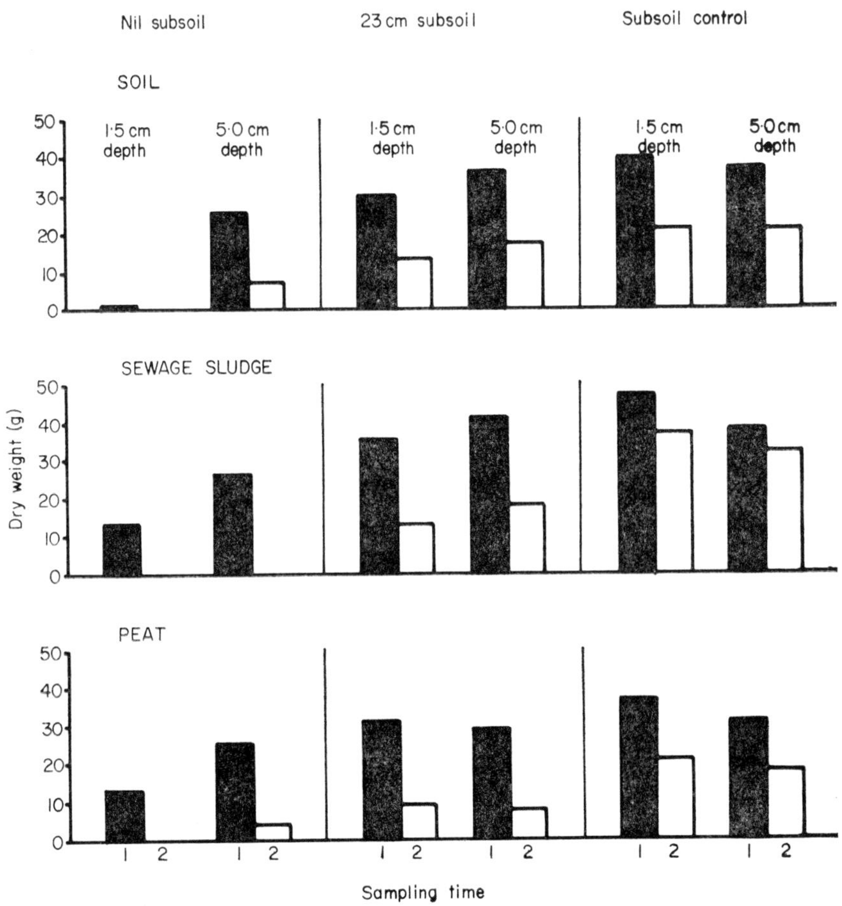

Figure 3. Growth production and recovery from cutting of grass established on chromate smelter waste amended with organic materials in the presence and absence of subsoil. Sampling times 1 and 2 were at 6 months and 16 months from planting respectively.

subsoil covering. A feature of the results is that whereas 5·0 cm depths of each of the three amendments without subsoil gave acceptable initial growth, acceptable recovery occurred only in the presence of subsoil. Although the recovery on subsoil amended waste was less than the controls in every case, the causative factor was limited moisture and nutrient availability rather than toxicity. At the end of the experiment, rooting was still active to within 2·5–3·0 cm of the subsoil/waste interface and there was no evidence of chlorosis or chromate toxicity.

Another important observation was that initial rooting in soil directly overlying waste did not reach the soil/waste interface, being restricted at 2·5 cm above it. This was attributed to chromate mobility and confirmed by analysis of the top 1 cm layer of soil which showed the presence of 0·13 mE/kg ammonium acetate extractable chromate. Soil overlying subsoil treated waste showed no increase over the control.

A further experiment was carried out simultaneously with the above in order to obtain information on the minimum depth of subsoil required for revegetation and whether chromate mobility could be induced by water stress. The effect of increasing subsoil depth was investigated in terms of growth productivity, recovery from cutting, and chromate contamination of the surface covering. The results showed that at 6 months from planting, growth was depressed at subsoil depths less than 10·0 cm but there was no increase in growth at levels greater than this, irrespective of watering regime. Chromate mobility was restricted to 2·5 and 5·0 cm above the level of the waste at the high and low watering regimes respectively. Analysis of recovery growth after 16 months from planting showed that recovery on 10·0 cm of subsoil was the same as the controls in the case of the high watering treatment but there was reduced recovery when watering was restricted. However, there was no visual evidence of toxicity or chlorosis and recovery was normal at higher rates of subsoil addition. Because the watering restriction was severe, it was concluded that dieback due to the toxicity of the underlying waste and chromate mobility phenomena was most improbable under British climatic conditions.

Discussion and conclusions

From the results of chemical substrate evaluations on chromate smelter waste, it was possible to determine the chemical nature of the toxicity factors operative and to obtain useful information on the relative importance with respect to growth inhibition of the various toxins present. Although toxicity was multifactorial, it was found that calcium chromate and calcium hydroxide were the principal causes of toxicity. This enabled a hypothetical treatment to be devised which was then tested under controlled experimental conditions and found to be valid.

In devising a practical reclamation procedure from the results described, it was necessary to take account of any constraints which might have been imposed by the nature of the engineering and other operations involved during reclamation. These were the practicality of spreading surface coverings of amendments in discrete layers without serious contamination from the underlying waste, the availability and costs of chemical and organic amendments, the feasibility of spreading the amendments evenly and at the rates required, and the stability, both physical and biological, of the treated waste. It was also considered likely that other factors could necessitate modifications of the technique during the actual reclamation work.

PROVISION OF THE SUBSOIL COVERING

As previously mentioned, the subsoil material tested was obtained from uncontaminated land surrounding the waste heaps. Since this land was acquired for spreading and levelling the waste, it was possible to execute the tip spreading and subsoil excavation operations simultaneously. Thus, the initial amendment treatment was of relatively low cost.

Because the spreading of the subsoil covering was carried out by large earth-moving machinery, it was impossible to control the depth applied to within fine limits. Therefore, it was necessary to provide a thickness greater than that indicated by the plant growth experiments so that the minimum depth would not fall below the critical level of about 10 cm. It was decided that a mean depth of 25 cm would be sufficient to cover variability in spreading.

In theory, the addition of subsoil would have satisfied the growth requirements on the site if the material could have been spread cleanly as a discrete surface covering and entirely free of contamination from the underlying toxic waste. In practice, contamination was unavoidable although it was possible to reduce it to a minimum. The chief source of contamination was chromate waste adhering to the tracks and wheels of earth-moving machinery which then came into contact with the subsoil during the excavation, spreading, and grading operations. It was also considered that the final cultivations necessary for seed-bed preparation and seed incorporation could result in further contamination. Because the waste was shown to be toxic in trace amounts, it was necessary to provide further amendments of ferrous sulphate and sewage sludge in order to achieve an acceptable cover of vegetation.

APPLICATION OF FERROUS SULPHATE

Ferrous sulphate was available as a chemical waste in the form of the heptahydrate ($FeSO_4.7H_2O$). The rate applied was 15,000–20,000 kg/ha although levels greater than this were spread on localized areas of high toxicity. Following spreading, the material was lightly incorporated into the surface of the subsoil covering prior to amelioration with sewage sludge.

If extreme overdoses of ferrous sulphate are used and followed by immediate planting, there is the possibility of toxicity and induced nutrient deficiency due to over-acidification. However, this is unlikely if the substrates contain lime. In practice, a period of natural leaching was allowed to occur in order to remove possible excesses of ferrous sulphate from the surface of the substrate and achieve chromate reduction and pH neutralization throughout the rooting medium.

AMELIORATION WITH SEWAGE SLUDGE

Sewage sludge was applied to the surface of the amended waste at the rate of 100,000 kg/ha following ferrous sulphate incorporation. This amendment was lightly incorporated into the surface material, deep cultivations being avoided in view of the possibility of contamination from the underlying waste.

ECONOMICS OF THE METHOD

A summary of the reclamation technique devised is presented in Table 5. The chief advantage of this method is that three waste problems are solved simultaneously, viz. chromate smelter waste, ferrous sulphate residues from

Table 5. Outline of revegetation technique for chromate smelter waste.

Amendment	Rate	Method	Effect
Porous subsoil	25–30 cm depth	Spread on surface of waste	Dilution medium and restriction of toxin mobility
Ferrous sulphate	Up to 40 t/ha $FeSO_4.7H_2O$	Surface incorporation by discing or harrowing	Chemical reduction of chromate and decrease in pH
Organic matter as peat or sewage sludge	100 t/ha or greater	Surface incorporation by discing	Reduces level of plant-available chromate and other toxins. Source of NPK

titanium extraction, and sewage sludge. Consequently, the costs of reclamation were reduced to a minimum by application of this method. The major costs involved were incurred in haulage and placement of amendment materials and these were relatively low in comparison with other reclamation schemes.

APPLICATION OF THE METHOD TO OTHER PROBLEMS

Research on the causes of toxicity and techniques of revegetation of chemical wastes has been seriously neglected. Therefore, any chemical waste or material affected by chemical contamination should be analysed on the lines of the techniques described here. This should identify the chemical nature of the factors inhibiting growth and enable a possible revegetation technique to be devised based on chemical treatments, organic matter additions, and/or dilution techniques with inert or other media. It will then be necessary to test the effects of the amendment materials proposed, both separately and in combination if necessary, so that their interactions with the toxic waste factors can be assessed in terms of growth improvement. A reasonable knowledge of the chemical and physical interactions of the amendments with the waste will allow the technique to be usefully modified to suit the conditions prevailing in the substrate at the time of reclamation. It may even be possible to modify the technique to solve other reclamation problems.

Two wastes which have been recently investigated on the lines described above are alkali waste from the obsolete Leblanc process for sodium carbonate, and blast furnace slag. Both wastes were inhibitory to vegetation but little or no information was available concerning the factors responsible for toxicity. Investigations carried out showed that in both cases the chemical factors involved were similar to some of those encountered in chromate smelter waste and it was thus possible to devise suitable corrective treatments on the basis of the technique evolved for chromate waste reclamation.

Alkali waste from the Leblanc process for sodium carbonate

The toxicity of this waste was found to be caused by high levels of calcium hydroxide which resulted in pH values of up to 12·7. As in chromate smelter waste, sulphate levels were high but not apparently involved in relation to the toxicity observed. Prolonged exposure of the waste removed the hydroxide and toxicity completely but because the time factor required was in the order of tens of years, this factor could not be utilized in reclamation.

Although ferrous sulphate had some effect in reducing the high pH and alleviating the toxicity in the undiluted waste, the levels of application required were found to be far too high for such treatment to be a practical feasibility. Clearly, it was impractical to establish vegetation direct into the waste and top-coverings of suitable amendments were obviously required. Such treatment could be reinforced by ferrous sulphate incorporation if necessary, followed by the sowing of *Festuca rubra* which is relatively tolerant of the high pH status of the substrate.

Blast furnace slag

Investigations on blast furnace slag from the Furness area of Lancashire showed that the factors limiting vegetation establishment were alkalinity and nutrient deficiency, the pH of unweathered material being in the range 10·2–10·6. Material exposed for a number of years had a pH of 7·5–8·0 but vegetation was still absent because of nutrient deficiency.

Ferrous sulphate had little effect on lowering the pH or reducing toxicity, presumably because of its reactions with calcium carbonate as well as calcium hydroxide in the waste. However, it was found that the alkalinity of the waste changed very quickly on exposure to natural leaching and weathering, the pH falling from 10·5 to 8·0 in about 8 months. This enabled a treatment to be devised which utilized the ameliorating effect of exposure, the waste being treated and planted when the pH had fallen to 9·0. High levels of phosphate were applied in combination with slow-release sources of nitrogen, the waste being sown to the high pH tolerant *Festuca rubra*. Slow-release nitrogenous fertilizers such as sulphur-coated ureas were considered to be better than more soluble forms of nitrogen on account of the porosity of the waste and the unavailability of nitrogen following leaching into the underlying material of high pH. The success of this treatment was verified under field conditions, enabling planting to be carried out direct into the waste.

Summary

A method has been described for the detection and identification of substrate toxicity factors in chromate smelter waste. Such information is vital for devising revegetation techniques for use on chemically contaminated sites. In many cases, the techniques evolved from this approach may be modified to utilize other waste materials and so achieve reclamation at very lost cost. Before use, revegetation methods based on the results of analytical information should be tested experimentally under glasshouse conditions and later by field trials if time permits.

In the case of chromate smelter waste, a wide variety of contaminants was found to be present. Some of the toxins may be encountered in other materials such as alkali waste and blast furnace slag. Therefore, the technique evolved for revegetating chromate waste may be adopted, with suitable modifications, for treating other wastes.

References

BREEZE V. (1973) Land reclamation and river pollution problems in the Croal Valley caused by waste from chromate manufacture. *J. appl. Ecol.* 10, 513–25.

GAFAFER W.M. (1953) Health of workers in chromate producing industry. *Amer. Pub. Health Service Pub.* No. 192, U.S. Government Printing Office, Washington.

GEMMELL R.P. (1971) *The Ecological Behaviour of Species—populations of Grasses Susceptible and Tolerant to Heavy Metal Toxicity*. Ph.D. thesis, University of Wales.

GEMMELL R.P. (1972) Use of waste materials for revegetation of chromate smelter waste. *Nature, Lond.* 240, 569–71.

GEMMELL R.P. (1973) Revegetation of derelict land polluted by a chromate smelter I: Chemical factors causing substrate toxicity in chromate smelter waste. *Environ. Pollut.* 5, 181–97.

HIMES F.L. & BARBER S.A. (1957) Chelating ability of soil organic matter. *Soil Sci. Soc. Amer. Proc.* **21**, 368–73.

LUCAS R.E. (1948) Chemical and physical behaviour of copper in organic soils. *Soil Sci.* 66, 119–29.

PITCAIRN C.E.R. (1969) *An Ecological Study of the Factors Influencing Revegetation of Industrial Waste Heaps Contaminated With Heavy Metals*. Ph.D. thesis, University of Wales.

PROCTOR J. (1969) *Studies in Serpentine Plant Ecology*. D.Phil. thesis, University of Oxford.

ROBINSON W.O., EDGINTON G.A. & BYERS H.G. (1935) Chemical studies of infertile soils derived from rocks high in magnesium and generally high in chromium and nickel. *U.S. Dept. Agric. Tech. Bull.* 471.

STREET H.E. & GOODMAN G.T. (1967) Techniques of revegetation in the Lower Swansea Valley. In *The Lower Swansea Valley Project* (Ed. by K.J. Hilton), Longmans Green, London.

WESTON R.L., GADGIL P.D., SALTER B.R. & GOODMAN G.T. (1965) Problems of revegetation in the Lower Swansea Valley, an area of extensive industrial dereliction. In *Ecology and the Industrial Society* (Ed. G.T. Goodman, R.W. Edwards & J.M. Lambert), Blackwell, Oxford.

SECTION FOUR

Planning for resource renewal

Population, pollution and natural resources—an inevitable clash?

R. C. CURNOW *Science Policy Research Unit, University of Sussex, Brighton, U.K.*

Introduction

Two recent publications (Forrester 1971; Meadows *et al.* 1972) have stimulated speculation about the future of the world. Both publications reach quite firm conclusions; that if present trends continue the world is heading for disaster well before the year 2100, either from shortage of natural resources, from the effects of pollution created by industrial and agricultural technologies, from lack of food or from a combination of these causes. What makes their Malthusian conclusions different from similar previous statements is that they are based on computer models of the world. The models are used to show that even the most 'optimistic' advances in technology would not be able to prevent the disasters foreseen and which, they say (Forrester 1971; Meadows *et al.* 1972), can only be avoided by adopting rather dramatic social policies.

The results have been widely quoted, for example, in the *Blueprint for Survival* (Ecologist 1972), by Gabor (1972) and by Mansholt in his letter to the European Commission. All have argued for the need to achieve some kind of equilibrium society at an early date. At the same time, the work has received strong criticism as in a recent study by the World Bank and by Beckerman in his television exposition *Growth is Good For You.* Nevertheless, the work of Meadows and his associates (Meadows *et al.* 1972) has had a particular impact in the Netherlands with sales of *Limits to Growth* reaching nearly half a million copies, and reputed sales of 2·6 million worldwide.

A team at the Science Policy Research Unit (S.P.R.U.) has entered this controversy (Cole *et al.* 1973). In order to establish the status of the models and to examine the claims of their authors they have investigated the models, their assumptions and the data upon which they are based. They have also examined some of the ideological background to the models in the context of the current environmental debate. They conclude that, whilst one cannot afford to be complacent, there seems, in their opinion, little reason to believe the results of the models. The results do not appear to be insensitive to the inclusion of plausible changes in technology as their authors have claimed, or to other structural changes; neither does the interpretation of the available data used in the models always appear to be reasonable.

The Science Policy Research Unit team believe that the results presented by Forrester (1971) and Meadows *et al.* (1972) could have undesirable consequences, in that the unnecessarily pessimistic views of the world's future that they present may instil a sense of fatalism. This, it may be argued, could lead to the neglect of important current problems, for example the plight of the under-developed countries.

Notwithstanding this danger, it is felt that many of the problems raised by the MIT work are important, and that the debate which they have stimulated should be very constructive, provided that the utility of models is understood. Since computer models will almost certainly become a more regular feature of societal planning, there is clearly an urgent need for society at large to appreciate both the advantages and limitations of these models and, following from this, the use to which they should be put in any debate.

Accordingly, as additional project work in the Unit has increasingly involved examination of the applicability of dynamic modelling to complex socio-economic issues, we are concerned with obtaining an overview of the modelling process in general, and of the dynamic modelling of world problems in particular.

Definitions of a model

Following Ackoff (1962) a 'model' may be defined in such a way as to incorporate the three different usages found in everyday language, i.e. as a noun (implying a representation), as an adjective (meaning an ideal), and as a verb (meaning to demonstrate or describe). A model will thus be defined as *an idealized representation describing some phenomena whose behaviour is to be highlighted.*

Attention is drawn to two aspects of models—their necessary degree of simplification and its attendant danger of over-simplification, and the purposive character of a model in its conscious choice of phenomena to be highlighted. Both these aspects involve the operational nature of modelling, as might be expected from a definition drawn from the discipline of operational research.

Before giving a structured account of the modelling process as tentatively viewed from interpretation of the control engineer's viewpoint, it may be useful to note the definition of a model as given by a philosopher of science (Hesse 1965). 'A model is intended as a factual description if it exhibits a positive analogy and no negative analogy in all respects hitherto tested, and if it has surplus content which is in principle capable of test.' In this case attention is drawn to the requirement of surplus content, i.e. the model must be capable of showing behaviour which is equally capable of being shown by the real world being modelled, although there is no necessity for the real world yet to have done so.

The modelling process

In looking at modelling from a very personal interpretation of the control engineer's viewpoint, the Science Policy Research Unit is constructing a model of that process of modelling and its links with the real world. It follows, therefore, that this model will be a simplified representation, idealized in that it will separate activities which normally overlap to some degree, and biased in that it chooses to emphasize some of those activities.

Figure 1 is a flow diagram illustrating activities, their sequence relationship, and some of the more obvious control loops in the total process.

The process begins with the establishment of the purpose of the proposed model. Here we are primarily concerned with the utility of modelling as an aid to the decision-making process; in this regard, a model may, by improving understanding of the major interactions between the pertinent parameters, indicate the likely repercussions of any proposed policy, and may also improve communication between experts in particular fields and policy makers. Alternatively, a model may be constructed for more academic purposes, e.g. as a teaching aid, to improve interdisciplinary communication, or, by indicating potentially important relationships for which theory and data are of a low standard, to act as a springboard for research necessary to improve understanding of a system's behaviour. If the problem under consideration is that of optimizing specific entities rather than that of obtaining a description of a system on which various policies are to be tested, a normative approach, in which an objective function is maximized subject to certain constraints, can be used.

The process continues with observations on the real world, in order to obtain an overall assessment of the major sub-systems whose behaviour has an important influence on the problems being considered. This is followed by an analysis of the internal structure of these sub-systems, which culminates in a set of theories, from which the final model is to be synthesized. Normally, of course, the modeller will have various theories of the behaviour of his sub-systems already available; however, his task in the synthesis stage may be very difficult, in that the theories pertaining to a particular sub-system, and those referring to different sub-systems, may be incompatible, and the degrees to which they have been separately verified may differ widely. Similar problems arise when the structure decided upon is calibrated; the data available may be scarce or of poor quality, as is frequently the case in socio-economic systems where the data have been collected at different times and for different purposes.

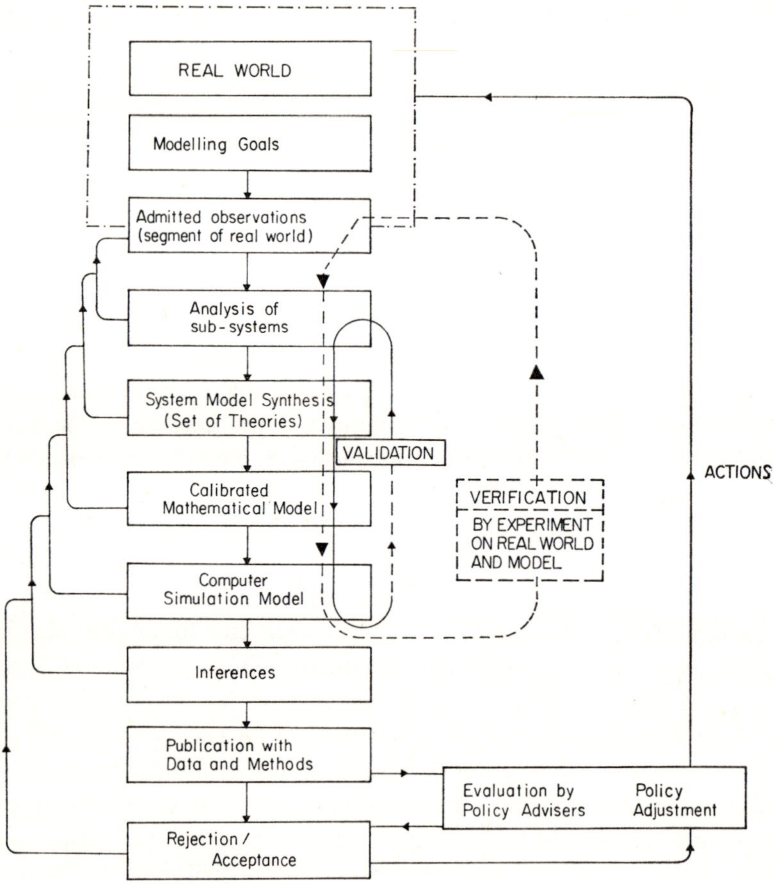

Figure 1. Computer simulation modelling—a model.

When the calibrated model has been obtained, a simulation language must be selected. Criteria used here include ease of programming, core requirements and required accuracy.

Note that the process of *validation* is here used to describe the process of ensuring that the model well represents the theoretical structures it is intended to represent following the systems analysis-synthesis steps.

This process involves the testing of the model structure, e.g. checking that a known fixed response is obtained from a specific input, and ascertaining the sensitivity of the model to small changes in the input parameters. The process of *verification*, which can be considered complete after finding *no falsification* under test, is in general only applicable when the 'real world' simulated is capable of test. Thus, a high degree of verification is possible if one is considering a computer simulation model of a diesel engine: even most social scientists would agree that verification is much less possible with models of socio-economic phenomena. Nevertheless, verification in both cases where claimed is tantamount to a statement of *not yet falsified though tested to some degree*. However, because verification by experiment on *model and real world* is certainly not possible if one is discussing world models, the verification loop is shown here as a potential rather than actual loop.

If the computer simulation models being considered are intended as in any sense predictive or projective, e.g. of future behaviour of the real world being modelled, then a further issue needs to be emphasized. In extrapolating the interacting trends of a model which might accurately represent behaviour to date in the real world, it is probable that the model will predict a form of behaviour which has not yet been exhibited by the real world. In this case, there is an onus on the modeller to show that no negative analogy exists in the separate futures of model and real world (i.e. he needs to demonstrate that he has not omitted plausible mechanisms which may be important in the future, but which have not yet been active and need not have been represented in a model of the world to date).

For the model constructed as a decision-making aid, the implications of various policies can now be tested, and the results evaluated by policy-makers who, if they have sufficient confidence in the model, will use its results as indicative evidence when taking decisions which affect the real world. This confidence, we feel, is only likely to be achieved if the policy makers have maintained contact with the modeller throughout, and are aware of the structure and limitations of the model. It seems likely that a conversion to an interactive formulation, if this has not been done already, will help the decision maker to become familiar with the behaviour of the model.

Of course, we are concerned with dynamic modelling, since it is clear that most problems of interest are drawn from dynamic interacting systems.

Two possibilities now exist for continuing this analysis. The different types and levels of critical evaluation given in Cole *et al.* (1973) can now be mapped on to this structure of the modelling process, or alternatively a general appreciation of the feasibility, utility and problems of dynamic modelling can be likewise developed and mapped on to this structure.

As mentioned before, this latter process is still under examination at Sussex, although a feasibility study of dynamic modelling for socio-economic issues has been completed for the European Economic Commission. A more general appreciation of dynamic modelling in relation to world problems is now being completed for the Research Councils of the United Kingdom, and should be available early in 1974.

Comments on the MIT world models

Reverting, therefore, to Cole *et al.* (1973) it is useful to look both at the critical evaluations of *Limits to Growth* (Meadows *et al.* 1972) therein and their place in a controlled process of modelling, with reference to the structure proposed above.

World 3 was constructed at the request of the *Club of Rome* as *Phase One* of their project on the *Predicament of Mankind.* The intent of this project, as stated in Meadows *et al.* (1972), is to examine the complex of problems troubling men of all nations. Clearly no single model can possibly represent anything more than a small number of these issues, but it can reasonably be asked:

> 'A world model, World 3, draws attention to certain issues and problems, i.e. it highlights the problem of finite limits. Whilst readily agreeing that if man is restricted to terrestrial limits, the finite limits of the Earth will eventually constrain any continued growth as currently defined, are these necessarily the most important or the most urgent problems?'

Clearly the choice or priority ranking of problems is a matter of human values. At Sussex we would argue that, leaving aside such issues as the current level of nuclear armament, the problems of disparities between and within societies confronts the world sooner and with greater problems than those raised by Forrester (1971) and Meadows *et al.* (1972).

Having chosen to represent physical limits, we may now ask whether an adequate set of state variables has been chosen, and whether the major interactions between these variables have been correctly identified. As discussed by Cole *et al.* (1973), it is felt that it is impossible to discuss the consequences of physical limits without an adequate representation of possible economic, social and political reactions. Although this is a question of belief and values, the implied assumption that the human race would continue to behave as it has previously behaved in the face of catastrophe seems at least questionable.

Moving on to the calibration and simulation stages, we note that, because of the complexity and non-linearity of *World 3*, no transparent analytical representation is possible. The model is constructed directly in the format appropriate to the simulation language used.

The question of the level of aggregation is important here; the best approach, especially when data are scarce, is to build a highly aggregated model in the first instance, and then to determine experimentally any advantages gained from disaggregation. *World 3* therefore cannot be dismissed as a valueless exercise because the variables used are highly aggregated, but it is necessary to draw attention to the difficulties of calibration arising from the need to attach numbers to a variable, such as pollution, which represents a combination of diverse phenomena.

Consider now the validation of *World 3*. Perhaps the only means of doing this is to examine its sensitivity. It is claimed (Meadows *et al.* 1972; Forrester 1971) that if present trends continue, the world faces catastrophic collapse before A.D. 2100. In other words, the implication is that simulation of the future gives rise uniquely to catastrophe, unless Draconian policies are implemented very soon. The question can be asked 'Are the models put forward, and accepting their structure, and their calibration, uniquely projecting a catastrophe?'

Figure 2 shows the standard runs of *World 2* and *World 3* (i.e. Forrester 1971; Meadows *et al.* 1972 respectively).

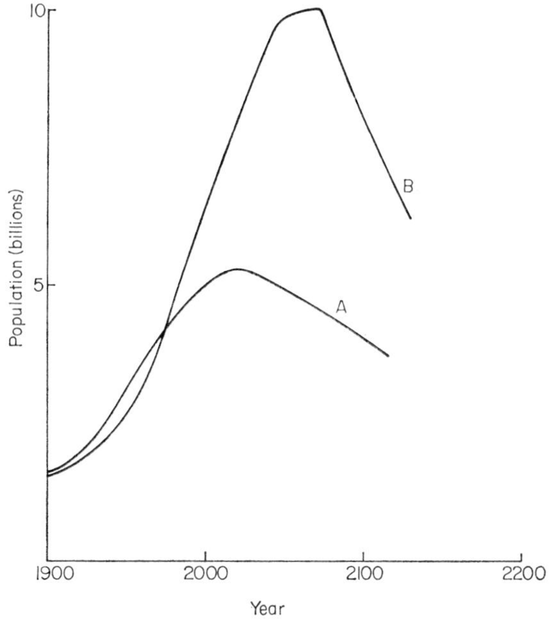

Figure 2. Standard runs of A: World 2 (Forrester 1971); and B: World 3 (Meadows *et al.* 1972) for population.

Figure 3 shows a run on *World 3* which involved less than 5% perturbation to the initialization values of *World 3*. This work was carried out by Dr Hugo Scolnik of Fundacion Bariloche in the Argentine, and involved optimization procedures around the initialization point. In all cases the perturbations were within the range of uncertainty of estimation admitted implicitly in the as yet unpublished *Technical Report* supporting *Limits to Growth*.

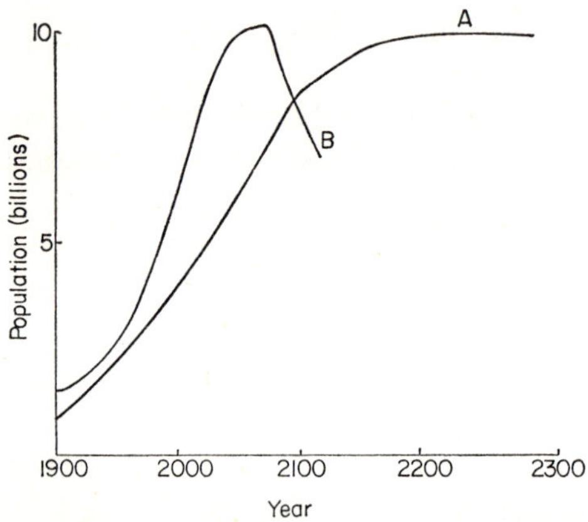

Figure 3. A comparison of A: the Bariloche World 3 population run with that of B: Meadows *et al.* (1972).

Although not all would agree, it would appear that a projection showing acceptable population growth to the year 2300 is phenomenologically a different outcome from a collapse well prior to the year 2100. It is of passing interest to note that the discrepancies between *World 2* and *World 3* appear not to have drawn comment, possibly owing to the fact that the graphs in Meadows *et al.* (1972) do not carry vertical scales. At this first level of argument, therefore, it can reasonably be asserted that the statement that *World 3* uniquely forecasts an early world catastrophe is not supportable.

To the counter argument that Fig. 3 shows a collapse, albeit later, it must be responded that the so-called equilibrium states involving rather dramatic social policies put forward by Forrester (1971) and Meadows *et al.* (1972) themselves show collapse if the runs are continued for several hundreds of years in the future.

At the next level of argument, it can be asked 'Do the models well represent those portions of the real world which they purport to represent, and if equally plausible assumptions are put forward, are the conclusions affected?' The bulk of Cole *et al.* (1973) is concerned with questions of this type, and Figs 4, 5, 6, 7 & 8 show runs which are designed to help answer this question. The standard run of *World 3* is included on each Figure for comparison. It should be made clear at this stage that SPRU does not necessarily believe that any of these runs will represent the actual evolution of the world, but rather, as argued by Cole *et al.* (1973), they show that a set of assumptions equally as plausible as those of Meadows *et al.* (1972) and can give rise to rather different conclusions. In other words, it would appear that the modelling stages of systems analysis and synthesis were not under adequate control in the evolution of *World 3*.

As mentioned before, verification of the Model cannot be carried out because of the impossibility of experimenting on the real world.

Following the MIT attempt at modelling the global system, many other research groups throughout the world have engaged in a similar activity. Not all these however are as pessimistic with regard to physical constraints as those of Meadows *et al.* (1972) and Forrester (1971). For example, the Fundacion Bariloche in Argentina have designed a powerful counter-initiative to the Malthusian view. The main goal of their project (which incorporates a physical model) is to show that if many present day *political* constraints were removed it is possible for the Third World countries to achieve a satisfactory living standard within the physical limits of the world.

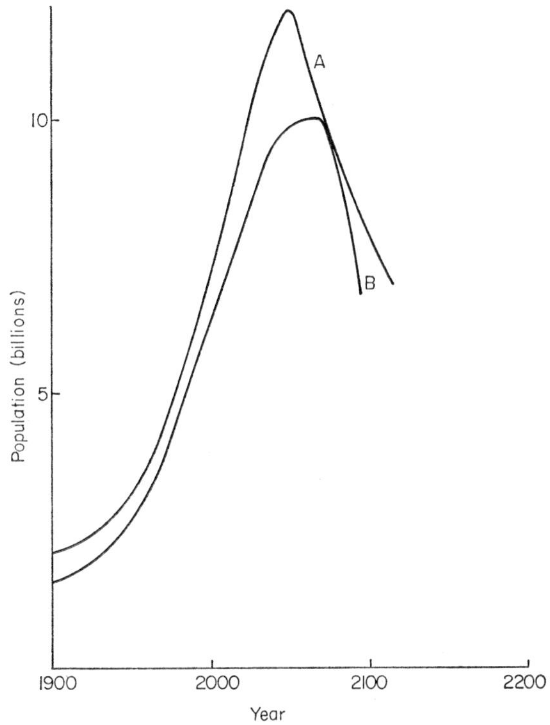

Figure 4. World 3 with changed resource costs (A), and Meadows World 3 (B).

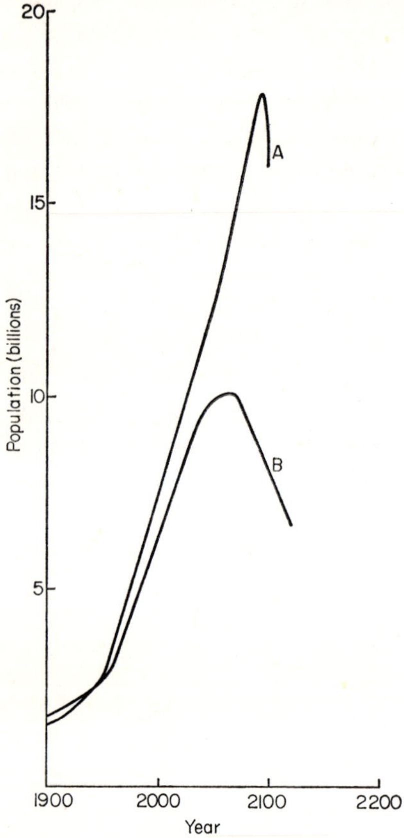

Figure 5. World 3 with changed resource costs and industrial pollution assumptions (A), and Meadows World 3 (B).

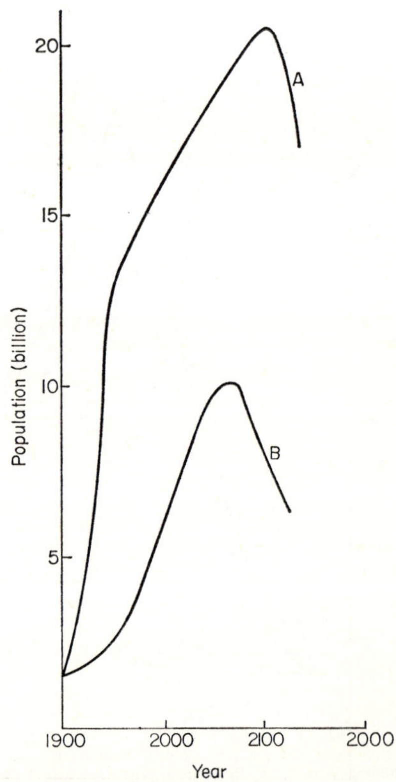

Figure 6. World 3 with changed resource costs and pollution assumptions (A), and Meadows World 3 (B).

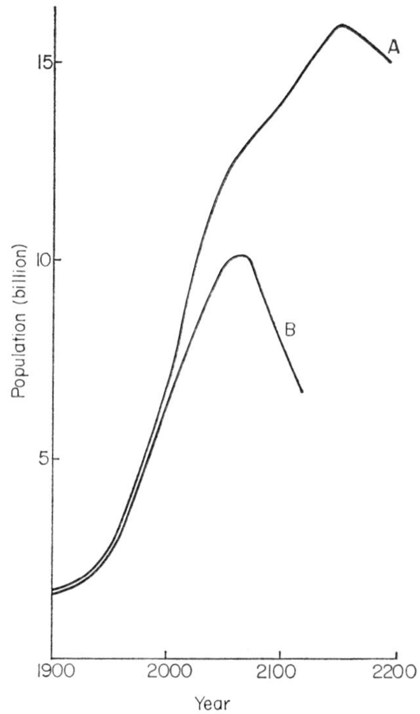

Figure 7. World 3 with changed resource, pollution and agricultural assumptions (A), and Meadows World 3 (B)

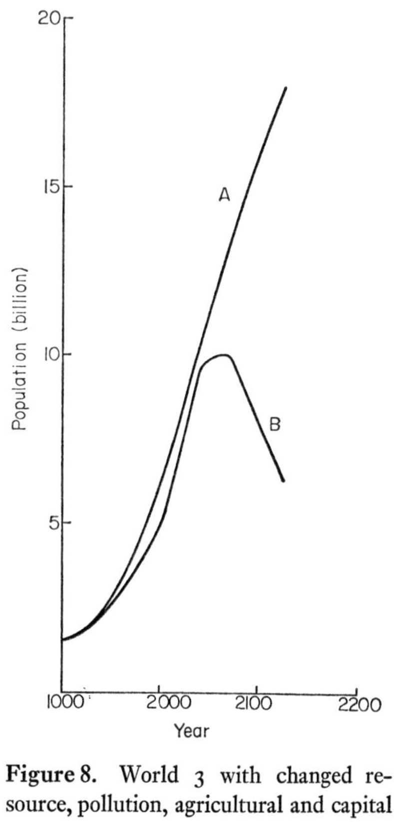

Figure 8. World 3 with changed resource, pollution, agricultural and capital distribution assumptions (A), and Meadows World 3 (B).

The details of the pollution sub-system in worlds 2 and 3

Since a full description of these sub-systems is given in Cole *et al.* (1973) in *The Pollution Subsystem*, and the arguments for their validation and verification examined there, it is sufficient in this paper to draw attention to the main critical assumptions (given below).

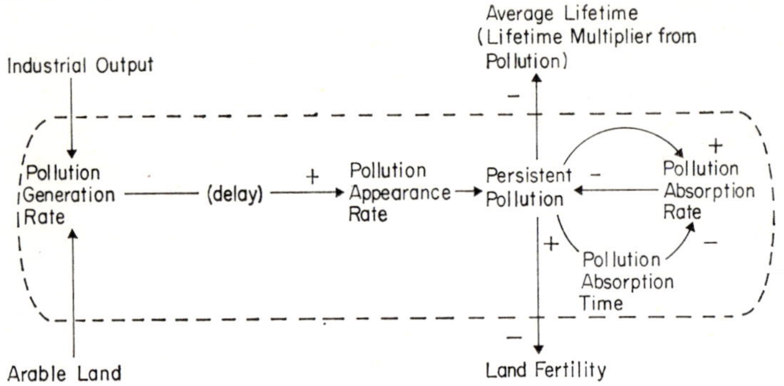

1. No statistical data quoted is based on a run of greater than 10–20 years.
2. No evidence is given for the assumption that absorption times are increasing.
3. The supporting Technical Report is inconsistent in that most of the 'pollutants' it quotes do not meet its own definition of persistence equivalent to a half-life of more than one year and an appearance time likewise.
4. There is no disaggregation between types of pollutants and their locality of appearance.
5. The interaction between pollution and other important sub-systems, such as land fertility and expectation of human life is based upon a use of data which could exaggerate these effects by orders of magnitude.
6. As the model is currently constructed, addition of technical change intended to counter pollution would have the reverse effect!

It will be seen that these models are simplistic to a high degree, and that much further research work is required, probably at a highly disaggregated level, before the conclusions or indeed any modelling representation can be regarded as acceptable in any scientific sense. Nevertheless, it also seems clear that it is possible that some of the current pollutants (which potentially include all discharge into the environment as a result of human activity) may be sufficiently critical that urgent action may not wait upon a full understanding.

Full development

It is pertinent to ask whether systems analysis can contribute to both the tactical and strategic problems of dealing with the environmental problems. In the chapter on Environmentalism in Cole *et al.* (1973), Sinclair has put forward a simple 4-point scale of various levels of risk to human life. These are:

1. in which the amenities and aesthetic qualities of life are violated;
2. in which there is injury or death to individuals from environmental contamination;
3. in which whole species are threatened with extinction from disturbances of ecological inter-relationships; and
4. in which fundamental cycles in the biologic pyramid and its natural environment are distorted or destroyed to such a degree that life for whole series of living forms becomes impossible over wide areas and possibly over the globe as a whole.

Developing this concept further, a possible structure of approach to environmental problems which would help distinguish between tactical and strategic issues could be as outlined below.

(a) Even if all discharge from human activity into the bio-sphere stopped today, any single pollutant vector i.e. point position in the 4-dimensions of risk outlined above would still evolve, due to the dynamic characteristics of its passage through the eco-systems.

(b) The dynamic characteristics of the changing pattern of pollutant discharge must also be researched.

(c) Even without pollutants, the bio-sphere is evolving and is in no sense a stable set of stable equilibria.

(d) Issues a, b, and c interact dynamically.

(e) The scales of human risk which are used in the 4-dimension risk structure evolve too, reflecting the varying social distributions of risk acceptable in different societies and cultures with different economic and social needs.

It is clear that the common and cooperative efforts of many types of scientists will be needed to approach these problems. In this view, ecologists have an important role to play.

Acknowledgment

Figures 2–8 are redrawn from Cole *et al.* (1973).

Summary

The pollution mechanisms included in the so-called World Models of Forrester (1971) and Meadows *et al.* (1972) are reviewed and found wanting in some critical aspects. Whilst not underestimating the potential gravity of many pollution issues, a plea is made for a more balanced approach to world problems in general and to the role of modelling in particular.

A possible taxonomy of pollutant effects in relation to risk to human life has been put forward and this concept is further developed to indicate how the complex problems in the current environmental 'crisis' can be approached.

References

Ackoff R.L., Gupta S.K. & Minas J.S. (1962) *Scientific Method—Optimizing Applied Research Decisions.* Wiley, New York.

Cole H.S.D., Freeman C., Jahoda Marie & Pavitt K.L.R. (1973) *Thinking About the Future.* Sussex University Press, London.

Forrester J.W. (1971) *World Dynamics.* Wright-Allen, Cambridge, Mass.

Gabor H. (1972) The new responsibilities of science. *Sci. Policy* 1, 1–8.

Hesse M. (1965) The role of models in scientific theory. In *Philosophical Problems o, Natural Science* (Ed. by D. Shapere). Collier-Macmillan, London.

Meadows D.H., Meadows D.L., Randers J. & Behrens III W.W. (1972) *The Limits to Growth.* Earth Island, London.

Reclamation of derelict land at Stoke-on-Trent

C. R. V. TANDY *Land Use Consultants, London, U.K.*

Introduction

The 'official' estimate of the amount of derelict land in the United Kingdom is in the order of 90,000 acres (36,450 ha), but the definition on which this estimate is based is a narrow one, and the true total is believed to be nearer 300,000 acres (121,500 ha). What is more disturbing is that, in spite of reclamation work being done by County and Local Authorities, the total of dereliction is still increasing by about 3,000 acres per year.

Since the Aberfan disaster, coal mining has been thought of as the culprit in this, but coal mining is actually responsible for only about 30% of the total. Of course, its 2,000 tips each containing millions of tons of waste material are the most *visible* part of the problem and more noticeable than quarries, holes, industrial wasteland and other waste tips which make up the remainder of the dereliction.

Governmental policy—supported by both parties—has been to pay grants of 50%, 75% in the 'intermediate areas', and 80% or more in 'Development Areas', to local authorities for the reclamation of land officially designated as derelict. The National Coal Board has also offered assistance to local authorities with a dereliction problem caused by past coal mining.

The problem at Stoke-on-Trent

In this programme, Stoke-on-Trent has a unique place. The City, which started with the handicap of possessing a larger area of derelict land within its boundary than any other British Borough, now has a record of one of the biggest continuous reclamation programmes in the country. The programme has benefited by derelict land grants from the Department of the

Environment, but the factor which made it more than a series of disjointed projects was the assistance which the National Coal Board gave to the city to prepare a feasibility study for incorporating the reclamation projects into a total open space system for the city.

The participation came about initially through the foresight of the City Council, which set up a Reconstruction Committee, comprised of officers from the City Architect's and Planning, City Engineer's, and Parks Departments, supported by the Treasurer, District Valuer and, under the chairmanship of the Deputy Town Clerk, invited representatives of the (then) Ministry of Housing Regional Office, the National Coal Board Area HQ and Regional Opencast Executive to join it. This Committee could commence work only after derelict land was acquired. It took some while to get a 'rolling programme' established, and work commenced with one small site of about 5 acres.

The National Coal Board Opencast Executive had, at that time, recently engaged the services of Land Use Consultants as their advisers on planning, land use, and the design of restoration and reclamation projects, and offered the services of these Consultants to the City of Stoke as part of their aid programme. Later, LUC became Consultants to the NCB and the city, jointly.

After briefing by officers of the Planning Department, LUC made an appraisal of the derelict land areas, their potential for development, and the open space needs of the City, and made a preliminary presentation to the City Council which successfully caught the imagination of the Councillors, and resulted in the immense enthusiasm for the work which followed. The programme, which started with a 15-year time scale, tentatively accepted by the Councillors, was later condensed to a little over 5 years, pursued with great vigour, and repeatedly extended.

To gain this enthusiasm, it was necessary to find one or more 'themes' large enough to knit together the diverse derelict sites into a recognizable pattern which could stir public imagination. LUC produced two concepts big enough to fill this bill. The first was the idea of a central forest park, large enough to act as a focal centre. Of course, there had been forest parks associated with towns before, but never one which was in the centre of an urban area. The other idea was the use of derelict railway lines and mineral lines, of which there were many, to form pedestrian greenways which would link reclaimed sites to the existing parks and to the urban centres.

A phrase which I coined at this time was later quoted by Lord Robens and well circulated: 'Stoke was originally six towns linked by industry; it is now six towns separated by dereliction.' The two complementary concepts of forest park and pedestrian greenways were to be the means to link again the six urban centres, with a pattern large enough to overcome the separation that dereliction had caused.

To catch the public imagination, it was fortunately possible to commence work on the forest park on a small piece of land, originally owned by the Duchy of Lancaster, adjacent to the new shopping centre in Hanley. When this small site had been reclaimed and laid out as public open space, with a sitting area at the entrance, plans, notices and sketches were exhibited telling the public of the future proposals. The City Planning Department also put on an exhibition in an empty shop window in Hanley centre which explained the programme, and invited public participation.

The designs prepared for the reclamation projects are simple, natural layouts, mainly of grass with mass tree planting, paths and seats. There are open areas for play, but no sophisticated types of recreation or entertainment. Detailing is robustly appropriate to the use of the area without being either sophisticatedly urban, nor sentimentally rustic. It is important that these new reclaimed open spaces should neither compete with, nor copy, the municipal parks. For this reason, there are none of the familiar attractions of urban parks. The only flowers are 'wild' flowers in meadows; trees and shrubs are common species appropriate to the region; 'fairways' are mown to make walking easy, but most of the areas are of longer grass, and colonizing species of common herbs.

The dereliction in Stoke was the result of decaying industries of several kinds, but predominantly coal mining with shale tips, and fireclay, ball and brick clay extraction. There are abandoned pottery kilns and brickworks, petro-chemical industries, steel and iron works, derelict railways, and the resulting wastes of iron slag, broken crockery and moulds, subsidence and flooding.

Reclamation has been done by contracts run by LUC, by the Planning Office, the Engineer and the Parks Departments of Stoke, and all projects were aided by Government grants from the Department of the Environment. As stated earlier, the work originally planned for a 15-year programme was so speeded up that the original schemes are virtually complete after 5 years. There have been many new sites, and there are 38 projects in progress by the City at present. The work has been spurred on by having such enthusiastic work behind it from the Reconstruction Committee, which is still in being.

Methods used

Clearly, each site is different, and each is a 'one-off' job, so one cannot generalize on methods used, but there have not been exceptionally difficult or unusual problems in the reclamation itself. The key to the work is having a large programme and a lot of imagination. Materials have been selected and disposed of appropriately: Weathered shale kept for top layers as it has soil-forming properties; burnt shale set aside for roads and paths; soil conserved where any exists; domestic and industrial rubbish mixed with rock waste to ameliorate it; soft wet tailings and clinker lumps buried, and wet areas used for water features.

Table 1 gives results of analyses made by the Biology Department, University of Keele, on the tip in the Central Forest Park as an example of the prevailing conditions.

Table 1. Mean values of spoil from the Central Forest Park before and after treatment.

	Shale tip		Tip after sludge treatment	
	N. facing	S. facing	N. facing	S. facing
pH	7·4	4·9	6·5	6·5
Available potassium (meq/100 g)	0·18	0·12	0·36	0·19
Available calcium (meq/100 g)	2·3	1·5	4·3	5·2

The contrast between the north and south facing slopes is worthy of note, but it is not solely a matter of aspect, for the south facing slope has retained most of its original vegetation, and was strongly acid through leaching, while the north facing slope had been regraded. The effect of sludge treatment was surprisingly small.

Very little spoil-moving outside the boundaries of a site has been necessary; usually, it has been re-shaped by 'cut and fill'. The shale, fortunately, is not highly acid, rainfall is above normal so that the grass can be established directly on to the shale, or with an inch or so of soil only. Sewage sludge has been liberally used (it is free) but it can be *too* rich, and causes a maintenance problem, particularly with nettle seeds which appear to be water-borne. Areas for seeding are cultivated to as fine a tilth as possible, given a pre-seed fertilizer, followed by a winter and spring feed of slow-release inorganic fertilizer of standard commercial mix (8:12:8).

Planting and seeding methods

We have been determined at Stoke to introduce natural climax vegetation patterns, and *not* create more municipal parks of cultivated species. Almost every site has a proportion of forestry—planted by commercial forestry methods with 2+2 transplants. At first we started with pollution-tolerant and pioneer species (Red Oak, Alder, Sycamore, Wych Elm) and avoided all conifers. We have found, however, that pollution levels are dropping, and in later stages, wherever conditions are good enough, we use English Oak, Ash, Birch, Willow, Corsican Pine, and hope to get an Oak-Birch climax which is characteristic of this part of Staffordshire.

Plate 1. Stoke-on-Trent: Westport Lake — reclaimed from a coal mining subsidence flash and now used as a swimming and boating lake with a very popular sandy beach.

Plate 2. Stoke-on-Trent: Westport Lake—the same beach in popular use during summer.

Plate 3. Stoke-on-Trent: Westport Lake—the Western fringe of the lake is set aside as a wet-land local nature reserve.

Plate 4. Stoke-on-Trent: Hanley Forest Park—a view from one of the tips, showing a 'pitch and put' course and illustrating the contrast between rough meadow grassland and mown fairways.

Plate 5. Stoke-on-Trent: Hanley Forest Park—showing tips, paths and cycle ways made from red shale, grassing and forest planting with some standard trees included for immediate effect.

Plate 6. Stoke-on-Trent: one of the pedestrian greenways created from derelict railway land. The locomotive fire box commemorates the assistance given by British Rail in this project.

Of course, this is basic. In addition, we use a number of 'edge' species to give variety, including Holly, White Poplar, Bird Cherry, Field Maple, *Robinia*, and we do use a number of large standard trees for effect at key points. As under-storey, and in shrub plantations, we choose from a wide range of common shrubs, according to soil and climatic conditions. In doing this, we are hoping to establish the 'intermediate' plant communities which are disappearing elsewhere, because parks horticulture is largely mown grass and trees; the countryside is forest or agricultural grasses and crops. Consequently, we are planting Laurel, Buckthorn, Snowberry, Thorns, Bramble, Whitebeam, Rowan, Wild Service (*Sorbus torminalis*), Guelder, Scotch and Ramanas Roses, Dogwood, Elder and Spindle.

In grass areas, we have to use some commercial mixtures for playing pitches and fairways, but elsewhere we introduce seeds of wild species whenever and wherever we can get them. We have members of local naturalist trusts, university and other students, and our own staff, collecting seeds. We buy from seed merchants who stock them, and we get screening-rejects from grass seed merchants if we can. Some of these are planted in trial plots, some used generally (Clover and Burnett); some wild plants colonize themselves, e.g. Rose-bay Willow-herb; some we have to remove because they are 'notifiable weeds'; but some of our staff have a sneaking liking for Docks which seem to be a good nurse for other plants, and detract vandals from young trees. Unfortunately, they will take over a complete area unless severely controlled by spot-weed-killing.

Common Midland plants used in the mixtures, include:

Yorkshire Fog	Trefoils	Chicory
Yarrow	Red Clover	Mustard
Vetch	White Clover	Medick

Ryegrass and Cocksfoot are avoided as far as possible.

The following plants would be welcomed, but it has not proved possible to obtain seed in quantity:

Poppy	Speedwell	Cow Parsley
Fumitory	Toadflax	Primrose
Senna	Marigold	Cowslip
Fennel	Cornflower	Mullein
Spurge	Hogweed	Thyme
Scabious	Lupin	

Some pools at Westport Lake have been cleaned out and planted as a nature reserve. The vegetation has been monitored by Keele University. Species of *Juncus*, *Epilobium*, *Lycopus*, *Myosotis*, *Typha* and others have become established; *Phragmites communis* has failed (to our surprise) but

8 species have been recorded for the first time. One maintenance worry is that *Glyceria maxima* is attempting to take over from every other species.

There is, of course, a danger that if one overdoes the introduction of wild plants and weeds, and fails to maintain intelligently, the site may begin to look as if it had not been reclaimed at all. In Stoke, we have not been aiming for a wholly 'wild' environment—this would be out of place in an urban situation—but a continuation of the typical Midlands countryside.

Of course, maintenance is the key to success. Initially, the long grass was cut only twice a year and left lying, but bad weather turned it into a soggy mat, and raking cost more than extra mowing. It was then mowed 4 times a year. We mow regularly (9 to 14 times a year) all 'fairways', site boundaries adjoining other properties, and the edges of 'wild' areas to give an appearance of being cared-for.

To gain public respect, the open spaces must *appear* to be well-managed, and we have found that the contrast between the closely mown strips and the rough grass areas helps to show this control, and to identify the purpose of the various spaces.

There was prejudice to be overcome from those who could not envisage any form of public open space that was not as well maintained and manicured as are the municipal parks in Stoke-on-Trent. Our work was accused of being unkept, untidy, and badly maintained. At one time there was a risk that all our wild plants would be lost to gang-mowers, but we are beginning to overcome this prejudice—perhaps because there are now skylarks back in the centre of the town, three kestrels live on Sneyd tip, and the Central Forest Park is the local habitat for cowslips!

Appendix

Seed mixes used in reclamation at Stoke-on-Trent

A. RECREATION AREAS	
Lolium perenne N.Z.	34%
Phleum pratense S.50	17%
Festuca pratensis S.53	29%
Festuca rubra S.59	13%
Trifolium repens S.100	3%
Trifolium hybridum	4%

B. FORESTRY AND PARKLAND AREAS	
Agrostis tenuis	10%
Festuca rubra S.59	33%
Festuca tenuifolia	33%
Trifolium pratense S.123	24%

C. MOORLAND AREAS	
As B (above)	50%
Mixed screenings (including largely Yorkshire Fog, but excluding Ryegrass & Cocksfoot)	50%

D. SPECIAL SEEDED AREAS	
Yarrow	33⅓%
Common Vetch	33⅓%
Trefoil Clover	33⅓%

E. SPECIAL SEEDED AREAS

Burnett*	40%	
Sainfoin (common milled)	40%	
Broad-leaved Red Clover	20%	N

* Where Burnett is unobtainable, Black Medick and the Trefoils may be substituted.

F. SPECIAL SEEDED AREAS

Chicory	75%
Wild White Clover	25%

Reshaping and draining derelict landscapes

M. F. DOWNING and B. HACKETT
Landscape Architecture Staff, University of Newcastle, Newcastle upon Tyne, U.K.

The basic approach to the design of landscape modification which the authors have always followed has been that of ensuring that after modification the landscape can function in a physiological sense. To give an example of this, the grading of the landform would be devised in such a manner that the level of resistance to erosion is always greater than the forces developed by the run-off of surface water, or by the wind. This state of stability can be brought about in several ways, acting separately or, as is preferable, in combination. Derelict land reclamation in Great Britain is important because it brings into use land which, if left derelict, would reduce the country's stock of good agricultural land. This consideration is central to the programme of reclamation in this country. It is this shortage of land which explains the great lengths, and the great expense, to which reclamation has gone here. The use of land will of course have a great bearing on the sort of slopes that can be achieved. Allowing for this, and assuming that the use is one which does not demand critical gradients, the slopes can be designed to be sufficiently gentle as to avoid the washing away of surface material providing artificial means are available to cope with the run-off from less gentle slopes.

The use of special techniques can lead to the speedy development of vegetation cover as a counter-erosive measure, and the installation of clay or concrete pipes and channels provides additional control of the movement of water.

Those familiar with the whole programme of restoring derelict land in Britain, will know that the money available does not cover the costs of what has come to be regarded as the convention in landscape work. This would probably include adequate topsoil provision, and perhaps turfing rather than seeding to achieve a quick vegetation cover. Nor could this be expected on the scale of operations of much reclamation work, though for some end uses it may seem very desirable. Even with the development of cheap methods the costs of reclamation are very high, and far in excess of the agricultural value of the land. Thus it is necessary as a designer to attempt to emulate the action of natural forces over a period of thousands of years. This means looking at a derelict site as though it were the landscape left after the retreat of the ice after the last Ice Age. At that time there would have been large areas of material with a variety of topographical features, and a natural drainage pattern based on gravity would not have been developed. As the centuries passed, gradually a stable pattern of streams and rivers would develop to drain the area, and this stability would be aided by the development of a vegetation cover. We believe that if this period of development can be telescoped into a short period of years the results of reclaiming derelict landscape will be successful, and comparatively easily achieved.

If we adopt this approach and the objective is a healthy landscape in an ecological sense, it is obviously necessary to have an understanding of ecological principles. It is highly improbable that any one could produce a scheme whose every detail will function precisely according to the design forecast. Thus the dual principles of flexibility and evolution, rather than finality and immovability, are proper to fulfilling the objective.

Reverting to the emergence of the natural drainage pattern after the Ice Age, it would appear that some features of the pattern found their place purely as a matter of chance. For example, the direction of an emergent drainage channel could have been determined by the position of a large boulder in the boulder clay, and from the particular direction taken, the landform would have been modified as a result of this initial chance event. This fact can be used as an argument for a degree of flexibility in designing landform and drainage patterns for reclaiming derelict landscapes.

A question that is bound to arise is the degree to which it is desirable to try to revert to the landform and drainage pattern that existed before dereliction occurred. Sometimes there is pressure to create a new landform which will suit a particular use, such as a sportsground or some form of agriculture. In the former example a series of almost flat tables would be appropriate, while in the latter case a gentle south facing slope might be arranged where previously the site had a northerly aspect. Clearly, if the design is to be judged on the basis of ecological principles it could be criticized on the grounds that nature will eventually produce a reversion to the original pattern unless specific measures are taken by human endeavour to counter this.

We have found some divergence of opinion among ecologists over the question whether certain man-made changes will be permanent changes to the basic landscape, and further discussion of the point of how far a dramatic change can be compatible with ecological balance and stability might be interesting.

In the experience of the authors, in reclaiming derelict landscapes there seem to be convincing arguments both for creating or preserving artificial landforms, and conversely for designing the new landform to be so in sympathy with the surrounding land that the newcomer visiting the site cannot find it. It would seem that by taking note of the original and the surrounding topography, and bearing in mind the point already made about flexibility, it is possible to produce a design which has a good chance of success from an ecological point of view.

A practical consideration must be that many of the angles of slope will need to be gentler than those of the surrounding topography because the surface material is likely to be very easily eroded in the initial stages.

A topic about which we need to know more for the design of new landforms is the detailed variations in microclimate brought about by differences in the configuration of landform. This is important not only because of the beneficial environments that could be created for human use but also because of the way in which these variations could affect the development of the vegetation cover. For almost a year, we have had in Newcastle a limited project involving a physicist with specialist expertise in microclimatology, as well as landscape design staff, looking at this. The information we hope to be able to provide from our pilot project will almost certainly be very scanty, but we hope that its significance will be such that it will lead to further work in this field.

There are many difficulties in designing landforms that will be successful in an ecological sense. We are all by now familiar with the fact that north facing slopes, which are usually regarded as more lowly rated in the habitat sense than south facing slopes, prove to be the best habitats for the pioneer vegetation when material like colliery spoil is very freely drained and lacks organic content. Even in the climate of the north-east of England, extremely high temperature and lack of moisture can cause problems for developing young plants. On the other hand, in the future such south facing slopes in the northern temperate zone will presumably develop, with the creation of a congenial soil, into a much more favourable habitat.

The cover of vegetation, of course, often forms the link between the landform and the surface drainage. With the comparatively unstable material of colliery waste, one could imagine that if no vegetation ever developed the site would very quickly erode to form a plane. But with the advent of vegetation the erosion process is stopped at a stage dependent on a combination of slope, vegetation and the rates of run-off. As previously stated a design approach which attempts to speed up the natural processes would appear sensible. Obviously, it is necessary to work back from the possible drainage outlets to ensure drainage by gravity, and also to take into account any run-off from adjoining land or from streams.

Some mention has already been made of the use of artificial aids to the drainage process (required to retain the shape of the land to suit some particular use) against erosive forces which would, before maximum stability is achieved with the development of the vegetation cover, wear it down to a more even profile. This has often resulted in the need to place 'cut off' drains along the tops of steep embankments, as well as at their bases, and to conduct the water down the slopes either by means of concrete flumes, or more successfully in sealed pipe systems of salt glazed ware. Preference should be shown for a system which is proof against sudden extreme conditions. Flumes for example have not always been reliable and when bypassed by flood water, not only does the water cause erosion of the bank but this can also lead to the disturbance of the foundations of the flume itself with consequent interference of its operation. All other things being equal the flatter areas which can be graded to gentle slopes between these embankments can be relied upon to develop a cover of grass more quickly.

Because of the fragile nature of reclaimed land in the period immediately after regrading and before the development of the vegetation cover a number of devices must be adopted to maintain the stability of sites. The use of contour ditches is one such device. These divert the flow of water from the surface before the volume becomes too great, being dug out parallel to each other at perhaps only a few yards apart, and varying in size from a single plough furrow turned down hill to a properly constructed ditch several feet deep. After they have served their purpose these ditches may either be filled in and grassed over, or can be piped or filled with rubble to form a conventional drain. Being excavated almost along the lines of the contours these drains have a minimum fall and conduct surface water at a reasonable rate to outfalls either in the natural stream system or an artificial system of pipes or ditches. They have the added advantage in many conditions of retaining water which would otherwise be quickly discharged into the main drains, and which consequently is enabled to percolate into the soil. Colliery shale materials are commonly extremely dry and but for this assistance would tend to remain so. Another device, though one which it must be admitted is less certain in performance, is the provision of check dams to form steps in the courses of main drains where these would otherwise be too steep. The advantage of these can be that like contour drains they hold water which is slowly allowed to percolate into the soil and improve its potential for plant growth. Some ditches may be planted deliberately with suitable species to grow in them and slow down the intermittent flow of flood water. The disadvantage of these check dams in the unstable material of colliery sites is that they can result in scouring below the dams, and even laterally, and in the deposition of silt immediately above them. These problems result in the need to reinforce the bed and sides of the stream or ditch and this may result in greater outlay than desired.

The foregoing presumes a large scale commitment to moving shale to form new landforms, the scale of which in this country may surprise foreign observers. The approach to the design of new landforms has been the subject of considerable debate in recent years, particularly concerning the creation or retention of artificial shapes in the landscape. Among designers there are those who assert that artificial forms should be adopted consciously to create monuments, while others suggest that this is not appropriate in any landscape which topographically is dominated by natural forms. While the debate continues many millions of tons of material have been transported. The criteria for regrading must be to try and obtain the optimum results for the minimum earth-moving outlay. Regrading can however be brought into play as an aid to the revegetation of a site. With knowledge of the materials of the site which can frequently be very variable, arrangements can be made to bury the intractable, using the most suitable as top dressing and conceivably excavating seams of suitable surface material previously hidden. This is where the value of a careful survey of the site conditions is clearly shown.

Unless the pressure is particularly heavy on the designer for achieving a tailor-made landscape for some particular function, there is a responsibility to produce a design that will least inhibit future freedom of use of the site. The design should also weld the site, both visually and functionally into the surrounding landscape. In the absence of a specific brief a design might be created which includes a number of relatively level areas of reasonable dimensions, though different one from another, and a range of slope gradients and orientations within acceptable limits, to allow for flexibility in future requirements. Obviously each reclamation site is a different and unique situation and while such utilitarian matters as surface water run-off, gradients, cut and fill balance in the grading, sizes of ditches and streams, and the junction with the adjoining boundary can all be arrived at by computation, or by the application of formulae, the success of the design will depend ultimately on the manipulation of the many variables by the designer.

Spoil heaps in the Ruhr District and their integration into the landscape as a measure to improve environmental conditions

GERHARD PETSCH *Siedlungsverband Ruhrkohlenbezirk, Essen, Federal Republic of Germany.*

Introduction

The natural landscape in the Ruhr district, and also partly in areas west of the River Rhine, is characterized by its lavish variety and by the existence of industrial spoil heaps. Most of these result from the hard-coal mining industry or they consist of industrial spoils and waste materials. Others are controlled refuse tips. The largest and most significant number of spoil heaps were produced by the mining industry. These *artificial mountains* consist of strata associated with the coal seams, of mineral substances washed out during the exploitation of the coal or of residual rock materials. The ground materials range from slate and sandstone to the conglomerates of these materials. The grain sizes are diverse and range from flat, round and granular shapes to portions similar to fine-grained sand particles.

Transportation and tipping are effected by means of plant-owned railroad systems or with an 'endless chain' of trucks. All the heaps are located in the vicinity of mining operations. The location of spoil heaps was determined by the shortest possible transport routes, by the situation of coal washing plants, of the coal dressing and sorting area or by the extension of the site owned by the coal mine concerned.

Within the framework of rationalization of the mining industry's operations, spoil is now deposited on centrally located heaps. These have acquired new dimensions and thus new concepts for the integration of these heaps into the landscape are required. The Land Planning Community SVR (Ruhr Regional Planning Authority) has endeavoured ever since 1932 to co-ordinate and harmonize these operations and, in the course of regional planning, have determined colliery spoil heap locations. Success has been achieved to varying extents and requires mutual understanding by parties with different interests. In order to gain an idea of the area covered by spoil heaps it is interesting to point out that with an assumed width of 100 metres, the total spoil site would extend for 130 kilometres!

Structural changes in the coal mining industry, with resulting pit closures, and the current increase in road construction operations, have produced antagonisms in the siting and construction of heaps. On the one hand, the policy of closing down uneconomic collieries leave a few large-scale coal mines with accordingly large and rapidly growing tipping areas. These will have a massive impact on the naturally evolved landscape. On the other hand, substantial quantities of spoil material will be removed from partly reclaimed heaps on which vegetation was established in a way that constitutes an ideal and exemplary approach to integration of spoil heaps in the surrounding natural landscape, and this for the sole purpose of using that material for road construction projects. Old heap locations are restored to their original surface character and these surfaces are often treated and used for the extension of existing industrial estates.

In the near future, the rate of deposition will decline in volume but this will be a short-term development. It is thought that quantities deposited in concentrated heap areas will subsequently increase rapidly again. In the Ruhr district the planning of tip locations not only takes into account individual mining operations but also the development requirements of areas within the region.

The Ruhr development programme

The Ruhr Development Programme was made effective by the Government of the Federal Land of Northrhine-Westphalia for the period 1968–73. It includes Item 8, Section 1: *Vegetational Establishment on Spoil Heaps, Regional Green Areas, Conservation of Woodland.* The Government of the Federal Land regulates the establishment and landscaping of tips by regulations on spoil heaps in the Section on The Supervision of Mining Activities (4 September 1967). The significant fact is that this regulation predominantly refers to those tips that are subject to the German Law on Mining Operations with respect to underground mining activities. With a hard-coal output in the Ruhr district of some 91 million tons per year, an additional 50% (roughly 45 million tons) of stone and rock material are extracted.

Some 80% of this material is subsequently employed for backfill and packing operations, for road construction or similar projects. The remaining 20% must be deposited on spoil heaps. The actual quantity of spoil materials deposited is rated at 10 million tons per year. In addition there are 2 million tons per year of residual materials from the steel producing industry and 2·5 million tons per year of ash material from the power generating industry.

Table 1.

Location	No.	Tip existing	Area remaining and being added to ha	Existing capacity ha	Remaining capacity (and being added to) (Mio m³)	Present ha	Heap removal (Mio m³)	Heap removal ha	Planned (Mio m³)	revegetated surface ha
Gelsenkirchen	10	139·50	101·30	26·45	40·10	45·00	5·19	6·00	0·88	22·72
Gladbeck	8	64·45	86·80	8·21	20·29	3·90	0·50	2·25	0·18	13·50
Bochum	14	43·73	37·57	9·16	8·74	11·02	0·62	—	—	6·79
Duisburg	1	9·82	—	1·30	—	9·82	1·30	—	—	—
Ennepe–Ruhr–Kreis	2	4·40	1·00	0·38	0·09	3·40	0·29	—	—	—
Wanne–Eickel	5	15·60	15·60	1·41	1·41	—	—	—	—	3·50
Bottrop	9	88·99	72·40	15·88	26·84	14·34	2·79	—	—	16·60
Dinslaken	4	109·00	154·00	18·50	36·70	—	—	—	—	18·20
Unna	12	119·95	150·54	14·38	23·42	11·00	2·00	—	—	11·25
Wattenscheid	3	14·70	5·60	1·32	—	9·10	0·62	—	—	—
Recklinghausen (Stadt)	5	20·10	19·10	3·72	3·69	—	—	1·00	0·03	18·00
Recklinghausen (Kreis)	22	237·53	281·83	32·67	79·31	5·50	0·27	1·10	9·07	63·71
Herne	4	19·04	17·18	2·86	4·04	—	—	6·46	0·93	5·72
Lünen	7	71·61	64·47	5·36	9·60	—	—	12·50	1·30	8·00
Oberhausen	4	29·32	15·40	6·70	2·70	13·92	4·00	—	—	9·65
Dortmund	17	138·25	190·40	15·22	20·12	22·20	1·51	18·40	1·95	23·60
Essen	25	109·27	72·97	9·09	10·12	38·30	1·31	13·92	0·63	27·29
Moers	6	39·10	160·80	23·80	61·15	—	—	—	—	47·78
Mülheim	1	5·00	—	·10	—	5·00	0·10	—	—	—
Castrop Rauxel	3	20·04	20·04	2·14	7·90	—	—	—	—	2·40
Total	162	1,296·40	1,467·00	198·65	356·22	192·50	20·50	61·63	5·97	298·71

In addition to technical considerations, care is also taken to ensure that a vegetational establishment plan will be worked out for every single spoil heap. This plan has the character of a landscaping plan and becomes part of the operation plan. It must be approved by the mining supervision authority. The landscaping plan will include the technique of, and the period of time during which, afforestation is to be performed to integrate the spoil heap into the surrounding landscape. The plan will also specify the volumes authorized for reintegration. To this end it is imperative to design the plan in close co-operation with the owner of the extracting operations and to examine questions of financial feasibility. The landscaping plans can be established by independent landscaping architects or by specialized institutions by way of a consulting procedure. The landscaping plans must be designed and arranged in such a way that *partial stages* of the plan can be realized during the course of the overall project. For certain recultivation measures supporting funds are made available by the Federal Land or, in the Ruhr district, by the SVR (Ruhr Regional Planning Authority). However, these financial aids are only granted if their investment will contribute towards reaching an overall integration of heaps and contribute to visual improvement in surrounding areas. It is a guiding principle that causal responsibility will be coupled to the duty of reclamation.

Table 1 gives details of the various tips in the Ruhr district and shows the stage of vegetational establishment reached to date (1973).

Current tipping and reclamation

When studying and evaluating these data it must be pointed out that in case of coal mines being closed down, although tipping operations will cease the future shaping and design of these tips will only be continued when it has been established that supplementation deposits of refuse materials will take place. This means that simultaneously additional volumes will be created for the deposit of such materials. This policy helps to meet the ever increasing demand for new deposit space for industrial and domestic waste materials. This is particularly important in the Ruhr district. An essential consideration in this connection is the attitude adopted by the mining supervision authorities whose decision will be based on the prerequisite that approval has been obtained for *joint* and *combined* desposits of colliery spoil materials, industrial waste and domestic refuse materials. Joint deposits of these three types of waste materials will not be possible in all cases (sometimes for reasons of water protection and water conservation). However, this possibility is always investigated to gain extensions of deposit areas.

Future tipping and reclamation

With regard to new heaps, the principle will be that these surfaces must be integrated into the environment in such a way that if a sudden discontinuation of projected tipping operations occurs, complete sectional areas will remain available for vegetational establishment and integration into the original natural landscape. Despite all biological attempts at reintegration, tips are 'technical configurations' that require a long-term process of reintegration into the natural scenery. Surface shaping and design must be developed to ensure that mechanical stability of the tips will be given top priority in order to prevent catastrophes caused by uncontrolled movements of the materials deposited on heaps.

Flat top heaps have successfully been established in the Ruhr district. These specially shaped heaps have frequently been terraced by means of transverse paths in such a way that the heap will be stable and also offer optimum conditions for drainage. However, it is still debatable whether terraced shaping as now proposed in operational planning procedures is actually the most favourable shape of tip, particularly for plant establishment in the course of revegetation.

In practice it has become evident that tips with a naturally established tipping angle, without transverse paths interrupting the slope, give high initial growth rates. In future this aspect will require more intense and accurate investigation to determine exact interrelationships between tipping shape, ground material and initial plant growth. A final decision must not only take into account biological considerations but also that, in changing the previously characteristic tip shape, no additional land will be needed as this is a scarce commodity in the Ruhr region. It is certain, however, that in future the shaping of tips must be geared to a greater extent to original and natural features of an area than has currently been accomplished by the 'cuboid' shape of tips exclusively designed to meet the technical requirements of the tipping process.

Tip design and plant growth

Many heaps in the Ruhr district offer real opportunities for comparing rates of plant growth. However, every tip shows different location features and thus has different microclimatic conditions. Thus nearly all call for separate and specific evaluation. In the age of generalization, particularly in the field of landscape conservancy and afforestation procedures, it is likely to be dangerous to rely too heavily on general concepts of vegetational establishment, particularly concepts based on quick successes being achieved rather than those based on an understanding of long-term biological processes. The 'computerized afforestation pattern' is a concept of limited value. Whether preference must be given to the 'pre-forest' stage or to the establishment of a 'real forest structure' that will develop within the framework of biological conversions, must be decided strictly on environmental considerations. It is actually feasible to combine both aspects on one tip. Many biological disciplines, if applied to vegetational establishment, are capable of yielding new findings which may be interpreted both from a plant sociology standpoint and from a practical point of view. It is important that all those involved in the task of reclamation have biological understanding and apply it. Biotechnical measures, as they are predominantly offered today in the form of applications of plastic materials coupled simultaneously or subsequently with grass seeding, must be regarded as limited environmental protection measures only. Such applications have however the advantage of quickly producing vegetation on tips and preventing erosion of dust as a result of wind. Attempts to create vegetation in this way are still being carried out but these tests need to be evaluated separately and over long periods.

Conclusions

However, in my opinion, three fundamental principles must be recognized:
(i) Recultivation is a measure that must be given the same priority and support in the social sphere as the maintenance and continuation of profit-oriented, capacity production operations for commercial products; financial restriction on the re-establishment of a balanced biological structure is, and continues to be, an act of social hostility committed in the name of a false thriftiness. This principle applies equally to a planning authority forced to think in terms of economy, as well as to legislative bodies bearing the responsibility for the common good of the general community.
(ii) Substantial investments must be made in order to integrate spoil heaps into the natural scenery. From a long-term point of view these investments can only be justified if regulations are established to ensure that spoil heaps on which vegetation has been established (i.e. spoil heaps that serve a genuine function in the climatic and biological field of the urban landscape) are legally protected and will not be jeopardized by operational ultra-short-term decisions.
(iii) In the planning stage procedures of vegetational establishment must be chosen that are secure, reliable and quick acting and can stand comparison in terms of the expenditure and labour required and success achieved. Such a comparison must not exclusively be calculated from the point of view of the producer but must also make allowance for its effect on the living area.

What may be a very expensive and even an almost impracticable requirement for the producer can become the lowest possible basis of securing the sound maintenance of a complete living environment.

On the one hand no demands and requirements must be raised that are impossible to meet, whereas on the other hand one cannot demand that all requirements and demands be based on scientifically exact knowledge and information since scientifically-based knowledge on vegetational establishment still lags behind the practical experience acquired. Whoever expects landscapes and vegetational establishment on spoil heaps to have the genuine character of idylls is establishing false standards that are eventually harmful to reclamation objectives.

All reclamation measures are a 'bundled instrument' for the design of an industrial landscape which keeps some of its natural character and provides a higher quality living environment. In the future spoil heaps will continue to be inevitable. This is why the systematic evaluation and the application of all previous results is an activity that is not only of immediate and short-term importance. This task will find ever increasing importance in the era of environment protection and control.

The hard-coal mining industry in the Ruhr district has voluntarily, and in close cooperation with the communities concerned, made achievements which today—in an era of changes in the mining industry—is not given the recognition that it deserves. What will have to be done in future is therefore to evaluate this experience because it offers the opportunity of studying financial and manpower expenditures in a unique way.

At a low point in our national economic development, capacities and performance reserves were made available for vegetational establishment on spoil heaps which set exemplary standards for other parts of Europe, even at a time when other tasks were held in higher esteem by the majority of people. If today we are reaping a part of this success in the form of green areas, for which the basis was sown in the 1950s, this shows quite clearly that the long-term character of certain measures and activities must be taken into account in future, particularly as the pace of industrial production will quicken and industry will be forced to prepare for day-to-day reorientation.

The partial recovery of the metal polluted River Rheidol

A. NEVILLE JONES and W. ROSCOE HOWELLS
South West Wales River Authority, Llanelli, Wales, U.K.

Introduction

The fissures associated with the faults of the Lower Palaeozoic rocks of mid Wales have been infilled with quartz and baryte and in many areas this has been accompanied by ore minerals such as galena (lead sulphide) and sphalerite (zinc sulphide). These ores have been reputedly mined since pre-Roman times but the greatest mining activity was during the period 1750 to 1900. The industry had declined by the end of the last century due to foreign competition but some mines reopened for a short time during the 1914–18 war. The last mines to close in the Rheidol catchment were the Melindwr Mine and Rhiwfron Mine in 1922 (Carpenter 1924a) apart from some minor activity at Erw Tomau Mine which finished in 1927.

Lead mining in Wales has been described in detail by Lewis (1967) whereas the geology of the mining district is described by Jones (1922). The River Rheidol was in the area most affected by mining and about 43 mines were worked in the catchment. An outline map of the river with the locations of the mines is given in Fig. 1. Ore was separated from the crushed rock by water flotation and the water which contained dissolved metal salts and suspended solids were then discharged into streams. The degradation of the environment as a result of this mining activity can be summarized as outlined below.

1. Oxidation of the ore during extraction and weathering of spoil heaps produced lead sulphate and zinc sulphate which are soluble and toxic to many freshwater organisms (Carpenter 1924b).
2. The suspended solids of ore and rock produced by crushing and flotation were discharged into the river causing siltation which adversely affected the benthos. The spoil heaped on the banks of the streams was washed into the rivers resulting in an unstable river bed unsuitable for the establishment of vegetation (Rees 1937; Jones 1950).

3. There was an effect on the marine environment in the area effected by the freshwater discharge and zinc concentration, for example, of certain littoral organisms has been shown to be higher in this area (Ireland 1973).
4. Soils of the area were affected by deposition of ore particles by the river in the flood plain and by aerial pollution from smelting and windblown slime (sediment) in the vicinity of the mines. In such areas crops were adversely affected and there were occasional reports of poisoning of stock (Griffith 1918; Alloway & Davies 1971).

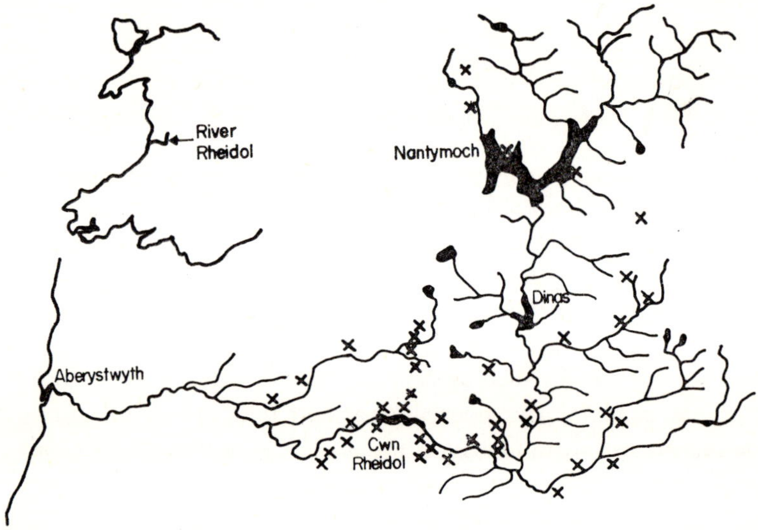

Figure 1. Outline map of River Rheidol showing location of mine workings (after Jones 1922 and Carpenter 1924).

5. When the mines were closed no attempt was made to landscape the area affected and they were left in a derelict condition. Because of the toxic nature and nutrient deficiency of the spoil heaps they did not become colonized with vegetation (Newton 1944).

Not all the legacy was on the debit side, however, and several lakes were created to supply water for the flotation process. These lakes are now used for angling and can be considered an asset to the environment.

The following account is primarily concerned with water pollution and its ecological effects. The recovery of the river since the decline of the mining industry is described to a stage where the fish population has virtually fully recovered yet the water quality is not yet regarded with confidence as entirely suitable for human consumption.

Water quality

Grimble (1904) refers to the River Rheidol and the neighbouring Ystwyth as 'two sewers used almost solely for the lawless and selfish benefit of industries which could well afford to take proper means to effectually prevent the poisoning of the waters'. The only treatment known at that time, however, was settlement which could not have affected the metal salts in solution. Settlement lagoons were installed at some mines but these 'slime beds' were probably intended more for the reclamation of ore than the prevention of pollution.

There appears to be no record of water quality for the period when mining was at its peak. The first analyses are quoted by Carpenter (1924b) for the period when the mining industry was in its final stage of decline. She records a lead level of 0·2 to 0·5 mg/l in the lower reaches and after 1922 there was an apparent improvement with the lead not normally determinable, and even in times of flood not exceeding 0·1 mg/l apart from one occasion in March 1924 when a lead level of 0·4 mg/l was recorded. This was the last record of so high a degree of pollution. James *et al.* (1932) using more 'critical' methods of analysis found lead levels of 0·02 to 0·17 mg/l and zinc levels of nil to 0·3 mg/l in the main river. No records of water quality are known from 1932 until 1955 when preliminary work started on a hydro-electric scheme. Since then analytical work has been undertaken on a large scale by the Central Electricity Generating Board and the South West Wales River Authority (formerly the River Board). A comparison of lead and zinc content for the lower reaches only is given in Table 1. This tends to show

Table 1. Historical comparison of lead and zinc estimates for the lower reaches of the River Rheidol.

		Range in mg/l	
		Pb	Zn
1919–21	(Carpenter 1924b)	0·2 –0·5	
1922–23	(Carpenter 1924b)	nil–0·1*	
1931–32	(James *et al.* 1932)	0·02–0·1	0·14–0·30
1971–72	(River Authority)	nil–0·04	0·20–0·83

* Plus one record of 0·4 mg/l in a flood in 1924.

that there was a pronounced improvement in lead content after the final closure of the mines in 1922 and that there has been an apparent further improvement since 1931–2, so that nowadays lead is not often detectable in the main river. Methods for zinc analysis were not developed at the time of the study made by Carpenter (1924b) and all that can be said about the zinc content is that there has apparently been no improvement from 1931–2 to the present day. Lead was the main metal extracted by the mining industry and whereas sphalerite is commonly found on the spoil heaps at the present time galena is only rarely found.

Direct comparisons in the quality of water in the catchment between one year and another are, however, of doubtful validity. Very low levels of these metals are being measured and the results of a sampling programme can be affected by a number of factors, e.g. the weather conditions, the location of the sampling station, the river flow, the form in which the trace metal may exist in the water, the nature of the sampling devices and containers and the stability of the samples. It is clear that a good deal more work requires to be done to standardize sampling techniques. Comparison of today's results with those published forty years ago should be made with great caution, particularly as analytical methods have changed. The current method is atomic absorption spectrophotometry.

The results of water analysis of the lower reaches of the River Rheidol are given in Table 2. As in all Cardiganshire rivers the water is slightly acid

Table 2. Water analysis for the lower reaches of the River Rheidol (at Penybont Bridge) based on seven samples taken in 1971–2.

	Mean	Range
pH	—	6·4–7·0
Conductivity (μmho/cm)	60	49–76
Biochemical Oxygen Demand	1·2	0·6–1·9
Dissolved Oxygen	11·5	10·3–13·6
Hardness as $CaCO_3$	24·8	13–36
Chlorides	6·3	5–9
Nitrates	0·47	0·38–0·64
Nitrites	—	nil–0·005
Ammoniacal Nitrogen	—	nil–0·17
Albuminoid Ammonia	—	nil–0·13
Iron (total)	0·95	0·46–1·90

(results expressed as mg/l except pH and Conductivity)

and very soft. This softness is a disadvantage as far as metal toxicity is concerned. The river shows no sign of organic pollution, the ammonia levels are normally low and the water is always well saturated with oxygen. The only town in the catchment is Aberystwyth which discharges its effluent to the sea near the estuary.

Levels of lead and zinc in the catchment are relatively low and results of sampling in 1971–2 are given in Table 3. The main sources of pollution in the middle and lower reaches are the Llywernog Mine on the Nant Llywernog, the Ystymtuen Mine on the Afon Tuen and the Goginan Mine on the Afon

Melindwr. The Cwm Rheidol Mine on the main river remains a very serious source of pollution and some deterioration in water quality downstream of the mine is thought to have taken place in recent years due to prospecting activities.

Table 3. Heavy metal content of River Rheidol at routine sampling stations based on seven samples taken in 1971–2.

	pH	Hardness as $CaCO_3$		Zn		Pb
	range	mean	range	mean	range	range
Main river						
1. Ponterwyd Bridge	6·3–7·3	28	8–45	0·10	0·02–0·21	nil–0·04
2. Below Cwm Rheidol Mine	6·5–7·2	27	16–33	0·45	0·15–0·81	nil–0·04
3. Below Capel Bangor	6·4–7·5	34	20–72	0·41	0·20–0·83	nil–0.04
Tributaries						
4. Nant Llywernog	6·6–7·7	37	19–53	1·08	0·40–1·28	nil–0·04
5. Afon Tuen	6·4–7·7	38	16–48	0·14	0·04–0·35	nil–0·04
6. Afon Mynach	5·4–7·2	41	13–68	0·21	0·05–0·43	nil–0·10
7. Afon Melindwr	6·6–7·0	44	36–52	0·81	0·60–1·10	nil–0·10

(results expressed as mg/l except pH)

The biological recovery

BENTHOS

Studies of the benthos provide one of the earliest examples of the value of invertebrates as indicators of pollution. Carpenter (1924b) started her pioneer work after the First World War and between 1919 and 1921 she was only able to find 14 species consisting almost entirely of insect larvae and a few crustacean species. In 1922 there was an apparent improvement and 29 species were found including trichopteran larvae and a planarian species. Laurie & Jones (1938) surveyed the river in 1931 and 1932 and recorded 103 species—a considerable improvement, including molluscs and fish which had previously been absent. They noted no further change in 1936. Jones (1949) made a further study of the Rheidol in 1947 and 1948 and recorded 191 species but only 130 of these occurred in the main stream. Thus the recovery of the benthos and increase in species diversity has been comprehensively documented in this series of papers by workers at the University College of Wales, Aberystwyth. A comprehensive review of the early work including botanical work and work on neighbouring rivers is provided by Newton (1944).

A survey by A. Jenkins (unpublished) in 1965–6 indicated that the most abundant species were as follows:

PLECOPTERA
Amphinemura sulcicollis (Stephens)
Leuctra inermis (Kempny)
Leuctra hippopus (Kempny)

EPHEMEROPTERA
Rithrogena semicoloratea (Curt.)
Baetis rhodani (Pict.)

DIPTERA
Simulium reptans L. (local)

TRICHOPTERA
Hydropsyche instabilis (Curt.)
Rhyacophila dorsalis (Curt.)

Only one species was found which had not previously been recorded namely the trichopteran *Glossosoma conformis* Neboiss. This suggests that there has been little, if any, change since 1947–8. Indeed the main recovery appears to have been prior to 1931–2 and some of the apparent later improvement may well be due to a more thorough examination and better identification to species.

FISH

Royal Commissions on River Pollution in 1874 and 1900 established that early in the nineteenth century the River Rheidol had been a healthy river abounding in fish. The destruction of the fishery appeared to have coincided with the introduction of fine grinding machinery by the mining industry. When Carpenter (1924b) surveyed the river in 1919 to 1921 she found the main river devoid of fish except for the occasional stray brown trout *Salmo trutta fario* from unpolluted tributaries or sewin (migratory trout) *Salmo trutta trutta* from the sea, and apparently these neither survived nor bred.

The first fish to become established in the main river were sticklebacks *Gasterosteus aculeatus* L. recorded by Carpenter (1925). An angling club called the Rheidol Protection Society was formed in 1930 when the trout fishery first showed signs of recovery and Laurie & Jones (1938) recorded that in 1932 brown trout were found as high up as Aberffrwd. Jones (1949) described the fish fauna in 1947–8 as decidedly poor and recorded stickleback, brown trout, sewin, eels *Anguilla anguilla* (L.) and some uncertain records of salmon *Salmo salar* L. In 1949 some minnows *Phoxinus phoxinus* (L.) made an appearance probably from introductions a few years previously from the River Dyfi (Jones 1950) but these fish have not survived. During this period the run of sewin was increasing each year but the salmon were much slower to colonize the river. Salmon were first established as present in 1952 when Mr R. L. Marston of the *Fishing Gazette* confirmed that a fish specimen from the River Rheidol was a salmon parr.

Restocking of the River Rheidol was advocated by the West Wales River Pollution Sub-Committee in 1935 and this received substantial local press coverage. It was stated that abundant fish food existed and that pollution was so slight and intermittent as to be harmless. Restocking was again advocated in 1945 when the Rheidol Protection Society combined with the Aberystwyth Angling Association. Subsequently the river was restocked with 2,000 brown trout in the spring of each of the years 1947, 1948 and 1949. Moreover, in 1950 about 50,000 fresh-water snails and 50,000 fresh-water shrimps were placed in the river. There does not appear to be any record of the success or failure of these projects.

Salmon restocking was advocated by Mr Cecil Hutchings in 1952 and a five-year-plan was adopted by the Angling Association. A total of 165,000 salmon ova was planted in the river between 1953 and 1956. Following the establishment of their salmon hatchery at Dolbantau on a tributary of the River Teify, the South West Wales River Board started a second salmon restocking project. Over one million salmon ova and unfed fry have been planted to date as shown in Table 4.

Table 4. Salmon statistics for the River Rheidol.

	Restocking with ova and fry	Rod catch from statutory returns
1952	—	0
1953	25,000	0
1954	40,000	0
1955	50,000	1
1956	50,000	1
1957	—	0
1958	—	2
1959	—	1
1960	107,000	4
1961	83,500	2
1962	74,000	16
1963	201,000	12
1964	252,000	12
1965	153,000	94
1966	—	134
1967	70,400	110
1968	63,000	34
1969	10,200	42
1970	—	81
1971	—	120
1972	—	123

The water bailiff's redd counts are not regarded as particularly accurate but the increase in redd counts from 5 in 1953 to over 300 in 1966 by the same experienced bailiff, reflects the increasing salmon population of the river. Similarly the increasing rod catch shown in Table 4 reflects an increase in salmon population. To what extent this increase is natural recolonization and to what extent it is enhanced by restocking is difficult to ascertain. There is some circumstantial evidence that the restocking had a substantial effect. The first restocking by the River Board took place in the 1960–1 winter when 107,000 salmon ova and fry were planted out followed by 83,500 in the winter of 1961–2. In 1965, which is four years after the first restocking

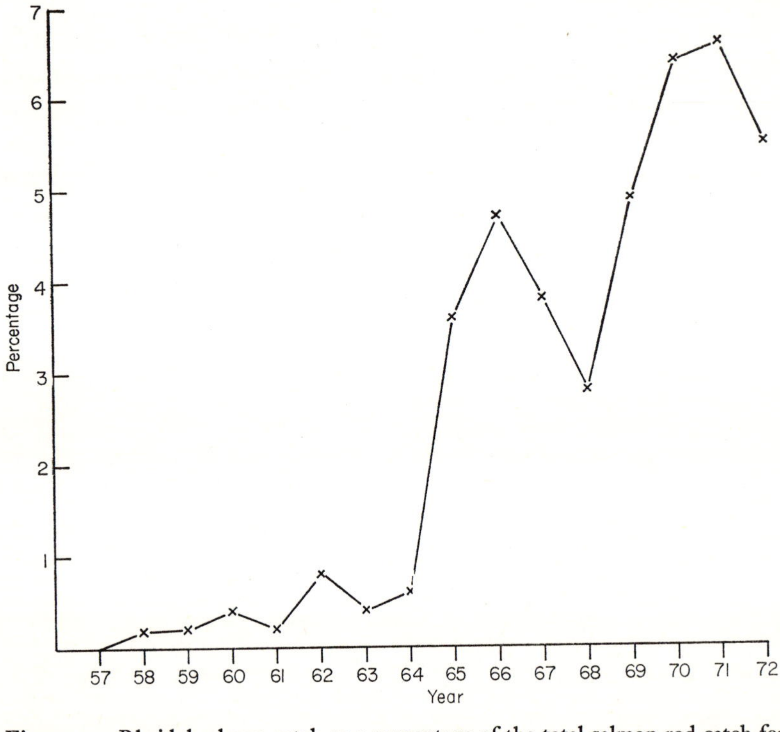

Figure 2. Rheidol salmon catch as a percentage of the total salmon rod catch for south west Wales.

and three years after the second restocking, the anglers' rod returns indicated a catch of 94 salmon consisting of small salmon (4-year-old) and grilse (3-year-old). This was a considerable increase over the 12 salmon reported for 1964. Salmon populations fluctuate considerably from year to year, and to allow for this the Rheidol salmon catch is expressed as a percentage of the total rod catch for all the rivers of south-west Wales in Fig. 2. This confirms the improving nature of the salmon populations, especially the 1965 increase. It does not seem likely that such a remarkable and sudden increase could have occurred without the boost given by the restocking project.

The statutory rod returns underestimate the rod catch because not all anglers make returns. Bailiffs' records indicate that 84 salmon were caught in 1969 and 208 in 1970 compared with 42 and 81 respectively listed on statutory returns. Thus the rod catch data in Table 4 is probably less than half of the true catch. Almost all the salmon are caught in the 12 miles of river below Cwm Rheidol reservoir and the yield of salmon is of the order of 20 per mile per annum which compares favourably with the best salmon rivers in south-west Wales. The poor salmon catch in 1968 is probably a reflection of the outbreak of Ulcerative Dermal Necrosis. The river has since recovered from this setback.

The hydroelectric scheme

BRIEF DESCRIPTION

The scheme was instituted by the Central Electricity Generating Board in 1958 and the works commissioned in 1961 and 1962. There are three stages and three reservoirs in this scheme which is outlined in Fig. 3. The upper

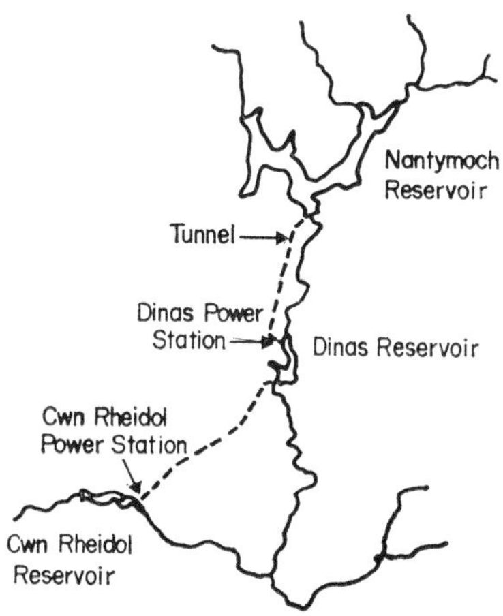

Figure 3. Plan of hydroelectric scheme

stage consists of Nantymoch reservoir and Dinas Power Station to which it is linked by the 4 km pressure tunnel. Nantymoch reservoir has an area of 275 hectares and enables the seasonal variation of rainfall to be regulated. Some streams outside the catchment are brought in by means of underground aqueducts. The total plant capacity of the upper stage is 13,000 kW.

The middle stage consists of the smaller Dinas reservoir which is linked to the main power station at Cwm Rheidol by a 4·4 km pressure tunnel. The total plant capacity of this stage is 42,000 kW.

The lower stage consists of Cwm Rheidol reservoir which was constructed to regulate the flow to the main river from the irregular generating periods. This stage has a plant capacity of 1 megawatt.

PROMOTION OF THE SCHEME

There was considerable apprehension about the effect on pollution and fisheries when the details of the hydroelectric scheme were announced in the early 1950's. The South West Wales River Board sought protection which was subsequently embodied in the North Wales Hydroelectric Act of 1955, and in a supplementary parliamentary undertaking. The implications of these protective clauses and undertakings were as outlined below.

1. The Generating Board appointed pollution consultants including Emeritus Professor Lily Newton who has maintained a close and active interest in the River Rheidol for over 40 years. The Board undertook treatment measures where necessary during construction of the scheme and where it was anticipated that zinc and lead bearing strata and spoil would be exposed and disturbed. This normally took the form of dressing with limestone.
2. The river flow between Dinas reservoir and Cwm Rheidol reservoir was reduced as a result of the scheme so that there was less dilution for the polluted mine water emanating from Cwm Rheidol Mine. In order to compensate for this a treatment plant was constructed in the form of a limestone filter which is described by Treharne (1962). In addition the flow of water through the mine workings was reduced by works which prevented surface water entering a mine shaft on the plateau above the mine. This latter work was carried out in co-operation with Cardiganshire County Council.
3. The Cwm Rheidol reservoir inundated a considerable amount of spawning gravel for salmon and sea trout. In order to compensate for this loss, the Generating Board incorporated a Borland Fish lock to allow migratory fish to pass the Cwm Rheidol dam. They also built a fish ladder at the opposite end of the reservoir to enable fish to pass Rheidol Falls and thereby utilize a stretch of river up to Gyfarllwyd Falls which had previously been inaccessible. These developments indicate foresight at a time when the migratory fish population was negligible.
4. A restocking fund was agreed for Nantymoch reservoir and Dinas reservoir and stocking with 15,000 brown trout took place in 1963 and 1964. The fishing was reasonably good initially but then deteriorated over a period of years. Two research studentships were awarded at University College of Wales, Aberystwyth, for the study of the ecology of Nantymoch (Billington

1972). Very few brown trout survive in the reservoir for more than four years and the possibility of fish being affected by metal pollution caused by run-off from mine spoil is being considered (E. Howells *pers. comm.*). The restocking fund was also used to build three trout rearing tanks at Cwm Rheidol Power Station. Each tank has a capacity of 1,000 yearlings and rainbow trout *Salmo gairdneri* Richardson and brown trout are transferred to Dinas reservoir on a monthly basis. This put and take fishery has proved to be very successful. The rearing tanks were affected, however, by intermittent metal pollution during high rainfall periods following dry weather when a proportion of the fish would die. Recirculation of water is now carried out during such periods in order to avoid using polluted stream water.

REGULATION OF FLOWS

The regulation of river flow below Cwm Rheidol Reservoir since 1962 has probably played an important part in improving water quality. The large oscillations in water quality at times of heavy rainfall following dry spells have been dampened sufficiently by the regulation to reduce the levels of toxic metals.

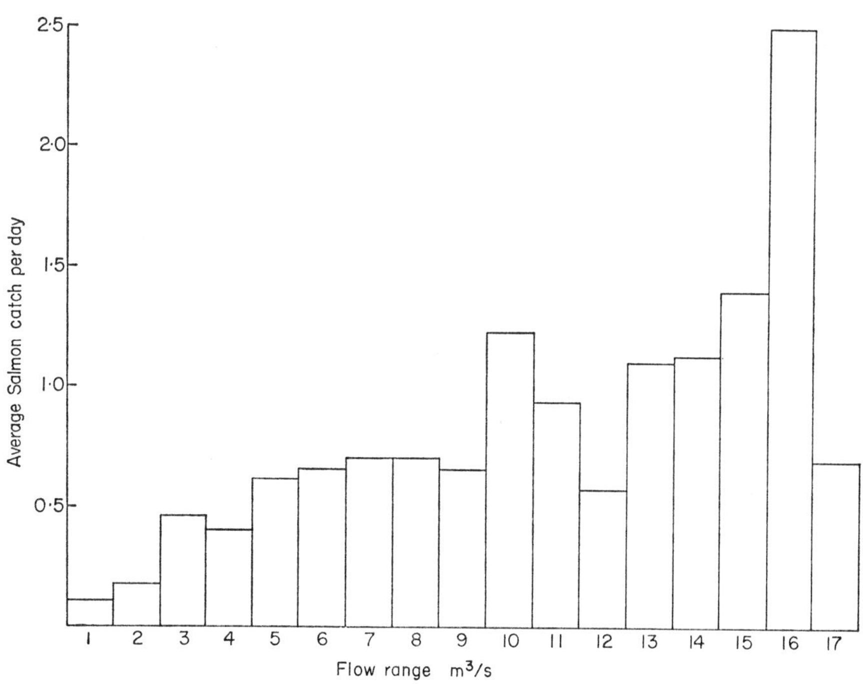

Figure 4. Relationship of salmon catch and flow for the River Rheidol 1966–1971.

Prior to regulation the river bed was unstable and supported hardly any rooted vegetation. The vegetation consisted of only small patches of algae and bryophytes on the more stable stones. After regulation the river bed became stabilized and vegetation such as water crowfoot *Ranunculus* sp. and water starwort *Callitriche* sp. became established. The regulated flows probably benefited the whole food chain in the river and may have resulted in an improved production of salmon parr. Indeed this may well have contributed substantially to the apparent success of the River Authority restocking project.

The compensation flow of 1·87 m^3/s is frequently augmented in the summer months by generating water. This often results in good angling conditions during dry weather when neighbouring rivers are experiencing very poor angling conditions. Conversely, but less frequently, a spate provides good angling conditions for neighbouring rivers which may not be experienced by the River Rheidol due to impoundment. A study of the relationship between river flow and angling success (River Authority Annual Report 1972–3) confirms that angling success for salmon increases with increased flow, as shown in Fig. 4. The average salmon catch is particularly poor at flows of less than 3 m^3/s which includes the flow range most frequently occurring.

Future exploitation of natural resources

MINERALS

In 1968 a number of companies started to show interest in prospecting for minerals in the Rheidol and neighbouring catchments. This development was viewed with alarm by the River Authority and the Generating Board. Any disturbance of ores or spoil or release of mine water could increase pollution. A close watch was therefore kept on these activities. An example of the dangers to the river of even preliminary prospecting occurred in October 1969 when a bulldozer levelling a site for a drilling rig caused the collapse of the front of a sealed off adit at Cwm Rheidol Mine. The 'blow out' of underground water contained 3,106 mg/l suspended solids, 1,500 mg/l iron, 230 mg/l soluble zinc and 148 mg/l soluble lead. Fish in the river were killed, and the river bed down to Cwm Rheidol reservoir was covered with yellow ochre. The company involved was convicted for the pollution but the consequences remain today. The Generating Board's limestone filter became blocked with ochre and was rendered ineffective. Renewal of the filter media allowed only of temporary relief and the filter became gradually silted up again. Typical analytical results for mine water entering and leaving the filter for before the 'blow out' and after are given in Table 5. The opening of the sealed adit has apparently increased the flow of water to the filter, and in allowing access of air, may have accelerated oxidation resulting in a higher concentration of metals in solution.

Table 5. Analysis of mine water entering and leaving the C.E.G.B. filter at Cwm Rheidol Mine.

	pH	Zn	Pb
23 May 1969			
entering	3·5	22·5	0·06
leaving	6·0	2·6	nil
5 June 1973			
entering	3·2	50·0	0·07
leaving	3·6	45·0	0·03

(results in mg/l except pH)

Mine water is exempt from the statutes (Rivers (Prevention of Pollution) Acts 1951–1961) if discharged in the same quality as when raised from underground. The River Authority applied to the Secretary of State for Wales for Orders revoking this exemption. A Public Enquiry was held in 1970 and subsequently the Secretary of State approved the application and made the appropriate Orders. This strengthens the River Authorities' powers to control pollution in the catchment. It was stated at a Public Enquiry that it was not the intention of the River Authority to stop mining in the area but only to ensure that discharges comply with reasonable standards so that the receiving waters are not polluted.

WATER SUPPLY

Routine water samples are taken by the River Authority at Nantymoch reservoir and its principal feeder streams. Typical results for reservoir water near the dam for the period May to November 1972 are given in Table 6. There appears to be no significant difference in metal content at different depths but zinc concentrations are higher near the submerged Bryn-ar-arf mine. The results in Table 6 indicate that the metals are usually within the World Health Organization International Drinking Water Standards of 1·0 mg/l for copper, 0·1 mg/l for lead and 5·0 mg/l for zinc. It is sensible, however, to utilize water which is less polluted than Nantymoch if available. Thus although Cardiganshire Water Board have an allocation of water in

Nantymoch reservoir, they prefer to abstract water from two unpolluted tributaries of the reservoir.

Table 6. Analysis of Nantymoch water based on six samples from May to November 1972.

	Mean	Range
pH	5·2	5·1–5·3
Cu	—	nil–0·04
Fe	0·37	0·28–0·54
Pb	—	nil–0·01
Zn	0·066	0·03–0·10
Cd	nil	nil

(results in mg/l except pH)

The Water Board is interested in the water quality of the lower reaches of the River Rheidol and has instigated a sampling programme in collaboration with the River Authority in order to assess its suitability for water supply. Interest has also been shown in abstracting water from the river gravels of the lower reaches but because of possible metal pollution (particularly the levels of lead), the matter is still under consideration.

A recent report by the Water Resources Board, the Wales and Midland Study (1971), proposes a considerable enlargement of Nantymoch reservoir and its joint operation for transferring water to regulate the River Severn and for the generation of electricity as at present. This proposal is currently the subject of careful consideration and investigation. Some of the aspects being studied are (a) the effect of reduced flows on the dilution available for mine effluents (b) the effect of reduced flows on salmon angling success and fish migration and (c) the possibility of a deterioration in the quality of water at Nantymoch due to an enlarged reservoir flooding more mines and spoil tips.

FISHERIES

Future development is possible along the following lines.

1. Dinas Reservoir has been developed as a successful 'put and take' fishery by the C.E.G.B. When the demand for more fishing is sufficient the next stage would be to similarly develop Nantymoch Reservoir. This would be a difficult prospect, however, because of the large size of the reservoir, its unfavourable water quality, and the very large fluctuations in water level.
2. The reservoir at Cwym Rheidol contains immature, and at times, mature migratory fish and not a great deal of angling is carried out there. There is

currently a proposal to develop the lake as a brown trout fishery through restocking. This will require careful consideration because of the possibility of high predation by the brown trout on the juvenile migratory fish resident in the reservoir and which pass through the reservoir.

3. The main river between Rheidol Falls fish ladder and Gyfarllwyd Falls is not yet fully utilized by migratory fish. There is a tendency for migratory fish to 'home' to the part of the river in which they originated and the colonization of this section of the river can be expected to continue gradually.

4. Gyfarllwyd Falls on the main river and Mynach Falls on the Afon Mynach both form obstructions to migratory fish. The provision of fish ladders is not practicable at the present time and in any case no work should be done which would detract from their aesthetic qualities. It would however, be possible to develop the water upstream by stocking with unfed fry. It would first be necessary to find out if smolts could survive the migration down the falls. If not then smolt traps could be installed upstream and the smolts transported manually in oxygenated tanks to sites downstream. A similar trapping and trucking scheme is operated in the upper reaches of the River Towy (Howells & Jones 1972).

The recreational use of the River Rheidol and its reservoirs is currently almost exclusively angling. There is an increasing interest in other water-based sports and if the population continues to increase one can foresee demands requiring the integration of these sports in the Rheidol catchment.

Concluding remarks

The main recovery of water quality after the cessation of mining activities was a natural process. Man's active role has been largely to control new activities in order to prevent any further degradation of the environment. The three major potential threats to the river have been the hydroelectric development, the recent mineral prospecting and the proposals for water supply. So far as the hydroelectric scheme is concerned the C.E.G.B. has shown a highly responsible attitude and have minimized the effect of their activities on the environment. Moreover, the controlled flows have probably benefited the river ecology whereas the creation of three reservoirs has provided a new amenity.

The abstraction of water or the transfer of water to other river systems from the upper reaches of a river is potentially harmful. In the case of the River Rheidol the loss of water might mean (a) less volume for diluting the more serious mine effluents such as at Cwm Rheidol Mine; (b) the reduced flow might adversely affect angling success as indicated by Fig. 4; (c) the reduced flow might cause a reduction of wetted area and result in a reduced biological production which could affect the juvenile migratory fish. From an ecological point of view it would be preferable to regulate flows and abstract water from above the tide (Iremonger 1970; Howells & Jones 1972).

At the time of writing there are no major exploratory works for metals being undertaken in the Rheidol catchment. The possibility remains however, that further interest will be shown should the world price of some of these metals continue to rise. So long as the human population increases and governments are committed to a growing economy then the demand for minerals will increase until the resource is exhausted. It is normally technically feasible to treat polluted mine water to a standard which will reasonably safeguard river systems and this should be one of the conditions imposed on those seeking to exploit the mineral resources of this country. The cost should be borne by the company concerned and the capital and running costs should be taken into account in assessing whether the operation is an economically sound enterprise. It would be undesirable to allow the activities of one section of the community to take place at the expense of another and every effort should be made to prevent the River Rheidol from degenerating to the sterile conditions of the nineteenth century.

References

ALLOWAY B.J. & DAVIES B.E. (1971) Trace element content of soils affected by base metal mining in Wales. *Geoderma* 5, 197–208.

BILLINGTON C.A. (1972) *Spatial Heterogeneity and Primary Productivity of the Algae of Nantymoch Reservoir*. Ph.D. thesis, Univ. of Wales.

CARPENTER K.E. (1924a) Notes on the history of Cardiganshire lead mines. *Aberystwyth Studies* 5.

CARPENTER K.E. (1924b) A study of the fauna of rivers polluted by lead mining in the Aberystwyth district of Cardiganshire. *Ann. appl. Biol.* 11, 1–23.

CARPENTER K.E. (1925) On the biological factors involved in the destruction of river fisheries by pollution due to lead mining. *Ann. appl. Biol.* 12, 1–13.

GRIFFITH J.J. (1918) Influence of mines upon land and livestock in Cardiganshire. *J. agr. Sci.*, Camb. 9, 365–95.

GRIMBLE A. (1904) *The Salmon Rivers of England and Wales*. Kegan Paul, London.

HOWELLS W.R. & JONES A.N. (1972) The River Towy regulating reservoir and fishery protection scheme. *J. Inst. Fish Mgmt.* 3, 5–19.

IRELAND M.P. (1973) Result of fluvial zinc pollution on the zinc content of littoral and sub-littoral organisms in Cardigan Bay, Wales. *Environ. Pollut.* 4, 27–35.

IREMONGER D.J. (1970) Rivers regulation and fisheries. *Salm. Trout Mag.* 191, 64–88.

JAMES T.C., LAURIE R.D. & NEWTON L. (1932) A study of the present condition of the river Rheidol. Standing Committee on River Pollution, West Wales Sub-committee. Rep. No. 419, 1–29.

JONES J.R.E. (1949) An ecological study of the river Rheidol, North Cardiganshire, Wales. *J. Anim. Ecol.* 18, 67–88.

Jones J.R.E. (1950) A further ecological study of the river Rheidol. The food of the common insects of the main stream. *J. Anim. Ecol.* 19, 159–74.

Jones O.T. (1922) *Lead and Zinc. The Mining District of North Cardiganshire and West Montgomeryshire.* Spec. Rep. Mineral Resources 20, Geol. Surv. of Great Britain.

Laurie R.D. & Jones J.R.E. (1938) The faunistic recovery of a lead polluted river in North Cardiganshire, Wales. *J. Anim. Ecol.* 7, 272–89.

Lewis W.J. (1967) *Lead Mining in Wales.* University of Wales Press, Cardiff.

Newton L. (1944) Pollution of the rivers of West Wales by lead and zinc mine effluent. *Ann. appl. Biol.* 31, 1–11.

Rees M.J. (1937) The micro-flora of the non-calcareous streams Rheidol and Melindwr with special reference to water pollution from lead mines in Cardiganshire. *J. Ecol.* 25, 385–407.

Treharne W.D. (1962) Pollution problems of the Afon Rheidol. *Wat. Waste Treat.* 8, 610–13.

Water Resources in Wales and the Midlands (1971) Water Resources Board, London, H.M.S.O.

Author Index

Bold figures refer to pages where full references appear

Subject Index